# Secure and Digitalized Future Mobility

This book discusses the recent advanced technologies in Intelligent Transportation Systems (ITS), with a view on how Unmanned Aerial Vehicles (UAVs) cooperate with future vehicles.

ITS technologies aim to achieve traffic efficiency and advance transportation safety and mobility. Known as aircrafts without onboard human operators, UAVs are used across the world for civilian, commercial, as well as military applications. Common deployment includes policing and surveillance, product deliveries, aerial photography, agriculture, and drone racing. As the air-ground cooperation enables more diverse usage, this book addresses the holistic aspects of the recent advanced technologies in ITS, including Information and Communication Technologies (ICT), cyber security, and service management from principle and engineering practice aspects. This is achieved by providing in-depth study on several major topics in the fields of telecommunications, transport services, cyber security, and so on.

The book will serve as a useful text for transportation, energy, and ICT societies from both academia and industrial sectors. Its broad scope of introductory knowledge, technical reviews, discussions, and technology advances will also benefit potential authors.

**Yue Cao** is currently the Professor at School of Cyber Science and Engineering, Wuhan University, China. Previously, he was a faculty member at Lancaster University, Northumbria University, UK. His research interests are under the umbrella of ICT-empowered Intelligent Transport Systems covering use cases for autonomous valet parking, E-Mobility, UAV.

**Omprakash Kaiwartya** is currently Senior Lecturer in the Department of Computer Science, Nottingham Trent University, UK. Previously, he was Research Associate at the Northumbria University, Newcastle, UK, and Postdoctoral Research Fellow at the University of Technology Malaysia (UTM), Malaysia. His research interest focuses on future smart technologies for diverse domain areas focusing on Transport, Healthcare, and Agriculture Production.

**Tiancheng Li** is currently the Professor at the School of Automation, Northwestern Polytechnical University, China. Prior to this, he had worked with University of Salamanca, Spain. He received the Excellent Doctoral Thesis Award of Shaanxi Province in 2017 and the Marie Sklodowska-Curie Individual Fellowship from EU in 2016.

# Secure and Digitalized Future Mobility

## Shaping the Ground and Air Vehicles Cooperation

Edited by

Yue Cao, Omprakash Kaiwartya, and Tiancheng Li

CRC Press is an imprint of the
Taylor & Francis Group, an **informa** business

First edition published 2023
by CRC Press
6000 Broken Sound Parkway NW, Suite 300, Boca Raton, FL 33487-2742

and by CRC Press
4 Park Square, Milton Park, Abingdon, Oxon, OX14 4RN



*CRC Press is an imprint of Taylor & Francis Group, LLC*



ISBN: 978-1-032-30753-4 (hbk)
ISBN: 978-1-032-30761-9 (pbk)
ISBN: 978-1-003-30650-4 (ebk)

DOI: 10.1201/b22998

Typeset in Minion
by codeMantra

# Contents

# Acknowledgement

This work is supported in part by the Wuhan Knowledge Innovation Program (2022010801010117)

# Contributors

**Geetika Aggarwal**
School of Science and Technology
Nottingham Trent University
Nottingham, United Kingdom

**Yue Cao**
School of Cyber Science and Engineering
Wuhan University
Wuhan, China

**Harish Chander**
Department of Computer Science and Applications
Sat Jinda Kalyana College
Kalanaur Kalan, India

**Zhuo Chen**
School of Cyber Science and Engineering
Wuhan University
Wuhan, China

**Jixing Cui**
School of Cyber Science and Engineering
Wuhan University
Wuhan, China

**Kai Da**
National Key Laboratory of Science and Technology on ATR
National University of Defense Technology
Changsha, China

**Piyush Dhawankar**
Department of Computer Science
University of York
York, United Kingdom

**Xueru Du**
School of Cyber Science and Engineering
Wuhan University
Wuhan, China

**Hongqi Fan**
National Key Laboratory of Science and Technology on ATR
National University of Defense Technology
Changsha, China

**Tianhao Guo**
College of Physics and Electronic Engineering
Shanxi University
Taiyuan, China

**Hassan-Jalil Hadi**
School of Cyber Science and Engineering
Wuhan University
Wuhan, China

**Chongwen Huang**
College of Information Science and Electronic Engineering
Zhejiang University
Hangzhou, China

**Kai Jiang**
School of Cyber Science and Engineering
Wuhan University
Wuhan, China

**Himanshu Joshi**
School of Computer and Systems Sciences
Jawaharlal Nehru University
New Delhi, India

**Rupak Kharel**
Department of Computer Science
University of Central Lancashire
Preston, United Kingdom

**Ahmad Mohammad Khasawneh**
Department of Cybersecurity
Amman Arab University
Amman, Jordan

**Aanchal Khatri**
Department of Computer Science and Applications
Sat Jinda kalyana College
Kalanaur Kalan, India

**Vishal Khatri**
Bhagwan Parshuram Institute of Technology
Guru Gobind Singh Indraprastha University
Delhi, India

**Kirshna Kumar**
National Informatics Centre, Ministry of Electronics and Information Technology
Government of India
New Delhi, India

**Sushil Kumar**
School of Computer and Systems Sciences
Jawaharlal Nehru University
New Delhi, India

**Sifan Li**
School of Cyber Science and Engineering
Wuhan University
Wuhan, China

**Tiancheng Li**
School of Automation
Northwestern Polytechnical University
Xi'an, China

**Endong Liu**
Department of Computer Science
University College London
London, United Kingdom

**Shuohan Liu**
School of Computing and Communication Systems
Lancaster University
Lancaster, United Kingdom

**Qiang Ni**
School of Computing and Communication Systems
Lancaster University
Lancaster, United Kingdom

**Yonglong Peng**
Keepv Technology (Nanjing) Co. Ltd
Nanjing, China

**Quoc-Viet Pham**
Korean Southeast Center for the 4th Industrial Revolution Leader Education
Pusan National University
Busan, Korea

**Shiv Prakash**
Department of Electronics and Communication
University of Allahabad
Allahabad, India

**Mukesh Prasad**
Faculty of Engineering and Information Technology, School of Computer Science
University of Technology Sydney
Sydney, Australia

**Kashif Naseer Qureshi**
Department of Computer Science
Bahria University
Islamabad, Pakistan

**Saim Shahid**
Department of Computer Science
Nottingham Trent University
Nottingham, United Kingdom

**Ajay Sikandar**
Department of Computer Science and Engineering
GL Bajaj Institute of Technology and Management
Greater Noida, India

**Qiang Tang**
SG Star Energy (Sichuan) Technology Co. Ltd
Chengdu, China

**Lei Wen**
College of Electronic Science and Technology
National University of Defense Technology
Changsha, China

**Xu Xia**
Telecom Research Institute
Beijing, China

**Yongjun Xu**
School of Communication and Information Engineering
Chongqing University of Posts and Telecommunications
Chongqing, China

**Qianqian Yang**
College of Information Science and Electronic Engineering
Zhejiang University
Hangzhou, China

**Zhaohui Yang**
Department of Electronic and Electrical Engineering
University College
London, United Kingdom

**Benru Yu**
School of Electronic and
Optical Engineering
Nanjing University Of Science and Technology
Nanjing, China

**Zhaoyang Zhang**
College of Information Science and Electronic Engineering
Zhejiang University
Hangzhou, China

**Huan Zhou**
College of Computer and Information Technology
China Three Gorges University
Yichang, China

CHAPTER 1

# The Overview of Air-Ground Cooperation


Zhuo Chen and Yue Cao

*Wuhan University*

Lei Wen

*National University of Defense Technology*

Zhaohui Yang

*University College London*


## CONTENTS



## LIST OF ABBREVIATIONS

**BCR** Battery consumption rate

**BS** Base station

**BnB** Branch and bound method

**CD** Centralized decision

**DD** Decentralized decision

**FSTSP** Flying sidekick traveling salesman problem

DOI: 10.1201/b22998-1

**GU** Ground unit
**HD** Hybrid decision
**IoT** Internet of things
**MAC** Multiple access control
**MEC** Mobile edge computing
**MILP** Mixed integer linear programming
**PDSTSP** Parallel UAV scheduling traveling salesman problem
**RM** Resupply multi-modal model
**RSU** Road side unit
**SCA** Successive convex approximation
**SM** Synchronized multi-modal model
**TDMA** Time division multiple access control
**UAV** Unmanned aerial vehicle
**UGV** Unmanned ground vehicle
**UMEC** UAV-MEC system
**UM** Unsynchronized multi-modal model

## 1.1 INTRODUCTION

Unmanned aerial vehicles (UAVs) are an emerging technology that can provide a wide range of applications and services in all areas of industry and human life. Due to their inherent flexibility, ease of use, and rapid deployment capability, UAVs have been widely used in commercial applications. Typical applications include air and ground detection, communications missions over predetermined areas, real-time monitoring, logistics distribution, etc. Collaborative systems that use multiple UAVs and vehicles are well suited to these tasks. The main driving force in this area is the optimal control and use of air and ground resources to achieve mission objectives reliably and efficiently.

A wide variety of unmanned ground vehicles (UGVs) are in use today. These vehicles are mainly used to replace humans in dangerous situations, such as handling explosives and bomb-banned vehicles, requiring extra strength or smaller size, or places where humans cannot easily go. They can be used in industries such as agriculture, mining, and construction. Military applications include surveillance, reconnaissance, and target acquisition. UGVs are very effective in naval operations, and they are important in helping marines fight. They can also be used for logistics operations on land and offshore.

In the UAV-UGV cooperation system, UGVs can more accurately locate or detect objects on the ground, compensating for the weaknesses of UAVs. In addition, the UGVs have greater payload capacity and powerful computing power, as well as the ability to charge UAVs and carry them on longer flights. For UAVs, they can fly at high speed and have a wide field of vision, which is good for guiding UGVs to avoid obstacles. This chapter discusses the problem of coordinated deployment of UAVs and UGVs.

## 1.2 UAV-UGV COOPERATION SYSTEM

### 1.2.1 UAV

UAV is operated by radio remote control equipment and self-provided program control device. UAV is actually a general designation of unmanned aerial vehicles. From a technical perspective, it can be divided into unmanned fixed-wing aircraft, unmanned vertical take-off and landing aircraft, unmanned airship, unmanned helicopter, unmanned multi-rotor aircraft, unmanned parachute, etc.

Compared with manned aircraft, the UAV has the advantages of small size, low cost, easy to use, low environmental requirements, strong survival ability, and so on. UAVs are often better suited for tasks that are too "dull, dirty or dangerous." According to the application field, UAVs can be divided into military and civilian. In terms of the military, UAVs are divided into reconnaissance aircraft and target aircraft. Due to the importance of unmanned aircraft in the future of air warfare, the world's major military countries are stepping up the development of unmanned aircraft. In terms of civil use, the application of UAV in aerial photography, agriculture, plant protection, express transportation, disaster rescue, wildlife observation, surveying and mapping, power inspection, and other fields has greatly expanded the use of UAV itself. Many countries are also actively expanding the industrial application and development of UAV technology.

The cooperative control and path planning of UAVs has always been a concern and has received widespread attention in the past decade. UAVs have been widely used in different fields, including surveillance, target detection, and search and rescue [1]. During the Euro 2020 football tournament, for example, UAVs were used for photography and video recording. In Ref. [2], the authors considered the application of forest mapping involving UAVs and UGVs. They developed algorithms to plan the routes of fuel-constrained UAVs and UGVs that can act as filling stations for the UAVs during missions.

In addition, a UAV-based logistics system is emerging, and Amazon has also announced a plan to provide a 30-minute delivery service using UAVs. Prime Air aims to use UAVs to deliver packages to customers in 30 minutes or less. Google has also announced projects with similar goals to Amazon's. Therefore, logistics systems based on UAVs will become popular in the near future.

UAVs could also be used to enhance space communications. In particular, a large number of small, low-cost UAVs can perform remote missions at high altitudes or in uninhabited areas with difficult terrain, enabling a UAV-based communications network. There has been some research on the efficiency and security of UAV communication networks. For example, Agogino et al. used evolutionary algorithms to improve the performance of UAV communication systems [3]. Zhang et al. implemented a UAV communication network based on dynamic time division multiple access control and multiple access control protocol, achieving a low collision rate and high channel utilization [4]. Olsson et al. demonstrated how to create a relay chain for measuring one target and how to create a relay tree for simultaneously monitoring multiple targets [5].

The current research has basically solved the problem of vehicle routing with limited communication. The authors [6] examined the problem of deploying teams of mobile agents to monitor multiple points of interest on a regular basis. Scholars [7] conducted an experimental study on the strategy of robot teams in experimental teams to maintain end-to-end communication links. In Ref. [8], a heuristic approach to changing the traditional vehicle routing problem through a simple communication model was studied.

### 1.2.2 UGV

UGVs can be used for many applications where the presence of a human operator is inconvenient, dangerous, or impossible. Typically, the UAV will have an array of sensors to observe the environment and will automatically make decisions about its behavior, or relay information to a human operator in a different location who will control the UAV remotely.

Similarly, the interest in UGVs has grown substantially over the past few decades, and several highly intelligent UGVs are emerging. A bionic robot created by Northeastern University can sense its environment like a real animal. The Spot, developed by Boston Dynamics, is a flexible quadruped robot that is rugged and customizable. It is well suited to unstructured environments, can navigate the terrain with unprecedented mobility, can climb stairs and traverse rough terrain, and automatically perform routine inspection tasks and data collection safely, accurately, and frequently. Formation control [9,10], regional search [11] and other cooperative control of multiple UGVs have been studied, and consensus control [12] has been widely carried out. In addition, various robot competitions such as RoboCup and the Robotics Competition organized by the American Institute of Aeronautics and Astronautics have greatly promoted the research of multi-UGV systems.

### 1.2.3 UAV and UGV Cooperation System

In practice, UAVs and UGVs are ideal for widespread deployment and use because of their low cost. However, both UAVs and UGVs have their own limitations, which to some extent significantly reduce the efficiency of their missions. On the other hand, it's also clear that UAVs and UGVs are highly complementary. The complementarity of UAVs and UGVs is mainly reflected in the following aspects. First, while sensors located on UAVs are typically limited by operational airspeed and altitude when capturing ground features, UGVs that can be deployed on the ground can accurately locate ground targets. Second, because UAVs are in the air, communication links between UAVs are less blocked by obstacles than between UGVs. Therefore, UAVs can be used as communication relays to achieve indirect communication links between UGVs at different locations. Finally, UAVs (especially small UAVs) are often limited by their short range due to energy limitations, whereas UGVs have a much larger payload capability. From what has been discussed above, UAVs and UGVs have strong complementarity in sensing, communication, payload capability, and other aspects, and have broad prospects for collaborative work.

UAVs and UGVs Cooperation Systems can be characterized as a group of UAVs and a group of UGVs operating in the same field, working together to achieve a common goal.

Given the huge heterogeneity and strong complementarity between UAVs and UGVs, the way they interact can be quite complex. Research to date has developed a variety of UAVs and UGVs Cooperation System for specific applications. To get a clear picture of these types, collaborative systems can be classified in three directions [13].

#### *1.2.3.1 Classification According to the Role*

Since UAVs and UGVs perform different functions in different missions, tasks can be classified according to the functional role the vehicle plays in the task. The collaborative system has four main functional modules: mobile sensor, mobile actuator, decision maker, and auxiliary facilities.

Based on the complementarity of UAVs and UGVs, a variety of UGV systems have been developed to assist UAVs as ancillary facilities. Ancillary facilities can provide energy, communications, computing, and other services to the main facility. The scenarios in which UAV and UGV play different functions in existing studies mainly include the following two scenarios.

1. *UGV as mobile actuator, UAV as decision maker, mobile sensor, or auxiliary:*
   In these systems, the UGVs act as a mobile actuator, and the UAV makes decisions for the UGV, provides environmental and mission information to the UGV, or acts as a communications relay to provide a communication link between a remote operation station and the UGV.
2. *UAV as motion sensor or actuator and UGV as auxiliary facility:*
   In these systems, the UAV acts as a motion sensor or actuator. However, in some cases, the low payload and short endurance of the UAV (especially small UAVs) greatly limit their efficiency in completing missions. As an alternative, UGVs can assist UAVs to accomplish their missions more efficiently by powering them or transporting them over a wide area.

#### *1.2.3.2 Classify According to Mission Objectives*

According to different task objectives, collaborative tasks can be divided into coupling objectives and decoupling objectives. Coupled objectives indicate that the movements of the two vehicles should be closely coordinated in order to achieve the mission objectives. Decoupling objectives indicate that, in some special cases, a task can be decomposed into multiple steps or sub-tasks, where different parts of the task can be accomplished by different types of vehicles. In this case, shared mission objectives can be decoupled into separate sub-objectives, each of which can only be achieved by UAVs or UGVs.

#### *1.2.3.3 Classify According to the Way Decisions Are Made*

Decision modes between UAVs and UGVs include centralized decision (CD), decentralized decision (DD), and hybrid decision (HD). Centralized decision-making means that the central controller collects mission-related information from UAVs and UGVs in a centralized manner and makes decisions for all vehicles. Although the centralized approach leads to a higher communication and computing load on the central controller, the optimal solution is theoretically guaranteed. With a decentralized approach, each UGV has autonomy and makes its own decisions based only on information from its neighbors. The decentralized approach is more robust and scalable than the centralized approach.

Both centralized and decentralized decisions have their own advantages and are applicable to specific scenarios. For small systems, a centralized approach may be a better choice, while a decentralized approach is appropriate for large systems. Mixed decision-making applies both centralized and decentralized methods. A hybrid approach can be implemented in a number of ways. The simplest way is for both the UAVs and UGVs to make their own decisions in a centralized manner, and then the UAVs leader and UGVs leader reach a consensus. The hybrid approach combines the benefits of centralized and decentralized approaches to achieve a better trade-off between solution quality and decision time costs.

## 1.3 COOPERATION SYSTEM APPLICATION

### 1.3.1 Cooperative Air and Ground Surveillance

The application of unmanned equipment in surveillance and exploration is increasingly prominent. Typical applications include air and ground mapping of predetermined areas for missions such as surveillance, target detection, tracking, and search and rescue operations. As shown in Figure 1.1, using multiple individuals to collaborate is well suited to such tasks. The main driving force in this field is the optimal control and utilization of unmanned equipment resources in order to achieve objectives reliably and effectively.

In Ref. [14], the authors proposed a method of cooperative target search, identification and location based on heterogeneous fixed-wing UAVs and UGVs. This article assumed

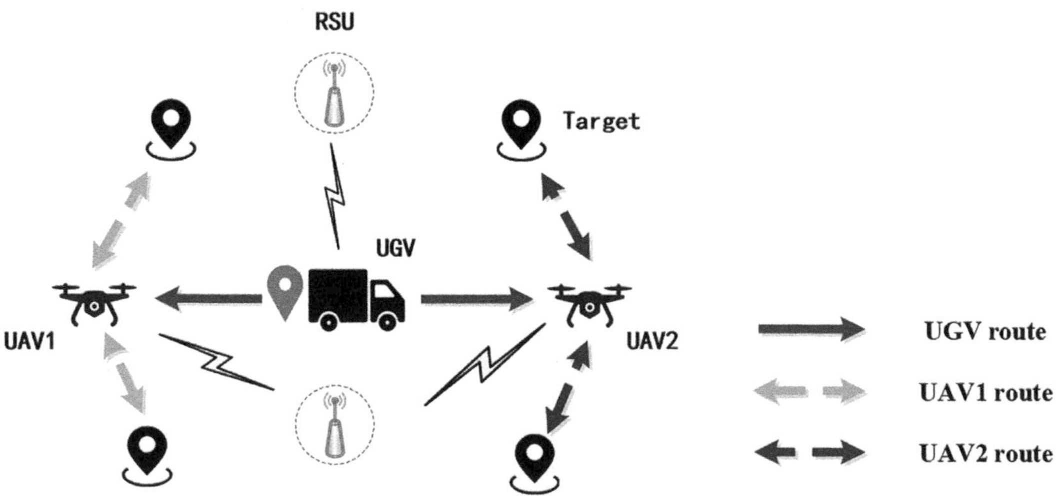


FIGURE 1.1 Cooperative air and ground surveillance.

standard solutions to low-level control of UAVs and UGVs and inexpensive off-the-shelf sensors for target detection. Their main contribution is a framework that is scalable to multiple vehicles and decentralized algorithms for control of each vehicle. The decentralized algorithms are transparent to the specificity of the team's composition and the behaviors of other members of the team. Unlike much of the literature on very difficult planning issues such as coverage and search and localization, their interest lies in: (1) reactive behaviors that are easily implemented; (2) independent of the number or the specificity of vehicles; and (3) offer guarantees for search and for localization.

In terms of assisted mobility, most of the research is toward the direction of using UGVs to transport UAVs. UGVs can carry heavier loads and save more energy while holding a fixed position. Therefore, it can be used to transport UAVs. Much less work has been done on collaboration for UAVs to transport ground vehicles due to their payload limits. The researchers [15] proposed a system in which the UAV could transport the UGV and deliver it to various surfaces in a multi-level structure. When used together, UAV and UGV can complement each other operationally. UAV can provide terrain data to the UGV from its aerial perspective to help them better navigate obstacles or find the best path.

Some researches focus more on path planning. The author of Ref. [16] described a collaborative path planning algorithm for tracking moving targets in an urban environment using UAVs and UGVs. The novel feature of this algorithm is that it considers the visual occlusion caused by the obstacles in the environment. The algorithm uses a dynamically occupied grid to model the target state, which is updated by sensor measurements using a Bayesian filter. Based on current and predicted target behavior, the path planning algorithm for a single vehicle (UAV/UGV) is first designed to maximize the sum of detection probabilities over a limited forward-looking range. The algorithm is then extended to multiple vehicle collaboration scenarios.

This work [17] presents the design and preliminary results of an experiment called pNEUMA (New Era of Urban traffic Monitoring with Aerial footage), which aims to create the most complete urban data set to study congestion. A fleet of drones hovered over Athens' central business district for 5 days, recording a congested traffic flow covering 1.3 km$^2$ of more than 100 km lanes, 100 busy intersections, more than 30 bus stops, and nearly a million tracks. The purpose of the experiment is to record traffic flow in multi-mode crowded conditions in an urban environment using UAVs, allowing in-depth investigation of key traffic phenomena. One of the aims of this work is to reveal the basic mechanism of large-scale network congestion pattern formation based on the complete data set collected by UAVs. The paper describes the design process of the experiment and various factors that need to be considered and optimized (such as UAV rules, number of UAVs, and the maximum duration of the flight). In this urban, multi-modal, busy environment, the analysis of the video allows different traffic phenomena to be tested on different research disciplines at both the micro and macro scales.

Other studies focus more on routing and scheduling algorithms. An efficient vehicle-assisted multi-UAV routing scheduling algorithm (VURA) was proposed in Ref. [18]. In VURA, they maintained and iteratively updated the memory containing the candidate UAV routes. VURA works by iteratively deriving solutions based on UAV routes selected from memory. In each iteration, VURA works together to optimize anchor selection, path planning, and trip allocation through nested optimization operations.

### 1.3.2 Cooperative Air and Ground Logistics

With the increasing improvement of people's living standards, the logistics and distribution industry has prospered under the influence of mobile Internet technology. As one of the symbols of modern economic life, it has become more and more data-oriented, intelligent, and standardized. Logistics transactions have gradually infiltrated into daily life, and its market development has entered a stable period, so the focus of logistics development should be transferred from mining incremental customers to improving distribution efficiency and enhancing customer service excellent experience. However, due to the rapid increase of logistics scale and the complexity of the environment and resource conditions that need to be considered, logistics operation and management become more difficult, and with the help of UAV logistics distribution links, it will be a great saving of manpower.

In recent years, a logistics model combining trucks and UAVs has emerged. The addition of UAVs in the distribution process can improve the efficiency of logistics distribution and better meet the growing demand, which will be a hotspot of future research. In this mode, the problem of low power and load of UAV can be effectively solved, the cost of traditional truck delivery can be effectively reduced, and the timeliness and accuracy of delivery can be enhanced. Collaborative delivery of UAVs and trucks will greatly improve the competitiveness of the supply chain, and more artificial intelligence technology will also be added to logistics delivery services in the future.

According to Ref. [19], the collaborative logistics operation mode of UAVs and trucks can be divided into three categories, Synchronized multi-modal (SM) model, Unsynchronized multi-modal (UM) model, and Resupply multi-modal (RM) model, as shown in Figure 1.2. In the UM model, traditional delivery trucks and UAVs move independently, serving customers separately. The SM model pairs UAVs with traditional delivery trucks.

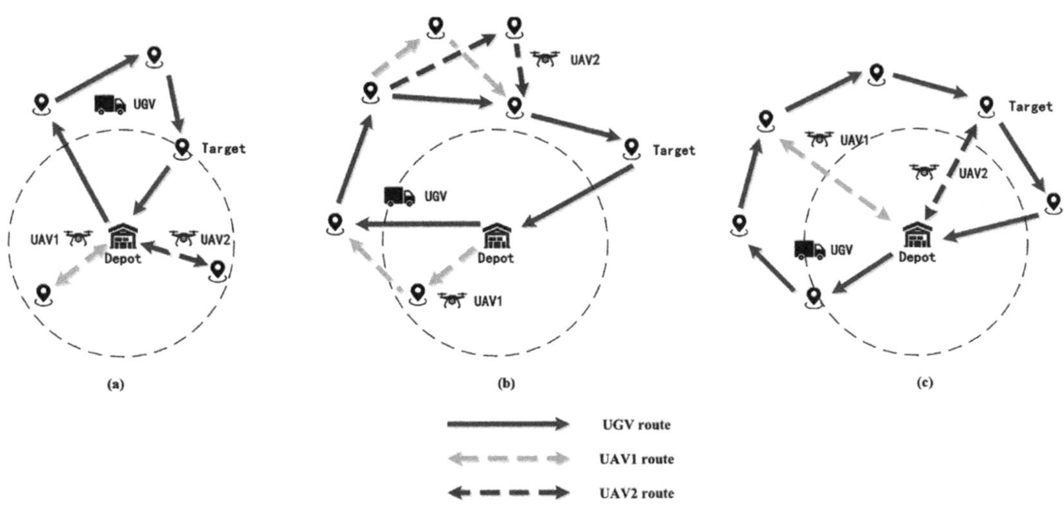


FIGURE 1.2 (a) UM model, (b) SM model, and (c) RM model.

Trucks serve as launching and landing vehicles to deliver UAVs close to customers, enabling them to serve customers within their flight range. RM model is that UAVs provide packages for trucks in real-time, ensuring same-day and even timely delivery service.

The research [20] proposed SM and UM models with a truck and a UAV, the so-called Flying Sidekick Traveling Salesman Problem (FSTSP), and Parallel UAV Scheduling Traveling Salesman Problem (PDSTSP). FSTSP tries to coordinate traditional delivery trucks with UAVs that can be launched from trucks. In the case that UAVs can't fly directly to customers from the distribution center due to their limited flight range, the solution to this problem can give full play to the advantages of UAVs. In situations where a large proportion of customers are located near distribution centers, the PDSTSP solution can provide customers with the best distribution of UAVs and delivery trucks. Due to the NP-hard nature of these problems, small-scale problems can only be optimally solved by a hybrid integer programming solver. They propose a mixed integer programming model and a simple and effective heuristic solution framework for solving large instances of FSTSP and PDSTSP.

The research [21] considered a similar FSTSP model with only a truck and a UAV. They allowed the truck to stay at the launch site or move to another customer location to retrieve the UAV. The problem is presented as a mixed integer linear programming problem model. They proposed two heuristic algorithms and two precise algorithms based on local search and dynamic programming to solve this problem. Heuristics firstly build truck routes and then use greedy partitioning and precision algorithms to build UAV's routes.

The research [22] extended the PDSTSP model by considering two different types of UAV missions: drop and pick. Once the UAV is finished, it can fly back to the warehouse to deliver the next package or directly to another customer to pick it up. It takes into account multiple warehouses, trucks, and UAVs, and assumes that pickup and delivery operations are limited by the customer's time window. The goal is to arrange pickup and drop-off services to minimize travel time. The problem is modeled as a parallel machine scheduling problem and solved by constrained programming. Vehicles traveling near the boundaries of the covered area may be more efficient in serving customers belonging to adjacent warehouses. This problem is uniquely modeled as unrelated parallel machine scheduling with sequence-dependent Settings, precedence relationships, and reentrant, which provides a framework for effectively considering these operational challenges. They proposed a constraint programming approach and tested it with problem instances from M-Truck, M-Drone, M-Depot, and hundreds of customers spread over an 8-square-mile area.

The research [23] introduced the problem of vehicle routing with UAV replenishment, which consists of a primary vehicle (truck) and an auxiliary vehicle. In this model, trucks make regular deliveries to customers, while UAVs redeliver packages for trucks. Once the truck has delivered all the packages, it stays in its last stop position until the UAV resupplies more packages. The goal is to maximize the number of customers served within the specified service time guarantee. Two solution strategies are proposed: the first is a limited resupply strategy, which allows UAVs to resupply trucks only when they have completed all of their previously assigned deliveries. The second approach is a flexible resupply strategy, where UAVs can resupply trucks whenever there is available space on board. The authors show that this RM model yields a considerable improvement over the truck-only model in terms of the number of customers served. However, if a tight service time window is guaranteed, the improvement is minimal. In addition, flexible replenishment strategies produce better results than restricted replenishment strategies.

The research [24] studies the design of a parcel delivery system based on UAVs, including the strategic planning of the system and the operational planning of a given area. The number of payloads affects the battery consumption rate (BCR), which can lead to disruptions in cargo shipments if underestimated during the planning phase, and unnecessarily higher costs if overestimated. Therefore, taking BCR as a function of payload in battle planning optimization, they proposed a reliable UAV package delivery scheme. The minimum set coverage method was used to model strategic planning, and the mixed integer linear programming problem is used to model operational planning. To improve the computation time of solving the combat planning model, they put forward the variable preprocessing algorithm, primal bound and dual bound generation method. The optimal solution provided the minimum number of UAVs and their flight paths to deliver packages while ensuring the safe return of the UAVs is related to battery levels. Experimental data showed that BCR has a linear relationship with payload, and customer access order affects the remaining cost.

### 1.3.3 Cooperative Air and Ground Communication

Low-altitude UAVs are widely used in different fields due to their versatility and high maneuverability. In terms of wireless communication, UAVs can be used as an aerial communication platform for auxiliary communication [25]. By installing a communication transceiver, it can provide communication services for ground targets with high traffic demand and overload. On the other hand, UAVs can also serve as aerial nodes for a variety of applications from cargo transportation to surveillance, which is often referred to as cellular-connected UAVs [26]. However, many of the existing works are limited to the role of UAVs in assisting cellular communications. In most cases today, UAVs are equipped with communications equipment or dedicated sensors for myriad applications, such as low-altitude surveillance, post-disaster rescue, logistics applications, and communications assistance. In addition, to support broadband wireless communication over a large geographical area, a group of UAVs formed FANET and established links with ground nodes for theoretical research and field experimental verification.

The work in Ref. [27] considered a UAV-enabled wireless access network, which acts as an aerial platform to communicate with a group of ground units (GUs) in a variety of practical ways of interest, including uplink data acquisition, downlink data transmission, and uplink and downlink data relay, as shown in Figure 1.3. Under this general framework, two UAV operation scenarios are considered: periodic operation, that is, UAV serves GUs periodically and repeatedly according to a certain trajectory; a one-time operation, where the UAV serves GUs for one flight and then leaves for another mission. In each scenario, the goal was to minimize UAV cycle flight time or mission completion time, while the target data needs of each GU were met through a joint UAV trajectory and communication resource allocation design method. An iterative algorithm for

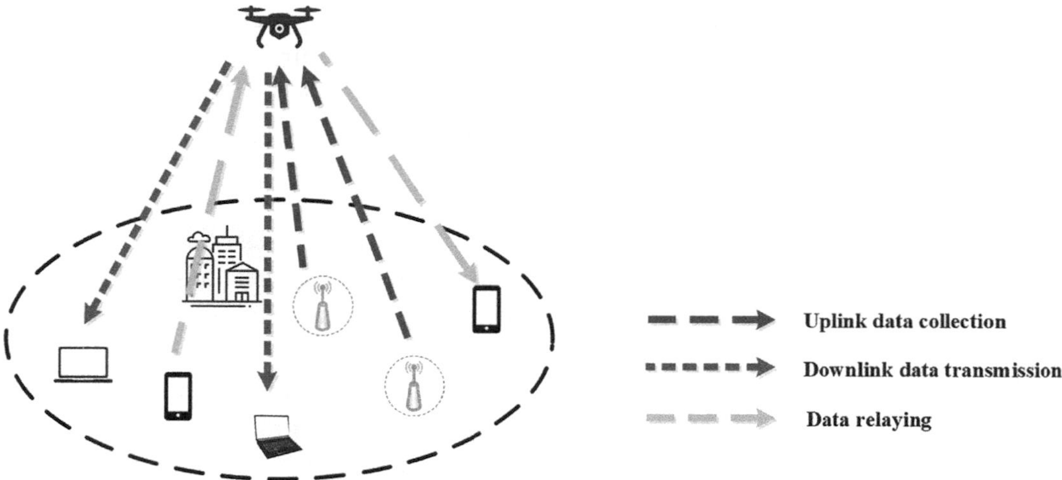


FIGURE 1.3 Support multi-mode communication UAV aerial platform.

solving efficient local optimal solutions is proposed by using successive convex optimization and block coordinate descent techniques. In addition, since the quality of the solution obtained by the proposed algorithm depends heavily on the initial trajectory of the UAV adopted, a new method is proposed to design the initial trajectory of two operation scenarios, using the existing results of the classical traveling salesman problem and the pickup problem. Numerical results show that the proposed trajectory initialization design has a significant performance gain compared with the benchmark initialization based on circular trajectory.

A wireless communication system was studied in Ref. [28]. The UAV supports data transmission of multiple ground nodes while maneuvering over the area of operation. The system considers the propulsion energy consumption of the UAV, so the UAV should operate at a certain speed and acceleration threshold. Through joint optimization of UAV trajectory, velocity, acceleration, and GNs uplink transmitting power, the minimum average rate maximization problem and energy efficiency maximization problem are obtained. Since these problems are generally non-convex, the successive convex approximation scheme is used. By using this method, the non-convex constraints are transformed into suitable convex approximations, and an iterative algorithm converging to the local optimum is proposed. The numerical results showed that the proposed algorithm outperforms the baseline scheme on both problems. In particular, the proposed algorithm achieved a gain of about 109% over the baseline scheme for EE maximization.

### 1.3.4 Other Cooperative Air and Ground Application

With the rapidly growing demand for pervasive communications and computing, ground vehicle networks alone cannot effectively and efficiently meet the demand. In the vision of the B5G/6G era, ground vehicle networks, airborne, and even space information communication technologies infrastructure are expected to be integrated to provide ubiquitous wireless connectivity and computing services anytime, anywhere. To facilitate B5G/6G application scenarios, Cloudlet or other mobile edge computing (MEC) facilities can be installed on UAVs to provide edge computing services from the air [29].

In the era of Internet of Things (IoT), MEC is considered to be a key enabling technology to support computing services of billions of IoT nodes [30]. Since most IoT nodes have limited power and computing compatibility, they can offload computing tasks to the edge of the network to extend battery life and improve computing efficiency. However, many IoT nodes operate in unattended, or challenging areas, such as forests, deserts, mountains, or underwater locations, to perform some computation-intensive applications. In these scenarios, the ground communication infrastructure is sparse and cannot provide reliable communication to nodes.

Therefore, UAVs are used to provide ubiquitous communications and computing support for IoT nodes. The UAV-Mounted Cloudlet can collect and process computing tasks from ground-based IoT nodes that cannot be connected to the edge of the ground. Because UAVs were fully controllable and operate at high altitudes, they could be sent to designated locations to provide efficient on-demand communications and computing services to IoT nodes in a fast and flexible manner [31]. Computing tasks generated by mobile users could be offloaded to nearby edge servers, such as small base stations (BSs) and Wi-Fi access points, to reduce computing latency and computing energy costs for mobile devices. In addition, by pushing traffic, computing, and networking functions to the edge of the network, mobile users could enjoy less task offload time and less backhaul usage [30].

This paper [32] studied the service provisioning problem based on UAV-based MEC. Under the task delay requirements and various resource constraints, service configuration, UAV movement trajectory, task scheduling, and computing resource allocation were jointly optimized to minimize the overall energy consumption of all ground user equipment. The Service Provisioning for UAV-enabled mobile edge computing (SPUN) problem is a non-convex mixed integer nonlinear programming problem because of its non-convexity and complex coupling between mixed integer variables. To solve this challenging problem, they proposed two alternate optimization-based suboptimal solutions with different time complexities. In the first solution with higher complexity in the worst case, the joint service deployment and task scheduling subproblem and UAV trajectory subproblem were solved iteratively by the Branch and Bound method and successive convex approximation, respectively. The optimal solution of the computational resource allocation subproblem was effectively obtained in closed form. In order to avoid the high complexity caused by Branch and Bound method, in the second solution, they proposed a novel approximation algorithm based on relaxation and random rounding techniques to combine service location and task scheduling subproblems, while the other two subproblems solved the same as the first solution. A large number of simulation results showed that the proposed solution could significantly reduce terminal energy consumption compared with the three benchmarks.

However, UAVs have limited power. Therefore, energy consumption is an optimal target in the process of information interaction. Combined with UAVs position optimization algorithm and LSTM-based mission prediction algorithm, an energy efficiency optimization algorithm based on a three-layer computing unloading strategy was proposed in this paper [33]. Experimental results showed that the proposed algorithm and structure could dynamically plan the computational unloading strategy of the MEC network supporting UAVs according to the required delay, UAV height, and data size, thus effectively reducing the energy consumption of UAVs.

On the other hand, the eavesdropper will take less time to attack UAV air-to-ground communications than ground-based wireless communications systems, which presents a significant security challenge. Much work has UAVs acting as secondary nodes to exert interfering noise over the eavesdropper to improve the physical layer security of legitimate communication links. For example, in Ref. [34], a coordinated UAV was deployed to jam the eavesdropper on the ground while other UAVs are transmitting confidential signals to users on the ground. In Ref. [35], the UAV with the best channel state information was selected from multiple UAV relay nodes, and other UAVs are used as jammers. A secure UAV-MEC system was studied in Ref. [36], in which the UAV can access multiple ground BSs to unload part of the computing bits. The total energy consumption of local computation, communication transmission, and flight propulsion was minimized through joint optimization of UAV trajectory, computational task allocation, BSs selection, and transmission power allocation.

In addition to ensuring communication security, the work in Ref. [37] designed an energy-saving calculation and unloading scheme of UAVs with an energy acquisition function, UAV-MEC system, which adopted a full-duplex protocol to receive confidential data from UAVs and broadcast control instructions. In order to improve the energy efficiency of UAV unloading, computational communication resource allocation is optimized so that UAV can consume and harvest energy while minimizing energy consumption. The secret offload rate and computational delay constraints in the worst case were considered to further improve the reliability and security of the system. Since the objective optimization problem is non-convex, they transformed it into a convex problem analytically and derived semi-closed expressions of unloading time, unloading data size, and transmission power, respectively.

Different security mechanisms are designed to deal with different cyberattacks. These mechanisms must keep messages private, prevent external intruders from entering the network, and authenticate nodes. In addition, detection techniques should be used to monitor nodes for improper behavior.

The paper [38] proposed a zero-sum network interception game to solve the network physical security threat of UAV communication systems. This plan considered the scenario of a supplier and an attacker, where the supplier lookup to find the path for UAVs between two different point locations, while the attacker's goal was to find the optimal location for the UAV to attack, resulting in the network or physical damage along the path, to increase the travel time of the UAVs. Therefore, the game could ensure the network security of the drone delivery system. However, due to processing tasks and verification, it increased the delivery time.

To be able to detect any malicious anomalies targeting the network, in [39], a hierarchical intrusion detection scheme was proposed, which was implemented on both UAVs and ground BSs. The work aims to classify monitoring UAVs and threats to detect malicious activities by combining behavior-based detection and rule-based attack. Therefore, based on these two mechanisms, each UAV had the ability to act as an observer to hear all traffic in the neighborhood to detect any malicious activities. Moreover, a high level of security is demonstrated with a high detection rate.

## 1.4 CONCLUSION

Air-ground coordination is becoming a reality, leading to the emergence of numerous UAV-UGV Cooperation Systems. This chapter classifies the different applications, each with its own characteristics, characteristics, and competitive advantages. In fact, every UAV-UGV collaboration has many challenging problems and deficiencies that need further study. For example, each application should carefully consider the different constraints associated with using drones and how to resolve them. In addition, because drones must be deployed in a given area, consideration must be given to the surrounding environment and wireless interaction with other devices. UAV network is vulnerable to different malicious attacks due to its dynamic topology and open links, and its security is equally important. To sum up, Air-Ground coordination is still in the early stages of research, and more research is needed to reach a complete research stage.

## REFERENCES

1. P. B. Sujit and D. Ghose, "Search using multiple UAVs with flight time constraints," *IEEE Transactions on Aerospace and Electronic Systems*, vol. 40, pp. 491–509, 2004.
2. P. Maini and P. B. Sujit, "On cooperation between a fuel constrained UAV and a refueling UGV for large scale mapping applications," in *2015 International Conference on Unmanned Aircraft Systems (ICUAS)*, Denver, Colorado, USA, 2015, pp. 1370–1377.
3. A. Agogino, C. HolmesParker, and K. Tumer, "Evolving large scale UAV communication system," in *Proceedings of the 14th Annual Conference on Genetic and Evolutionary Computation*, Philadelphia, Pennsylvania, USA, 2012, pp. 1023–1030.
4. Z. Zhang, Y. Wang, Y. Yao, and Y. Wang, "Implementation of UAVs communication network based on dynamic TDMA MAC protocol," in *2010 3rd International Conference on Advanced Computer Theory and Engineering(ICACTE)*, Chengdu, China, 2010, pp. V6-551–V6-554.
5. P. M. Olsson, J. Kvarnström, P. Doherty, O. Burdakov, and K. Holmberg, , "Generating UAV communication networks for monitoring and surveillance," in *2010 11th International Conference on Control Automation Robotics & Vision*, 2010, Singapore, pp. 1070–1077.
6. A. Ghaffarkhah and Y. Mostofi, "Dynamic networked coverage of time-varying environments in the presence of fading communication channels," *ACM Transactions on Sensor Networks (TOSN)*, vol. 10, pp. 1–38, 2014.
7. M. A. Hsieh, A. Cowley, V. Kumar, and C. J. Taylor, "Maintaining network connectivity and performance in robot teams," *Journal of Field Robotics*, vol. 25, pp. 111–131, 2008.
8. C. Sabo, D. Kingston, and K. Cohen, "A formulation and heuristic approach to task allocation and routing of UAVs under limited communication," *Unmanned Systems*, vol. 2, pp. 1–17, 2014.
9. T. P. Nascimento, P. M. António, and G. S. C. André, "Multi-robot nonlinear model predictive formation control: Moving target and target absence," *Robotics & Autonomous Systems*, vol. 61, pp. 1502–1515, 2013.
10. L. E. Barnes, M. A. Fields, and K. P. Valavanis, "Swarm formation control utilizing elliptical surfaces and limiting functions," *IEEE Transactions on Systems, Man, and Cybernetics, Part B (Cybernetics)*, vol. 39, pp. 1434–1445, 2009.

11. T. P. Breckon, A. Gaszczak, J. Han, M. L. Eichner, and S. E. Barnes, "Multi-modal target detection for autonomous wide area search and surveillance," in *Emerging Technologies in Security and Defence; and Quantum Security II; and Unmanned Sensor Systems X*, Dresden, Germany, 2013, p. 889913.
12. L. Sheng, Y. Pan, and X. Gong, "Consensus formation control for a class of networked multiple mobile robot systems," *Journal of Control Science and Engineering*, vol. 2012, pp. 1, 2012.
13. J. Chen, X. Zhang, B. Xin, and H. Fang, "Coordination between unmanned aerial and ground vehicles: A taxonomy and optimization perspective," *IEEE Transactions on Cybernetics*, vol. 46, pp. 959–972, 2015.
14. B. Grocholsky, J. Keller, V. Kumar, and G. Pappas, "Cooperative air and ground surveillance," *IEEE Robotics & Automation Magazine*, vol. 13, pp. 16–25, 2006.
15. B. Hament and P. Oh, "Unmanned aerial and ground vehicle (UAV-UGV) system prototype for civil infrastructure missions," *2018 IEEE International Conference on Consumer Electronics (ICCE)*, 2018, pp. 1–4.
16. H. Yu, K. Meier, M. Argyle, and R. W. Beard, "Cooperative path planning for target tracking in urban environments using unmanned air and ground vehicles," *IEEE/ASME Transactions on Mechatronics*, vol. 20, pp. 541–552, 2014.
17. E. Barmpounakis and N. Geroliminis, "On the new era of urban traffic monitoring with massive drone data: The pNEUMA large-scale field experiment," *Transportation Research Part C: Emerging Technologies*, vol. 111, pp. 50–71, 2020.
18. M. Hu, W. Liu, K. Peng, X. Ma, W. Cheng, J. Liu, and B. Li, "Joint routing and scheduling for vehicle-assisted multidrone surveillance," *IEEE Internet of Things Journal*, vol. 6, pp. 1781–1790, 2018.
19. M. Moshref-Javadi and M. Winkenbach, "Applications and Research avenues for drone-based models in logistics: A classification and review," *Expert Systems with Applications*, vol. 177, p. 114854, 2021.
20. C. C. Murray and A. G. Chu, "The flying sidekick traveling salesman problem: Optimization of drone-assisted parcel delivery," *Transportation Research Part C: Emerging Technologies*, vol. 54, pp. 86–109, 2015.
21. N. Agatz, P. Bouman and M. Schmidt, "Optimization approaches for the traveling salesman problem with drone," *Transportation Science*, vol. 52, pp. 965–981, 2018.
22. A. M. Ham, "Integrated scheduling of m-truck, m-drone, and m-depot constrained by time-window, drop-pickup, and m-visit using constraint programming," *Transportation Research Part C: Emerging Technologies*, vol. 91, pp. 1–14, 2018.
23. I. Dayarian, M. Savelsbergh and J. Clarke, "Same-day delivery with drone resupply," *Transportation Science*, vol. 54, pp. 229–249, 2020.
24. M. Torabbeigi, G. J. Lim and S. J. Kim, "Drone delivery scheduling optimization considering payload-induced battery consumption rates," *Journal of Intelligent & Robotic Systems*, vol. 97, pp. 471–487, 2020.
25. M. Mozaffari, W. Saad, M. Bennis, and M. Debbah, "Wireless communication using unmanned aerial vehicles (UAVs): Optimal transport theory for hover time optimization," *IEEE Transactions on Wireless Communications*, vol. 16, pp. 8052–8066, 2017.
26. M. Mozaffari, A. T. Z. Kasgari, W. Saad, M. Bennis, and M. Debbah, "Beyond 5G with UAVs: Foundations of a 3D wireless cellular network," *IEEE Transactions on Wireless Communications*, vol. 18, pp. 357–372, 2018.
27. J. Zhang, Y. Zeng and R. Zhang, "UAV-enabled radio access network: Multi-mode communication and trajectory design," *IEEE Transactions on Signal Processing*, vol. 66, pp. 5269–5284, 2018.
28. S. Eom, H. Lee, J. Park, and I. Lee, "UAV-aided wireless communication designs with propulsion energy limitations," *IEEE Transactions on Vehicular Technology*, vol. 69, pp. 651–662, 2019.

29. Z. Tan, H. Qu, J. Zhao, S. Zhou, and W. Wang, "UAV-aided edge/fog computing in smart iot community for social augmented reality," *IEEE Internet of Things Journal*, vol. 7, pp. 4872–4884, 2020.
30. M. Li, N. Cheng, J. Gao, Y. Wang, L. Zhao, and X. Shen, "Energy-efficient UAV-assisted mobile edge computing: Resource allocation and trajectory optimization," *IEEE Transactions on Vehicular Technology*, vol. 69, pp. 3424–3438, 2020.
31. N. Cheng, W. Xu, W. Shi, Y. Zhou, N. Lu, H. Zhou, and X. Shen, "Air-ground integrated mobile edge networks: Architecture, challenges, and opportunities," *IEEE Communications Magazine*, vol. 56, pp. 26–32, 2018.
32. Y. Qu, H. Dai, H. Wang, C. Dong, F. Wu, S. Guo, and Q. Wu, "Service provisioning for UAV-enabled mobile edge computing," *IEEE Journal on Selected Areas in Communications*, vol. 39, pp. 3287–3305, 2021.
33. G. Wu, Y. Miao, Y. Zhang, and A. Barnawi, "Energy efficient for UAV-enabled mobile edge computing networks: Intelligent task prediction and offloading," *Computer Communications*, vol. 150, pp. 556–562, 2020.
34. X. Zhou, Q. Wu, S. Yan, F. Shu, and J. Li, "UAV-enabled secure communications: Joint trajectory and transmit power optimization," *IEEE Transactions on Vehicular Technology*, vol. 68, pp. 4069–4073, 2019.
35. T. Shen and H. Ochiai, "A UAV-aided selective relaying with cooperative jammers for secure wireless networks over rician fading channels," in *2019 IEEE 90th Vehicular Technology Conference (VTC2019-Fall)*, Honolulu, HI, USA, 2019, pp. 1–5.
36. Z. Lv, J. Hao and Y. Guo, "Energy minimization for MEC-enabled cellular-connected UAV: Trajectory optimization and resource scheduling," in *IEEE INFOCOM 2020-IEEE Conference on Computer Communications Workshops (INFOCOM WKSHPS)*, Toronto, ON, Canada, 2020, pp. 478–483.
37. X. Gu, G. Zhang, M. Wang, W. Duan, M. Wen, and P. Ho, "UAV-aided energy efficient edge computing networks: Security offloading optimization," *IEEE Internet of Things Journal*, vol. 9, no. 6, pp. 4245–4258, 2021.
38. A. Sanjab, W. Saad, and T. Başar, "Prospect theory for enhanced cyber-physical security of drone delivery systems: A network interdiction game," in *2017 IEEE International Conference on Communications (ICC)*, 2017, pp. 1–6.
39. H. Sedjelmaci, S. M. Senouci, and N. Ansari, "A Hierarchical Detection and Response System to Enhance Security Against Lethal Cyber-Attacks in UAV Networks," *IEEE Transactions on Systems, Man, and Cybernetics: Systems*, vol. 48, pp. 1594–1606, 2018.

CHAPTER 2

# Autonomous and Connected UAVs/Drones

Kirshna Kumar

*Government of India*

Sushil Kumar

*Jawaharlal Nehru University*

Rupak Kharel

*University of Central Lancashire*

## CONTENTS



## LIST OF ABBREVIATIONS

**BCT** Block chain technology
**CIA** Confidentiality integrity and availability
**CPS** Cyber physical security
**CSMA/CA** Carrier sense multiple access/collision avoidance
**FC** Fog computing
**IoT** Internet of things
**LWA** Light weight authentication

DOI: 10.1201/b22998-2

| | |
|---|---|
| **LoS** | Line of sight |
| **PDR** | Packet delivery ratio |
| **QoS** | Quality of service |
| **SCADA** | Supervisory control and data acquisition |
| **UAVs** | Unmanned aerial vehicles |
| **WSNs** | Wireless sensor networks |

## 2.1 INTRODUCTION

Various applications of Internet of Things (IoT) such as smart home, smart city, intelligent transportation, green mobility, smart agriculture and aerial communication using connected aerial vehicle such as UAVs and drone will improve all aspects of day-to-day life [1,2]. Flying robots or drones are some examples of Unmanned Aerial Vehicles (UAVs), equipped with processors, sensors and wireless connectivity having the characteristics of self-programmed and remotely administered via group or individual [3]. Characteristics such as image definition and high flexibility in sensor-equipped UAVs promote the usages of UAV-based IoT to make modern applications smarter [4]. The applications of UAVs are wide in various areas such as remote monitoring, smart logistics, 3D mapping, military, rescue operations, cinematography because of high mobility, dynamic configuration and low maintenance cost of UAVs [5,6]. The researchers have not fully explored the various research issues such as energy, flying communication and security vulnerabilities in UAV network despite of rapidly growing usage of UAVs in several areas over the last few years [7].

This chapter presents a comprehensive review on autonomous and connected aerial vehicles such as UAVs/Drones in IoT environments focusing on research issue-based studies. The issue-based techniques in aerial environment have been categorically described focusing on their limitations and technical contributions while considering various research metrics. Specifically, the contributions of this chapter can be presented into the following aspects:

- Firstly, the significance of autonomous and connected aerial vehicles in IoT environments is highlighted while focusing on various use case implementations.
- Secondly, following the issues-based taxonomy, a technique-based qualitative review is carried out on various research challenges raised in aerial environments while focusing on main functional components, weaknesses and strengths.
- Finally, open research challenges related to autonomous and connected aerial vehicles: UAVs and drones are identified as future research directions in the area.

The remaining of this chapter is structured as follows: Section 2.2 highlights the significance of autonomous and connected aerial vehicles in connected drone environments considering various use cases including in precision agriculture, medical drones for COVID-19 pandemic management and traffic drones for on-road traffic management and enabling traffic services. Section 2.3 revisits the related literature surveys on autonomous and connected aerial vehicles UAVs or drones. In Section 2.4, a comprehensive review on autonomous and connected UAVs in IoT environment is presented focusing on issues-based taxonomy and model-based discussion. Section 2.5 identifies open research issues and challenges related to connected vehicles under aerial environments, followed by conclusions described in Section 2.6.

## 2.2 SIGNIFICANCE OF UAV USAGES IN IoT USE CASES

In this section, the significance of UAV usages in IoT environments is described, while considering a number of use case implementations.

### 2.2.1 Case Study: Smart Agriculture

UAV technology contributes a significant part to smart agriculture, with various usages such as disease detection in crops, advanced irrigation system, smart use of fertilizers and pesticides, field-level phenotyping and agricultural monitoring. A novel technique is introduced to register the images of agricultural crops taken by UAV [8]. In this technique, three-dimensional clouds on the field are aligned, and then 3D models for the crop are reconstructed for the monitoring of growth parameters on a plant-level basis. Genetic Algorithm and Particle Swarm Optimization is combined for utilizing the limited resources related to machinery equipment in farming system consisting of multiple components/UAVs/agents [9]. Figure 2.1 shows a UAV-based smart irrigation system, with task automation using the data gathered by the UAVs.

Another agriculture use case of drones has been investigated for identifying pests in fruit trees so that better pesticide spraying can be carried out using group of drones in precision agriculture [10]. A deep learning-enabled pest identification method has been developed which is embedded in drones for real-time pest position identification in agriculture field. Specifically, drones take photograph of the field for identifying actual pest positions. These pest positions are then used to plan optimal pesticide spraying routes for drones. This is a very potential agriculture application of drones enabling a range of benefits to farmers as well as environmental benefits such as accurate pesticide spraying, early detection reduce pesticide probability and lower energy consumption with optimal

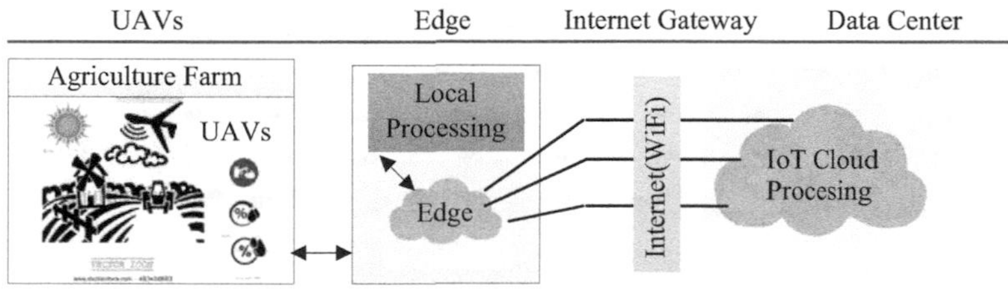


FIGURE 2.1 UAV-based smart irrigation system.

route planning. An example pesticide spraying drone's scenario is presented in Figure 2.2 where a group of drones are carrying out pest identification initially, and planning optimal route for pesticide spraying later. The pests' information from drones is then transmitted to nearest server which is connected to Internet for real-time data analysis and prediction in pesticide control in agriculture.

The technical architecture has been divided into two major parts including pests position data collection, and pesticide spraying route planning. In the initial flying

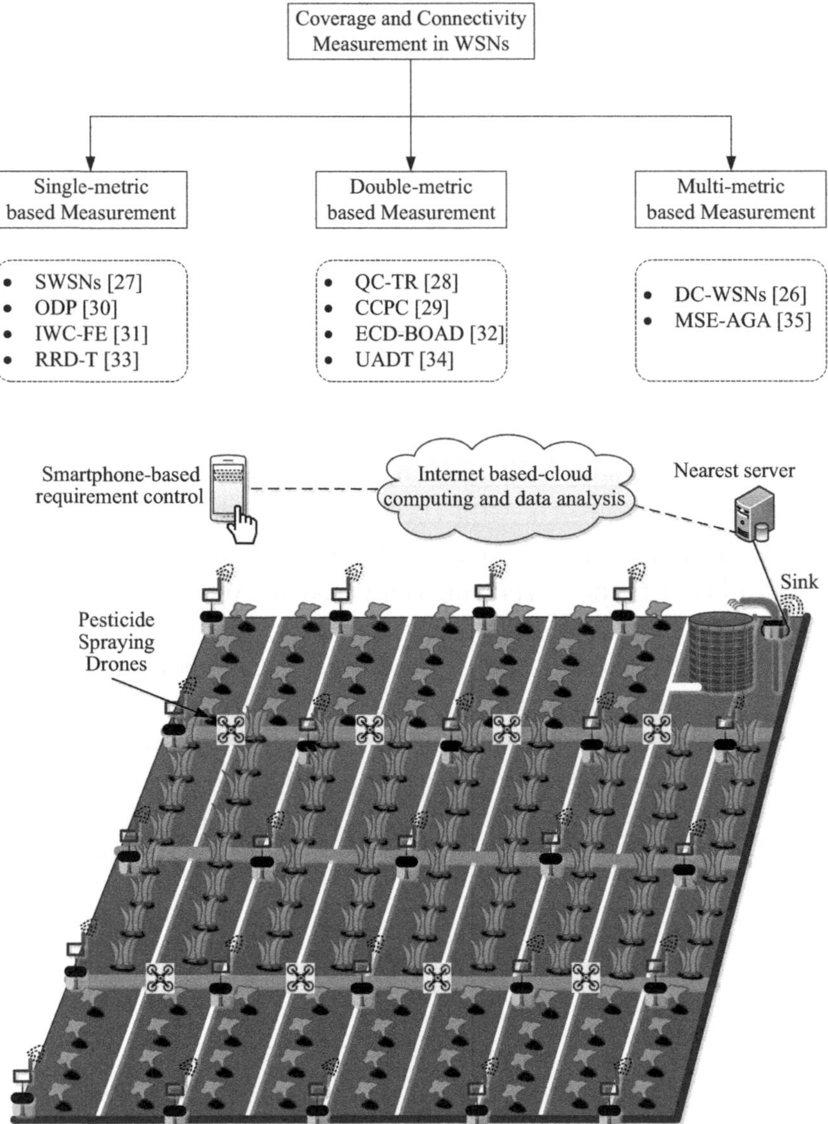


FIGURE 2.2 Pesticide spraying using drones in precision agriculture.

rounds, the drones basically take photographs of the fruit trees which are processed real time by an embedded deep learning-enabled platform in drones. The type of pests and position of pests in the agriculture field is then processed and distributed to the server, and cloud analysis tools. Once pests' identification and localization have been processed, an optimal pesticide spraying strategy is then developed at server side. The 3D geographical information of agriculture field is utilized for developing the accurate and optimal pesticide spraying path plan towards optimizing the cost and time in spraying apart from effectiveness in saving fruit trees from pests. The server communicates the spraying strategy to drones which performs the operation in agriculture field.

### 2.2.2 Case Study: COVID Disease Management

Drone-assisted COVID-19 health monitoring system has a prominent usage for monitoring respiratory and heart rates, temperature amongst crowded places such as workforces, potential risk groups, i.e., convention centres, seniors care facilities to handle pandemic situations. Wearable sensors are utilized for recording the observations in Body Area Networks through a push–pull data fetching technique. The substantial amount of data is gathered and stored in a stipulated time period, which provides help to take appropriate and timely action as per requirement [11]. In real-time drone-assisted COVID-19 health monitoring system implementation, observation proves the sanitization, patient identification and thermal image collection are possible while covering large area within a short time period through an aerial route.

Enabling COVID-19 vaccination and testing facilities in rural areas is another medical use case of drones [12]. Deep neural network technique has been used to provide real-time COVID-19 testing embedded as part of the medical drone's payload. An automated disinfection system is also attached along with medical sensors in the medical drone's payload. Intelligent path planning for test kit distribution, automatic drone battery recharge and automatic disinfection refilling is also attached as the medical drone's payload. Technically, four objectives were achieved towards developing this medical drone for COVID-19 pandemic handling. In objective 1, the proposed medical drone is enabled by two major modules including sensor-based user authentication and automatic spraying of disinfection. In objective 2, a deep learning-enabled COVID-19 detection module is developed for enabling the medical drone. A subsequent kit dispensing strategy is also embedded in case of high likelihood of positive symptoms. In objective 3, optimal travel path planning module for drone is developed considering rural delivery testing locations and battery recharge stations for drones. In objective 4, all the modules have been integrated as drone payload system and tested to enable the proposed medical drone. Overall, the COVID-19 centric drone use case can also be utilized for handling other types of medical emergency particularly in rural areas.

### 2.2.3 Case Study: Road Safety

The main applications of UAVs-based road safety system include risk assessment, accident investigation and road surveillance. In risk assessment, vehicle trajectories are analysed in detail through UAV-based videos for the identification of risky lane changes manoeuvres, potential conflicts, etc. [13]. Based on photographs and footages acquired from UAVs, accident investigation is performed by utilizing techniques and application of vision algorithms for reconstructing accident scenes.

## 2.3 LITERATURE SURVEY

A survey on various strategies and protocols suggested to attain green IoT utilizing UAV environment has been performed including challenges, opportunities and use cases [14]. The concept, discussion and focus of this survey are different from other reviews for sustainable and reliable smart world. A review on UAVs/Drones usage in various domains (military, civilian, etc.) is presented while analysing the exploitation of vulnerabilities inside communication links, hardware and smart devices to investigate cyber threats [15]. This review facilitates the ethical hackers to adopt and suggest novel technologies and techniques for improved detection and protection of UAV attacks. A survey of significant research related to the usage of UAV and IoT technologies in smart agriculture including scenarios such as pesticide usage, irrigation, monitoring of plant growth, weed management and disease management was provided [16]. The key principles of IoT technologies such as IoT sensors, networks and suggested techniques utilized in agriculture are described. A comprehensive review on blockchain-based security strategies and security challenges that arise in 5G-enabled UAV environment is presented [17]. A case study for implementing blockchain with UAVs for the security of industrial applications is presented with the critical research challenges. The routing techniques in the UAV environment including topology-based, hierarchical, stochastic, social network-based, deterministic and position based are reviewed while providing their comparative analysis and research challenges [18].

The main UAV challenges such as energy consumption, channel modelling and 3D deployment with potential applications are surveyed while describing several mathematical tools and analytical frameworks such as machine learning, transport theory, optimization theory, game theory and stochastic geometry [19]. In addition, the main guidelines are provided for analysing, designing and optimizing UAV-based IoT communication systems. A comprehensive survey regarding the limitation of existing techniques and potential usage of artificial intelligence in UAV-enabled networks for solving several problems related to drones/UAVs is presented [20]. Various aspects of cellular communication such as consumer UAVs types, issues related to interference and its solution, novel regulations suggested for the management of commercial UAV usage and Cyber Physical Security are reviewed in UAV environment [21]. An envisioned architecture for delivering the value-added IoT services based on UAVs in aerial networks is suggested and UAV-enabled applications and related issues are surveyed [22]. A survey on hand over management of connected drones and upcoming related research issues and their possible solutions in highly mobile heterogeneous networks using 6G technologies are provided [23].

## 2.4 ISSUES-BASED TECHNIQUES IN UAV SYSTEM

In this section, various issues-based techniques in the UAV system are qualitatively reviewed based on taxonomy shown in Figure 2.3. The techniques proposed by researchers have been categorized based on five issues faced in the UAV system. These five issues are named as data collection, computing, routing, security and energy efficiency. Edge and fog/cloud computing are sub-categorization of computing techniques. Existing routing techniques are sub-categorized in QoS provisioning and QoS non-provisioning. Blockchain non-assisted and Blockchain assisted are two more categorizations of security techniques in UAV system.

### 2.4.1 Data Collection Techniques

In this section, the existing data collection techniques are further classified into Wireless Sensor Network (WSN)-based and IoT-based data collection techniques.

#### *2.4.1.1 WSN-Based Data Collection Techniques*

Hybrid Medium Access Control Protocol [24] has been proposed in WSN-based UAV network for data collection while increasing the fairness among UAV nodes. This protocol utilizes CSMA/CA for the registration process of the sensor nodes and determines the transmission schedule of sensor node based on time slot scheme for data acquisition. A Compressive Data Gathering [25] based on projection has been suggested to acquire data in WSN-based UAV while aggregating data from a large number of sensor nodes to elected projection sensor nodes which acts as cluster heads. The primary focus of this technique is clustering of sensor nodes, construction of an optimal forwarding tree per cluster and collection of data from elected cluster head with shorter UAV trajectory distance.

#### *2.4.1.2 IoT-Based Data Collection Techniques*

Sink Rotating joint UAV Data Gathering [26] framework has been proposed while dividing the data collection process into two organic components: (1) S-SCADA (Supervisory Control And Data Acquisition) in-network data gathering, (2) utilize UAV to gather data between S-SCADA networks. A novel idea of great full coverage sub-graph having

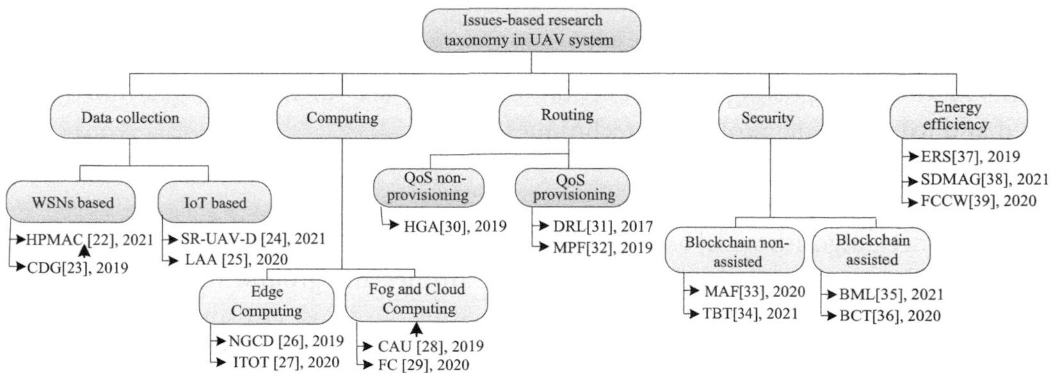


FIGURE 2.3 Issue-based research taxonomy in UAV system.

complete connectivity has been introduced while considering UAVs as mobile sink for generating candidate areas for data gathering in Lightweight Authority Authentication model [27]. Further, traditional K-centre problem has been solved while formulating as a multi-objective joint optimization having various constraints.

A summary of related literature studies on data collection and routing techniques for UAV system is described in Table 2.1, which comprises characteristics/protocol, issues, contributions, techniques, simulation tools utilized, metrics and research limitations of each protocol reviewed.

### 2.4.2 Computing Techniques

This section describes the categorization of existing data computing technique into edge and cloud/fog computing techniques.

#### *2.4.2.1 Edge Computing Techniques*

A non-cooperative game based cyber defence [28] mechanism has been developed for protecting the vehicle edge computing network from offloading threats. Various computation metrics such as computation overhead and node energy are taken into account. An Intelligent Task Offloading technique [29] has been suggested for UAV edge computing network having the capability of perceiving the network's environment in intelligent manner. On the basis of deep Monte Carlo tree search, offloading decision trajectories have been simulated to maximize the reward like lowest power consumption or delay.

#### *2.4.2.2 Cloud/Fog Computing Techniques*

A cloud-assisted UAV [30] protocol has been suggested to incorporate the computation ability of terrestrial cloud in UAV network. In this protocol, cloud-assisted system is analysed on demand need basis as well as UAV's control mechanism is impacted. A fog computing (FC) [31] based novel technique has been suggested while generating synthetic and large dataset for semantic and geometric urban usage. This technique consists of two key steps: modelling and rendering processed by UnrealEngine4 and CityEngine for obtaining synthetic aligned multi-model data. Still the technique has a scope of generating rich synthetic and more accurate data in future.

### 2.4.3 Routing Techniques

In this section, the existing data collection techniques are further classified into WSN-based and IoT-based data collection techniques.

#### *2.4.3.1 QoS Non-provisioning Routing*

A hybrid genetic algorithm [32] based scheduling and routing technique has been suggested permitting the communication between multiple drones and ground vehicle for efficient parcels delivery. Further, this algorithm comprises low visit cost crossover method, population initialization and hierarchical education algorithm for distributing fairly inside population, which avoids premature convergence and minimizes delay. But in this technique, there is no surety of providing optimized link lifetime.

TABLE 2.1 Summary of Data Collection and Routing Technique for UAV System

| Protocols | Characteristics | | | | | |
|---|---|---|---|---|---|---|
| | Issues | Contributions | Techniques | Simulators | Metrics | Limitations |
| HPMAC | Degraded fairness for data collection | Enhanced fairness and network performance | Hybrid medium access using CSMA/CA | Matlab | PDR<br>Delay<br>Fairness | Interference and energy consumption not evaluated |
| CDG | More energy consumption in data gathering | Energy efficient and robust solution | Projection-based compressive data gathering | Java-based simulator | Energy<br>Lifetime | Does not consider other QoS metrics |
| SR-UAV-DG | Increased maximum time cost of UAVs | Minimized maximum time cost in data collection | Fine grained trajectory plan technique | Matlab | Bandwidth<br>Time | Only consider to time metric |
| LAA | Issue of trust in mass data collection by UAVs | UAV-aided trustworthy data collection and traffic balancing | Light weight authority authentication in UAVs | Matlab | Trust | Limited operating efficiency |
| HGA | Issue of more delay and convergence | Avoid premature convergence and minimize delay | Hybrid genetic algorithm based scheduling and routing | | Delay | No surety of providing optimized link life time |
| DRL | Issue of energy consumption and reduced connectivity | Increased connectivity of drones and energy efficiency | Deep reinforcement learning | Network simulator 2 | Energy connectivity | Limited to centralized network |
| MPF | Issue of less connectivity and more delay | Increased link connectivity and reduced delay | Motion-assisted packet forwarding | Matlab | Delay<br>Connectivity | Does not consider energy efficiency and load balancing |

#### *2.4.3.2 QoS Provisioning Routing*

Connectivity of drone network and energy utilization is maximized with the introduction of a deep reinforcement learning-based technique for controlled 3D movement of more than one drones [33]. A reward function is provided based on QoS requirement and coverage fairness. When the value of reward function reduces drastically, the penalty is imposed for the disconnected network. For disconnected drone networks, a penalty is imposed with the reduction of the reward function value rapidly. Deep reinforcement learning is applicable for centralized network rather than for decentralized network. A motion-assisted packet forwarding technique [MPF] is proposed in the aerial networks with the utilization of two predictive heuristics for integrating delay tolerant, and location-based routing [34]. Motion-assisted packet forwarding technique only considers link connectivity, and delay, but does not focus on energy efficiency and load balancing.

### 2.4.4 Security Techniques

#### *2.4.4.1 Blockchain Non-assisted Techniques*

A multi-agent framework [35] has been suggested utilizing Artificial Immune System for the detection of the attacking UAV and selection of the safest route. In this framework, initially a route request packet is sent from source UAV to destination UAV for the detection of existing routes. Further, after receiving route reply packet, a self-protective technique is utilized while utilizing agents and knowledge base for the selection of the safest path and detection of attacking UAVs. A trust-based technique [36] has been proposed for securing the UAVs which determine the trust value assigned to sensors for checking the accuracy and correctness of the data. Trust value determination is based on direct, indirect, and final trust value determination in this technique.

#### *2.4.4.2 Blockchain-Assisted Techniques*

A blockchain and machine learning-based decentralized framework [37] has been proposed in UAV system for enhancing the performance of UAVs in terms of integrity in data storage. In this technique, blockchain is used for achieving decentralized predictive analytics so that machine learning models can be applied and shared in decentralized manner. Security of VC-based device data has been enhanced while suggesting blockchain technology (BCT) [38] assisted technique in which evaluation is performed in virtual UAV monitoring system. In this technique, data storage is done in an Ethereum-based blockchain for enabling seamless BCT transactions

### 2.4.5 Energy Efficiency Techniques

An energy efficiency based resource allocation technique [39] has been proposed to enhance quality of experience while storing UAVs offloading data having separate quality of experience requirement and relaying in multi-queueing environment. In this technique, satisfaction function is developed for energy efficiency and performance enhancement in UAVs. A Semi Dynamic Mobile Anchor Guiding approach [40] has been suggested to enhance the energy utilization in localization of large number of IoT devices and trajectory planning for drone networks. The proposed technique reduces energy consumption and data collection latency. A formation collision co-awareness technique [41] has been developed with the utilization of thin plate splines scheme for reducing deformation of swarm's formation. This technique utilizes a mapping function to minimize lag with the enhancement of energy efficiency.

## 2.5 OPEN RESEARCH CHALLENGES IN CONNECTED UAV SYSTEM

Many research issues related to the UAV system have been resolved and isolated by the researchers, but several research issues that still need attention are as follows:

- *Range and the Payload of Flights*
  Because of payload of a drone being between 2 and 4 kg, there is a possibility of failure for a drone's various components during flights which might result in a crash and may damage property or harm people. To overcome this type of payload-centric incidents various enhancements are required on drones for integrating enhanced battery technology, customizable apps for tablets and smartphones. This will result in improvements in-flight durations, reliability, better utilization capability of cameras and sensors required in various applications.

  Range and payload-centric drone development can also be benefitted from connected drones' communication where one drone can support the other drone for both enlarging communication range as well as sharing some payload [37]. This can be effectively implemented with more automation in drone operation such as on the flight payload transfer between drones, and automatic battery power sharing wirelessly. Developing use case centric drones is also needed for better payload and providing range guarantee. However, payload and range are also quite related to drone governmental regulations, and therefore, drone development needs legal advice before increasing payload support, and enabling long-range drone flights.

- *Resource Management for UAVs*
  The management of resources such as energy, bandwidth, UAV's flight time, transmit power and number of UAVs is also a big research issue in UAV-based aerial environment. A framework able to manage these resources among others dynamically is required. For instance, the challenge is to adjust the trajectory related to flying UAV serving ground users and the transmit power adaptively. Therefore, in this case, providing the optimal bandwidth allocation algorithms to capture the impact of mobility, the location of UAV, traffic distribution and LoS interference are the main issues.

  Resource management in drones would significantly benefit for battery capacity development R&D as it will ease in-flight resource management in drones. Greater battery capacity with smaller in size is one of the major requirements for attachable batteries in drones. Effective use of transmit power with controlled communication techniques can significantly enhance the battery resource utilization in drones. However, this controlled communication approach might increase computing resource requirement in drones. Therefore, a better balance between energy and computing resource utilization is needed for effectively enlarging battery life in drones.

This area of research is benefitted from the transmit power control centric wireless communication research.

- *Trajectory Optimization of UAVs*
  In a UAV-assisted aerial environment, UAV trajectory requires to be optimal in terms of main performance parameters including energy, throughput, delay and spectral efficiency as well as for type of UAVs and dynamic aspects. Despite numerous detailed studies about UAV trajectory optimization, still there exist various open research issues including mobility patterns of ground users-based trajectory optimization to maximize the coverage performance considering metrics such as energy consumption and user's delay.

  The drone trajectory optimization research is quite different from the traditional transport path planning research. Drone path optimization is a 3D travel path palling research where apart from distance to destination the height of travel path is important. It is also a real-time optimization research rather than distance-based static optimization considering probability of obstacles in 3D space, object detection and avoidance in real time. Therefore, trajectory optimization is quite complex research theme in drones-related R&D requiring image processing as well as travel path planning at the same time. Machine learning-oriented image processing research and heuristic optimization-based path palling research themes have the potential to benefit in drone trajectory optimization research and developments.

- *Performance Analysis*
  For the analysis of performance, there are several open issues that required to be focused. The performance of UAV-assisted aerial network consisting of both terrestrial and aerial users, and the base stations must be characterized in the form of capacity and coverage. Moreover, there is a requirement of fundamental performance analysis for capturing balanced trade-offs between energy and spectral efficiency inside UAV system while incorporating the UAV's mobility.

  Use case centric drone performance analysis is another research theme needing attention of researchers. Here, drone performance of individual components will be recorded in black box kind of module in drones which can be analysed real time at cloud servers. This can significantly improve the services enabled by drone networking scenarios as this performance analysis will enable guaranteed services to consumers for the service provider companies. Performance guarantee will also enable greater consumer confidence in drone services resulting in bigger market for drone-related business opportunity. Therefore, operation flight performance analysis of drones in real time has significant impact on enabling connected drones' services.

- *Security*
  Because of UAVs characteristics of carrying and transferring a lot of information through communication channels, they are much prone to several numbers and types of security threats. Even the expensive professional drones are not secure. Various attacks focus on affecting the security triad CIA including Confidentiality

Integrity and Availability. Such information exchange by UAVs through communication channels should be secured with the development of more advanced security techniques.

Towards enabling a wide range of drone use cases security of drone communication is a very potential research theme. Security of drone communication is quite different from the traditional security challenges considering the constraints in a drone networking environment. Here, lightweight security techniques get the preference over cryptographic security due to the computing resource-related constraints in drone networking scenarios. However, maintaining high level of security with lower volume of computing resource requirements is a quite challenging task in drone communication security research. Implementing some new security techniques such as BCT requires sincere efforts of researchers for drone communication considering connected drone networking environments.

## 2.6 CONCLUSIONS

Usage of connect aerial vehicles: UAVs or drones are evolving and expanding day by day. The researchers have not fully explored the various research issues such as energy, security vulnerabilities in UAV network in spite of rapidly increased usage of UAVs in several areas over the last few years. In this paper, a comprehensive taxonomy-based structure and a qualitatively study of associated publications on issues-based techniques for connected aerial vehicles in IoT environment have been described. Some potential drone use cases have been investigated such as in precision agriculture, COVID-19 pandemic handling and recovery management, and enabling connected road tariff environment using drone networking. Various research issues and approaches employed related to connected aerial vehicles have been analysed. Future, potential research directions have also been elaborated to support researchers' attention in drone networking and related themes. This chapter will enhance the overall understanding of UAVs or drones research directions in all relevant environments. We are currently working on different drone prototype development for specific use cases such as secure connected drone networking, tiny drone swarm for advertisement and battle grounds.

## REFERENCES

1. S. Garg, A. Singh, S. Batra, N. Kumar, and L.T. Yang, "UAV-empowered edge computing environment for cyberthreat detection in smart vehicles," *IEEE Network*, vol. 32, no. 3, pp. 42–51, 2018.
2. M. Wazid, A.K. Das, N. Kumar, A.V. Vasilakos, and J.J. Rodrigues, "Design and analysis of secure lightweight remote user authentication and key agreement scheme in Internet of drones deployment," *IEEE Internet of Things Journal*, vol. 6, no. 2, pp. 3572–3584, 2018.
3. A. Goodchild and J. Toy, "Delivery by drone: An evaluation of unmanned aerial vehicle technology in reducing $CO_2$ emissions in the delivery service industry," *Transportation Research Part D: Transport and Environment*, vol. 61, pp. 58–67, 2018.
4. O. Kaiwartya and S. Kumar, Enhanced caching for geo-cast routing in vehicular Ad Hoc network. In Durga Prasad Mohapatra and Srikanta Patnaik (eds.), *Intelligent Computing, Networking, and Informatics* (pp. 213–220). Springer, New Delhi, 2014.

5. C. Cambra, S. Sendra, J. Lloret, and L. Parra, "Ad hoc network for emergency rescue system based on unmanned aerial vehicles," *Network Protocols and Algorithms*, vol. 7, no. 4, pp. 72–89. 2015.
6. K. Kumar, S. Kumar, O. Kaiwartya, A. Sikandar, R. Kharel, and Mauri, J.L, "Internet of unmanned aerial vehicles: QoS provisioning in aerial ad-hoc networks," *Sensors*, vol. 20, p. 3160, 2020.
7. C.C. Baseca, J.R. Díaz, and J. Lloret, "Communication Ad Hoc protocol for intelligent video sensing using AR drones," *2013 IEEE 9th International Conference on Mobile Ad-hoc and Sensor Networks*, Dalian, 2013, pp. 449–453.
8. N. Chebrolu, T. Lbe, and C. Stachniss, "Robust long-term registration of UAV images of crop fields for precision agriculture," *IEEE Robotics and Automation Letters*, vol. 3, no. 4, pp. 3097–3104, 2018, doi:10.1109/LRA.2018.284960.
9. Z. Zhai, J.-F. Martnez Ortega, N. Lucas Martnez, and J. Rodrguez-Molina, "A mission planning approach for precision farming systems based on multiobjective optimization," *Sensors*, vol. 18, no. 6, 2018, doi:10.3390/s18061795.
10. C.-J. Chen, Y.-Y. Huang, Y.-S. Li, Y.-C. Chen, C.-Y. Chang, and Y.-M. Huang, "Identification of fruit tree pests with deep learning on embedded drone to achieve accurate pesticide spraying," *IEEE Access*, vol. 9, pp. 21986–21997, 2021, doi:10.1109/ACCESS.2021.3056082.
11. A. Kumar, K. Sharma, H. Singh, S.G. Naugriya, S.S. Gill, and R. Buyya, "A drone-based networked system and methods for combating coronavirus disease (COVID-19) pandemic", *Future Generation Computer Systems*, vol. 115, pp. 1–19, 2021.
12. N. Naren, et al., "IoMT and DNN-enabled drone-assisted Covid-19 screening and detection framework for rural areas," *IEEE Internet of Things Magazine*, vol. 4, no. 2, pp. 4–9, 2021, doi:10.1109/IOTM.0011.2100053.
13. A.Y. Chen, Y.L. Chiu, M.H. Hsieh, P.W. Lin, and O. Angah, "Conflict analytics through the vehicle safety space in mixed traffic flows using UAV image sequences," *Transportation Research Part C: Emerging Technologies*, vol. 119, p. 102744, 2020.
14. S.H. Alsamhi, F. Afghah, R. Sahal, A. Hawbani, M.A. Al-qaness, B. Lee, and M. Guizani, "Green internet of things using UAVs in B5G networks: A review of applications and strategies," *Ad Hoc Networks*, vol. 117, p. 102505, 2021.
15. J. P, Yaacoub, H. Noura, O. Salman, and A. Chehab, "Security analysis of drones systems: Attacks, limitations, and recommendations," *Internet of Things*, vol. 11, 100218, 2020.
16. A.D. Boursianis, M.S. Papadopoulou, P. Diamantoulakis, A. Liopa-Tsakalidi, P. Barouchas, G. Salahas, G. Karagiannidis, S. Wan, and S.K. Goudos, "Internet of things (IoT) and agricultural unmanned aerial vehicles (UAVs) in smart farming: A comprehensive review," *Internet of Things*, vol. 18, p. 100187, 2020.
17. P. Mehta, R. Gupta, and S. Tanwar, "Blockchain envisioned UAV networks: Challenges, solutions, and comparisons," *Computer Communications*, vol. 151, pp. 518–538, 2020.
18. M.Y. Arafat and S. Moh, "Routing protocols for unmanned aerial vehicle networks: A survey," *IEEE Access*, vol. 7, pp. 99694–99720, 2019.
19. M. Mozaffari, W. Saad, M. Bennis, Y.-H. Nam, and M. Debbah, "A tutorial on UAVs for wireless networks: Applications, challenges, and open problems," *IEEE Communications Surveys & Tutorials*, vol. 21, no. 3, pp. 2334–2360, 2019.
20. M.-A. Lahmeri, M.A. Kishk, and M.-S. Alouini, "Artificial intelligence for UAV-enabled wireless networks: A survey," *IEEE Open Journal of the Communications Society*, vol. 2, pp. 1015–1040, 2021.
21. A. Fotouhi, et al., "Survey on UAV cellular communications: Practical aspects, standardization advancements, regulation, and security challenges," *IEEE Communications Surveys & Tutorials*, vol. 21, no. 4, pp. 3417–3442, 2019.
22. N. Hossein Motlagh, T. Taleb, and O. Arouk, "Low-altitude unmanned aerial vehicles-based internet of things services: Comprehensive survey and future perspectives," *IEEE Internet of Things Journal*, vol. 3, no. 6, pp. 899–922, 2016.

23. J. Angjo, I. Shayea, M. Ergen, H. Mohamad, A. Alhammadi, and Y.I. Daradkeh, "Handover management of drones in future mobile networks: 6G technologies," *IEEE Access*, vol. 9, pp. 12803–12823, 2021.
24. M.R. Ramli, J.M. Lee, and D.S. Kim, "Hybrid MAC protocol for UAV-assisted data gathering in a wireless sensor network," *Internet of Things*, vol. 14, pp. 1–13, 2021.
25. D. Ebrahimi, S. Sharafeddine, P. Ho, and C. Assi, "UAV-aided projection-based compressive data gathering in wireless sensor networks," *IEEE Internet of Things Journal*, vol. 6, no. 2, pp. 1893–1905, 2019.
26. C. Luo, M.N. Satpute, D. Li, Y. Wang, W. Chen, and W. Wu, "Fine-grained trajectory optimization of multiple UAVs for efficient data gathering from WSNs," *IEEE/ACM Transactions on Networking*, vol. 29, no. 1, pp. 162–175, 2021.
27. M. Tao, X. Li, H. Yuan, and W. Wei, "UAV-Aided trustworthy data collection in federated-WSN-enabled IoT applications," *Information Sciences*, vol. 532, pp. 155–169, 2020, doi:10.1016/j.ins.2020.03.053.
28. H. Sedjelmaci, A. Boudguiga, I.B. Jemaa, and S.M. Senouci, "An efficient cyber defense framework for UAV-Edge computing network," *Ad Hoc Networks*, vol. 94, 2019, doi:10.1016/j.adhoc.2019.101970.
29. J. Chen, S. Chen, S. Luo, Q. Wang, B. Cao, X. Li, "An intelligent task offloading algorithm (iTOA) for UAV edge computing network," *Digital Communications and Networks*, vol. 6, no. 4, pp. 433–443, 2020.
30. F. Luo, C. Jiang, S. Yu, J. Wang, Y. Li, and Y. Ren, "Stability of cloud-based UAV systems supporting big data acquisition and processing," *IEEE Transactions on Cloud Computing*, vol. 7, no. 3, pp. 866–877, 2019.
31. Q. Gao, X. Shen, and W. Niu, "Large-scale synthetic urban dataset for aerial scene understanding," *IEEE Access*, vol. 8, pp. 42131–42140, 2020.
32. K. Peng, et al., "A hybrid genetic algorithm on routing and scheduling for vehicle-assisted multi-drone parcel delivery," *IEEE Access*, vol. 7, pp. 49191–49200, 2019.
33. M. Asadpour, K.A. Hummel, D. Giustiniano, and S. Draskovic, "Route or carry: Motion-driven packet forwarding in micro aerial vehicle networks," *IEEE Transactions on Mobile Computing*, vol. 16, no. 3, pp. 843–856, 2017.
34. P. Yang, X. Cao, X. Xi, W. Du, Z. Xiao, and D. Wu, "Three-dimensional continuous movement control of drone cells for energy efficient communication coverage," *IEEE Transactions on Vehicular Technology*, vol. 68, no. 7, pp. 6535–6546, 2019.
35. R. Fotohi, E. Nazemi, and F.S. Aliee, "An agent-based self-protective method to secure communication between UAVs in unmanned aerial vehicle networks," *Vehicular Communications*, vol. 26, 2020, doi:10.1016/j.vehcom.2020.100267.
36. Y. Su, "A trust based scheme to protect 5G UAV communication networks," *IEEE Open Journal of the Computer Society*, vol. 2, pp. 300–307, 2021.
37. A.A. Khan, M.M. Khan, K.M. Khan, J. Arshad, and F. Ahmad, "A blockchain-based decentralized machine learning framework for collaborative intrusion detection within UAVs," *Computer Networks*, vol. 196, 2021, doi:10.1016/j.comnet.2021.108217.
38. R. Ch, G. Srivastava, T.R. Gadekallu, P.K.R. Maddikunta, and S. Bhattacharya, "Security and privacy of UAV data using blockchain technology," *Journal of Information Security and Applications*, vol. 55, 2020, doi:10.1016/j.jisa.2020.102670.
39. A. Gao, Y. Hu, W. Liang, Y. Lin, L. Li, and X. Li, "A QoE-oriented scheduling scheme for energy-efficient computation offloading in UAV cloud system," *IEEE Access*, vol. 7, pp. 68656–68668, 2019.
40. S. Kouroshnezhad, A. Peiravi, M. S. Haghighi and A. Jolfaei, "Energy-efficient drone trajectory planning for the localization of 6G-enabled IoT devices," *IEEE Internet of Things Journal*, vol. 8, no. 7, pp. 5202–5210, 2021.
41. J.N. Yasin, et al., "Energy-efficient formation morphing for collision avoidance in a swarm of drones," *IEEE Access*, vol. 8, pp. 170681–170695, 2020.

23. I. Angjo, I. Shayea, M. Ergen, H. Mohamad, A. Alhammadi, and Y. I. Daradkeh, "Handover management of drones in future mobile networks: 6G technologies," *IEEE* [illegible] pp. 12803–12823, 2021.
24. M. [illegible] and D. S. [illegible] [illegible]
25. [illegible]
26. [illegible]

CHAPTER 3

# Multisensor Random Finite Set Information Fusion

## *Advances, Challenges, and Opportunities*

Tiancheng Li

*Northwestern Polytechnical University*

Kai Da and Hongqi Fan

*National University of Defense Technology*

Benru Yu

*Nanjing University of Science and Technology*

CONTENTS



## 3.1 INTRODUCTION

Multitarget tracking (MTT) refers to the joint estimation of both the number and states of multiple targets from noisy observations in the presence of random target birth/death and missing/false data. The random finite set (RFS) provides a new unified and fully probabilistic approach to MTT and information fusion [1], which has rapidly developed along

DOI: 10.1201/b22998-3

diverse pathways over the last two decades; see a comprehensive overview of both classic and RFS-based MTT approaches [2]. The cutting-edge focus of the MTT research is tangled with the wide distribution of sensor networks in many important commercial and industrial realms in which the cooperation between sensors gains in improving estimation accuracy, extending tracking coverage, and enhancing the viability/scalability of the local sensors [3,4]. In particular, the distributed network is more flexible and less prone to single-point failures than the centralized network. This has in turn significantly promoted the development of RFS-based tracking approaches. In this review, we focus on RFS-based multisensor MTT.

Existing multisensor MTT approaches can be classified into two major groups depending on the type of information shared between sensors: data-level and estimate-level, as shown in Figure 3.1. Briefly speaking, the majority of the former group is dedicated to computing the joint likelihood of measurements provided by all sensors, approaching the optimal multisensor Bayes posterior. Then, either raw observations or single-sensor likelihood functions are fused among sensors. Raw measurements might be heterogeneous, asynchronous, and communicatively extensive. Instead, the local multitarget likelihood functions produced by the raw measurements can be used for computing/approximating the joint likelihood function in order to obtain the multisensor multitarget posterior. By contrast, the multisensor density fusion handles posterior densities yielded by local sensors in tandem with the popular consensus approach, leading to a robust and conservative fusion approach to multisensor MTT. This fusion is, however, typically suboptimal due to the omitted cross-correlation between densities. To further

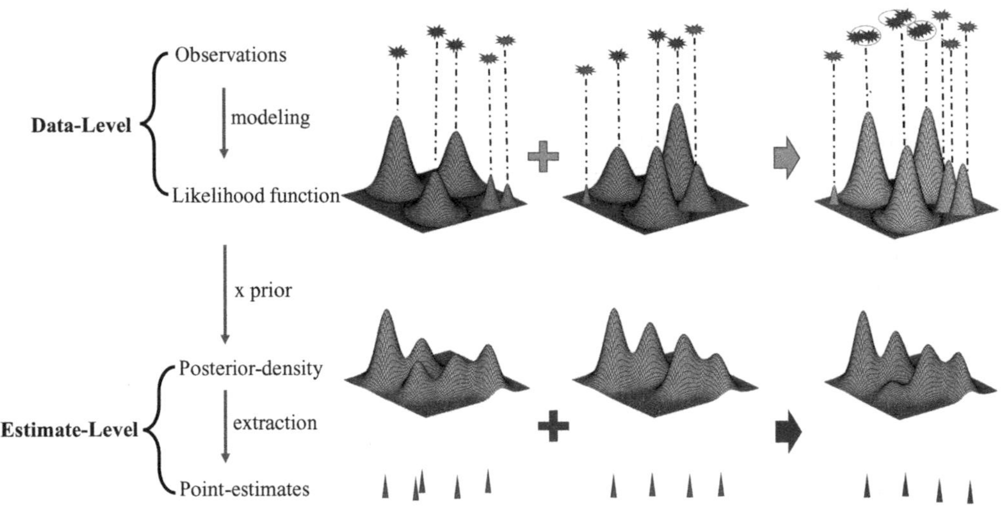


FIGURE 3.1 (a) Data-level fusion: either raw observations or single-sensor likelihood functions are fused among sensors. (b) Estimate-level fusion: either the local single-sensor posterior densities or the point estimates extracted from them are fused among sensors.

reduce the computation and communication cost, one may choose to exchange and fuse the point estimates among sensors. That is, only the filter point estimates are fused at the end which will not be feedback to each local filter that remains purely a single-sensor filter. All these issues constitute the main content of this chapter. We note that the major materials have been published in our recent review paper [5] except for some cutting-edge research findings such as the continuous-time trajectory estimation based on RFS addressed in Section 3.2.3.

The chapter is organized as follows. Various important RFS and benchmark RFS filters are reviewed in Section 3.2. Both heuristic and principled approximate multisensor measurement-fusion approaches are presented in Section 3.3. In Section 3.4, cutting-edge average-fusion, arithmetic and geometric average, and approaches for multitarget density fusion are detailed. Relevant topics and remaining open issues are presented in Section 3.5 to conclude the chapter.

## 3.2 RFS ESTIMATION

### 3.2.1 RFS Modeling

An RFS $X$ on $\mathbb{X}$ is a random variable taken from the space of finite subsets of $\mathbb{X}$, which can naturally characterize the false alarms, missed detections, and births and deaths of targets. A real-valued function $f(X)$ is a multitarget density function of $X$ if its unit is $u^{-|X|}$, where $u$ is the unit of $x \in X$. If $\int f(X)\delta X = 1$ is satisfied, then $f(X)$ is also called the multitarget probability density (MPD), where the set integral is defined with respect to any region $S$ by $\int_S f(X)\delta X = f(\varnothing) + \sum_{n=1}^{\infty} \frac{1}{n!} \int_{S^n} f(\{x_1,\ldots,x_n\}) dx_1 \ldots dx_n$. The probability hypothesis density (PHD) of $X$ is defined as $D(x) = \int f(\{x\} \cup X)\delta X$. The integral of PHD $\int_S D(x)dx$ in a region $S$ denotes the expected number of targets in that region if $X$ is employed to model the collection of target states. The cardinality distribution representing the probability that there are $n$ elements in $X$ is defined by $p(n) = \int_{|X|=n} f(X)\delta X$, where $|\cdot|$ denotes the cardinality of a set.

Given the target states at time $k-1$ denoted by an RFS $X_{k-1} = \{x'_1,\ldots,x'_{n'}\}$, each target $x' \in X_{k-1}$ may persist with a probability $p_S(x)$ or disappear with a probability $1 - p_S(x)$ at time $k$. That is, $T(x') = \varnothing$ with probability $1 - p_S(x)$ and $T(x') = \{x\}$ with probability $p_S(x)$, where $\{x\}$ is the state with distribution $\Phi_{k|k-1}(x \mid x')$. Then, the RFS motion model is given by $X_{k|k-1} = T(x'_1) \cup \ldots \cup T(x'_{n'}) \cup B_k \cup \Gamma_k$, where $B_k$ is the set of newly appeared targets and $\Gamma_k$ is the set of spawned targets at time $k$.

Consider the standard measurement model, i.e., each target generates at most a single measurement, and any measurement is generated by at most a single target. Given the target RFS $X_k = \{x_1,\ldots,x_n\}$, each target $x \in X_k$ generates a measurement with probability $p_D(x)$. That is, $\Upsilon(x_i) = \varnothing$ with probability $1 - p_D(x)$ and $\Upsilon(x_i) = \{z_i\}$ with probability $p_D(x)$, where $\{z_i\}$ is the measurement with distribution $g_k(z \mid x_i)$. The RFS-based measurement model is given by $Z_k = \Upsilon(x_1) \cup \ldots \cup \Upsilon(x_n) \cup C_k$, where $C_k$ is the set of measurements generated by clutter.

Usually, the above dynamic and measurement models are described by the Markov multitarget transition density $\Phi_{k|k-1}(X|X')$ and multitarget likelihood function $g_k(Z_k|X_k)$, respectively. Given the above models, the theoretically optimal multitarget Bayesian filter is given by the following prediction and correction steps:

$$f_{k|k-1}(X|Z_{1:k-1}) = \int \Phi_{k|k-1}(X|X') f_{k-1}(X'|Z_{1:k-1}) \delta X', \tag{3.1}$$

$$f_k(X|Z_{1:k}) = \frac{g_k(Z_k|X) f_{k|k-1}(X|Z_{1:k-1})}{\int g_k(Z_k|X) f_{k|k-1}(X|Z_{1:k-1}) \delta X}. \tag{3.2}$$

The multitarget state-extraction step, which may use a properly RFS-defined expected a posteriori (EAP) estimator or a maximum a posteriori (MAP) estimator, acts upon the updated posterior density to extract the target states at the end of the iteration.

### 3.2.2 Classic RFSs

The optimal multitarget Bayesian filter involving multitarget integrals is computationally intractable in practice and, consequently, a series of principled approximate filters have been proposed under different RFS processes. In what follows, we present these processes and filters related to them.

#### *3.2.2.1 Independent and Identically Distributed Cluster/Poisson RFS*

An RFS $X$ is considered as an independent and identically distributed cluster (IIDC) RFS if its cardinality is arbitrarily distributed according to $p(n)$, and for any finite cardinality, the elements are independently and identically distributed according to $s(x)$. The MPD and PHD associated with the IIDC RFS $X$ are given by

$$f^{\text{iidc}}(X) = n!\, p(n) \prod_{x \in X} s(x), \tag{3.3}$$

$$D^{\text{iidc}}(x) = \sum_{n=0}^{\infty} n p(n) s(x). \tag{3.4}$$

A Poisson RFS is a special IIDC RFS where the cardinality distribution $p(n)$ is Poisson with the mean of $\lambda$. The MPD and PHD associated with the Poisson RFS $X$ are given by

$$f^{\text{P}}(X) = \text{e}^{-\lambda} \prod_{x \in X} \lambda s(x), \tag{3.5}$$

$$D^{\text{P}}(x) = \lambda s(x). \tag{3.6}$$

The PHD filter [6–8] recursively propagates the PHD of a Poisson RFS modeling the multitarget state. The cardinalized PHD (CPHD) filter [9,10], by contrast, assumes that the multitarget state is formulated by an IIDC RFS and recursively propagates its PHD and cardinality distribution, thus yielding a better estimate of the number of targets but with higher complexity.

#### 3.2.2.2 (Multi-)Bernoulli RFS

A Bernoulli RFS $X$ either contains a single element with the probability of $r$ or is empty with the probability of $1-r$. The probability density of the Bernoulli RFS is given by

$$f^{\mathrm{b}}(X)=\begin{cases} 1-r, & X=\varnothing, \\ rs(x), & X=\{x\}, \\ 0, & |X|\geq 2. \end{cases} \tag{3.7}$$

A multi-Bernoulli (MB) RFS $X$ is the union of $n$ independent Bernoulli RFSs. The MPD and PHD of an MB RFS $X$ have respective forms as follows:

$$f^{\mathrm{mb}}(X)=\sum_{X_1\uplus\ldots\uplus X_n}\prod_{i=1}^{n}f_i^{\mathrm{b}}(X_i), \tag{3.8}$$

$$D^{\mathrm{mb}}(x)=\sum_{i=1}^{n}r_i s_i(x). \tag{3.9}$$

The Bernoulli filter [11,12] is Bayesian optimal for detecting and tracking at most one target in an arbitrary clutter background and detection profile. It recursively propagates the Bernoulli parameters of the spatial single-target density and existence probability. Similarly, the multitarget MB (MeMber) filters [13] recursively propagate the MB parameters rather than the MB distribution.

#### 3.2.2.3 Poisson MB Mixture RFS

A Poisson MB mixture (PMBM) RFS [14–16] is the convolution of a Poisson RFS and an MB mixture (MBM) RFS with MPD of the form:

$$f^{\mathrm{pmbm}}(X)=\sum_{Y\uplus W=X}f^{\mathrm{p}}(Y)f^{\mathrm{mbm}}(W), \tag{3.10}$$

where $f^{\mathrm{p}}(\cdot)$ is a Poisson density that represents all undetected targets. The MBM density $f^{\mathrm{mbm}}(\cdot)$ representing potentially detected targets is a weighted sum of MPDs of MB and has the form

$$f^{\mathrm{mbm}}(X)\propto\sum_{j\in\mathbb{J}}\sum_{X_1\uplus\ldots\uplus X_n=X}\prod_{i=1}^{n}w_{j,i}f_{j,i}^{\mathrm{b}}(X_i), \tag{3.11}$$

where $j$ is an index over all global hypotheses (components of the mixtures) and $\mathbb{J}$ is the index set of MBs in the MBM (with each term corresponding to a global association hypothesis).

As a conjugate prior under a standard dynamical model with Poisson birth, the PMBM filter admits closed-form Markov-Bayes recursion and achieves excellent performance in the case of low detection probability [17,18]. The MBM [16,19] and Poisson MB (PMB) [20] filters are special cases of the PMBM filter.

#### *3.2.2.4 Labeled RFS*

On the basis of the conventional RFS, a labeled RFS $X=\{(x_1,l_1),\ldots,(x_n,l_n)\}\subseteq\mathbb{X}\times\mathbb{L}$ serves to append distinct labels $l_1,\ldots,l_n\in\mathbb{L}$ to the elements in an RFS. The most known labeled RFS, namely the generalized labeled MB (GLMB) RFS, is a mixture of multitarget exponentials with probability density [21]

$$f^{\mathrm{gl}}(X)=\Delta(X)\sum_{c\in\mathbb{C}}w^{(c)}(\mathcal{L}(X))\prod_{(x,l)\in X}s^{(c)}(x,l), \tag{3.12}$$

where $\mathbb{C}$ is a discrete index set. $w^{(c)}(L)$ is the weight of the hypothesis that $|X|=n$ targets are presented with respective labels $l_1,\ldots,l_n\in L$ and respective probability distributions $s^{(c)}(x_1,l_1),\ldots,s^{(c)}(x_n,l_n)$, satisfying $\sum_{L\subseteq\mathbb{L}}\sum_{c\in\mathbb{C}}w^{(c)}(L)=1$ and $\int s^{(c)}(x,l)dx=1$.

A $\delta$-GLMB RFS is a special GLMB RFS with $\mathbb{C}=\mathcal{F}(\mathbb{L})\times\Xi$, $w^{(c)}(L)=w^{(I,\xi)}\delta_I(L)$, and $s^{(c)}(x,l)=s^{(I,\xi)}(x,l)=s^{(\xi)}(x,l)$. The pair $(I,\xi)$ represents the hypothesis that the set of tracks $I$ has a history $\xi$ of association maps. The MPD of a $\delta$-GLMB RFS $X$ is of the form

$$f^{\mathrm{dgl}}(X)=\Delta(X)\sum_{(I,\xi)\in\mathcal{F}(\mathbb{L})\times\Xi}w^{(I,\xi)}\delta_I[\mathcal{L}(X)]\left[s^{(\xi)}\right]^X. \tag{3.13}$$

The GLMB filter [21,22] supposes that the initial multitarget RFS in (3.1) is a GLMB RFS, and leads to an exact closed-form solution of the multitarget Bayesian recursion. Since a label is assigned to each state to distinguish individual targets, the GLMB filter provides the MPDs of the entire trajectories of the targets, rather than just the most recent states, thereby eliminating problems like track fragmentation [23]. Efficient implementations and novel approximations of the GLMB filter have been presented in [24,25] and [26,27], respectively.

### 3.2.3 Continuous-Time Trajectory Based on RFS

Despite the success of the labeled RFS, an insightful discussion on the optimality of linking discrete-time points is given in [28]. Another relevant approach models the trajectories of a random number of targets as an RFS to be estimated [29]. Compared to the point-state RFS, the trajectory RFS requires much higher computation.

Instead of estimating the *discrete-time state* of the target based on a sophistically designed state space model, we are actually more interested in estimating the *continuous-time trajectory* of the target in the context of target tracking, which contains more information than the discrete-time point set and is closer to be the nature of the target trajectory. To this end, [30,31] proposed to model the target movement by using a trajectory function of time (T-FoT) which reformulates the tracking as an online curve-fitting problem where the least squares approach plays a key role. At the core of the approach is a continuous-time T-FoT which is used in place of the stepwise Markov-jump model to describe the target motion, i.e., $\mathbf{x}_t = f(t)$. Then, the desired parametric T-FoT $F(t)$ as an estimate of $f(t)$ is obtained by minimizing the divergence between $F(t)$ and $f(t)$. To this end, the T-FoT can be more definitely specified as $F(t;C_k)$ of an engineering-friendly format, where $C_k$ are the parameters to be estimated which fully determine the T-FoT at time $k$. Further on, one may define a penalty factor $\Omega(C_k)$ as a measure of the disagreement of the fitting function with the a-priori model constraint, e.g., road constraint [32]. For instance, $\Omega(C_k) := \left\| F(t_0;C_k) - \mathbf{x}_0 \right\|$ measuring the mismatch between the fitting trajectory and the known state $\mathbf{x}_0$ that the target passes by at time $t_0$. This leads to a T-FoT optimization problem over a specific time-window $[k_1, k]$ (without false and missing data) as follows:

$$\hat{C}_k = \arg\min_{C} \sum_{t=k_1}^{k} \left\| \mathbf{z}_t - h_t\left(F(t;C), \overline{\mathbf{v}}_t\right) \right\| + \lambda\Omega(C), \tag{3.14}$$

where $\overline{\mathbf{v}}_t$ stands for the average bias of the observation/data and $\lambda > 0$ controls the trade-off between the data fitting error and the model fitting error.

Issues of false and missing data that are modeled by RFSs are further addressed in [33], based on the prerequisite that the target birth/initial position, velocity, and starting time are given in advance and that the target always exists in the scenario. These restrictive assumptions are relaxed in [34] which takes into account the random (in both the state space and time domain) birth, death, and even re-appearance of the target, improves the algorithm for better identifying false and missing measurements, and uses almost no a-priori model information but outperforms even the Bayesian optimal Bernoulli filter. An obviously valuable extension in this research direction is to underpin the T-FoT approach by the RFS theory for which a challenge is from the various lengths of the trajectories corresponding to a target RFS. That is, the elements (namely trajectories) of the trajectory set are of different, time-varying lengths. Fusion of the trajectory RFSs if they can be properly defined is an even more challenging issue.

## 3.3 MULTISENSOR MEASUREMENT FUSION

The optimal multisensor multitarget Bayesian corrector is given by

$$f_k\left(X \middle| Z_{1:k}^{1:s}\right) = \frac{g_k\left(Z_k^{1:s} \middle| X\right) f_{k|k-1}\left(X \middle| Z_{1:k-1}^{1:s}\right)}{\int g_k\left(Z_k^{1:s} \middle| X\right) f_{k|k-1}\left(X \middle| Z_{1:k-1}^{1:s}\right) \delta X}. \tag{3.15}$$

If the sensors observe the same targets and their measurements are conditionally independent, then

$$g_k\left(Z_k^{1:s} \mid X\right)=\prod_{i=1}^{s} g_k^i\left(Z_k^i \mid X\right). \tag{3.16}$$

Even though the above statistical independencies can be exploited for a drastic reduction of complexity, the multisensor MTT Bayes filter remains computationally intractable. In what follows, we introduce some heuristic methods followed by a theoretically satisfactory parallel combination approximate multisensor (PCAM) filter. In addition, by specifying the types of RFS, a variety of multisensor optimal RFS-based filters can be obtained. We note that in these approaches, the issue of measurement communication between sensors is ignored. A centralized network in which all sensors send their measurements directly to the fusion center is commonly assumed; however, when a distributed/decentralized sensor network is involved, the distributed flooding algorithm [35–37] can be applied by which each sensor serves as a fusion center to access the measurements of the other linked sensors at different distances.

### 3.3.1 Two Heuristic Approaches

The first heuristic approach is the widely used multisensor iterated-corrector (IC) filter [38,39] which iteratively applies the single-sensor corrector for each sensor measurement set. Such an approach has an obvious limitation in that the performance of the final result depends upon the sensors' processing order, which in turn depends upon the amount of information detected by sensors—the most informative sensor should be processed first. Thus, typically those sensors with large detection probabilities should be processed before others [40]. An improved IC-PHD filter is presented in Ref. [41] in which the probabilities of detection and misdetection are modified to reduce the influences of the order of the sensor updates. Additionally, in the IC-MeMber/GLMB filter, the component-truncation and resampling steps, which are important for constraining the computational cost of the filter, should be applied with caution in the Gaussian mixture (GM)/particle implementation, given that the components/particles weighted low in one sensor may actually correspond to a target weighted highly by other sensors.

As another heuristic approach, [37] converted the collection of measurements of different sensors into a set of proxy and homologous measurements (typically positions), overcome the problems of false and missing data and unknown statistics, facilitating linear PHD updating that amounts to standard single-sensor PHD filtering with no false or missing data. This simple scheme, however, demonstrates promising performance for certain measurement models and scenarios. In fact, these pseudo-position measurements themselves may provide high positioning accuracy [36] without modeling the dynamics of the targets.

### 3.3.2 PCAM Filter

The PCAM filter is a principled approximation of the multisensor Bayesian filter [42]. At the core of this filter is the assumption that the measurements of distinct sensors are conditionally independent, c.f. (3.16). The corrector of the multisensor multitarget Bayesian filter can be rewritten as a form of "Bayesian parallel combination" as

$$\begin{aligned} f_k\left(X \mid Z_{1:k}^{1:s}\right) &\propto g_k\left(Z_k^{1:s} \mid X\right) f_{k|k-1}(X) \\ &= g_k^1\left(Z_k^1 \mid X\right) \ldots g_k^s\left(Z_k^s \mid X\right) f_{k|k-1}(X) \\ &= \prod_{i=1:s} f_k\left(X \mid Z_{1:k-1}^{\{1:s\}\backslash i}, Z_{1:k}^i\right)\left[f_{k|k-1}(X)\right]^{1-s}, \end{aligned} \tag{3.17}$$

where $f_{k|k-1}(X) = f_{k|k-1}\left(X \mid Z_{1:k-1}^{1:s}\right)$ and $f_k\left(X \mid Z_{1:k-1}^{\{1:s\}\backslash i}, Z_{1:k}^i\right)$ is the MPD updated using the measurements of sensor $i$. Unlike the IC filter, the PCAM filter applies the correctors in parallel with the same predicted density, which does not depend on the sensor order.

Different PCAM RFS filters result from different assumptions about the prior $f_{k|k-1}(X)$, the singly updated posteriors $f_k\left(X \mid Z_{1:k-1}^{\{1:s\}\backslash i}, Z_{1:k}^i\right)$, and the sensor-clutter processes (e.g., PCAM PHD/CPHD filter [1, Chp.10]). However, [43] noted that the PCAM-PHD filter suffers a scale-unbalance problem in the particle implementation, which worsens as the number of sensors increases. Ref. [44] proposed a GM implementation of the PCAM PHD filter to mitigate the adverse effects caused by the update coefficient.

### 3.3.3 Multisensor Optimal RFS Filters

#### *3.3.3.1 Multisensor Bernoulli Filter*

The particle-based distributed Bayes-optimal Bernoulli filter is given by Ref. [45], which approximates the Bayes-optimal estimates of the target-presence probability and the target state. This is enabled by an extension of the likelihood-consensus method [46], which reaches consensus with regard to the likelihood function $g\left(Z^1,\ldots,Z^s \mid X\right)$ as in (3.16) with respect to the Bernoulli process. However, so far it is unclear how to extend the likelihood-consensus approach to other RFS MTT filters.

#### *3.3.3.2 Multisensor CPHD/PHD Filter*

The exact multisensor CPHD filter was derived by Ref. [47] by assuming that the target birth, clutter, and predicted multitarget processes are all IIDC RFSs. A multisensor PHD filter is a special case of the multisensor CPHD filter in which the target birth, clutter, and predicted multitarget processes are modeled by Poisson RFSs [38,48].

The PHD function of multisensor CPHD/PHD filter has the form

$$D_k(x) \propto \alpha_0 \prod_{i=1}^{s}\left(1 - p_D^i(x)\right) + \sum_{P \in \mathbb{P}} \alpha_P \left( \sum_{W \in P} \rho_W(x) \right), \tag{3.18}$$

where $P$ is the partitioning of all measurements from all sensors $Z_k^{1:s}$, $W \in P$ is a subset that contains at most one measurement from sensor $i$, i.e., $|W^i| \leq 1$, and $\mathbb{P}$ is the set of all such $P$; see Ref. [47] for the details.

It can be seen that the exact implementation of the multisensor CPHD/PHD filter needs to explore all partitionings, $P$, of the measurements, which is computationally infeasible. A reasonable alternative is to consider only those subsets that make significant contributions to the predicted PHD function, e.g., a greedy-approach-based implementation in [47]. Additionally, the update equations in the multisensor CPHD/PHD filter have similar forms to the single-sensor update for extended targets [49,50]. Therefore, the implementation of the multisensor CPHD/PHD filter can also benefit from those methods used to reduce the partitioning complexity for extended targets [51].

#### *3.3.3.3 Multisensor MeMber Filter*

The exact multisensor MeMber filter was derived by [52] by assuming that the target birth and predicted and posterior multitarget processes are all MBs and that the clutter process is Poisson. Unfortunately, the updated MPD $f_k(X)$ of the exact multisensor MeMber filter is not an MB, and so cannot be used in the next iteration directly. However, the updated PHD $D_k(x)$ is given by

$$D_k(x) = \sum_{P \in \mathbb{P}} \alpha_P \sum_{j=1}^{M_{k|k-1}} \rho_{W_j}^j(x) s_{k|k-1}^{(j)}(x), \tag{3.19}$$

the detailed parameters of which are introduced in Ref. [52]. This has a similar structure (i.e., a sum of weighted densities as (3.9)) to that of an MB process, even though $f_k(X)$ is not an MB. Thus, an MB approximation $\hat{f}_k(X)$ with the same PHD as that of $f_k(X)$ is obtained by constructing a Bernoulli component (BC) for each association between a predicted Bernoulli RFS and a partitioning [52].

In fact, the MB approximation with the same PHD $D_k(x)$ is not unique; other choices could be clustering the updated density into a number of BCs, or finding a higher-order approximation than PHD. The implementation of the multisensor MeMber filter suffers from the same problem of measurement partitioning (c.f.(3.19)) as does the multisensor CPHD/PHD filter, and similar solutions can be found in [47,52].

#### *3.3.3.4 Multisensor GLMB Filter*

The exact multisensor ($\delta$-)GLMB filter is also computationally intractable due to the assignment problem in each ($\delta$-)GLMB update. Ref. [53] developed a two-sensor $\delta$-GLMB filter using ranked-assignment algorithms to truncate the weighted sums of the multitarget exponentials; Ref. [27] proposed an approximate multisensor sequential-update approach for the multisensor $\delta$-GLMB filter. In each sequential step, an M$\delta$-GLMB density is used to define the principled approximation of the $\delta$-GLMB representing the true posterior in the sense of the multitarget Bayesian filter. Ref. [54] integrated the prediction and correction into a single step as follows:

$$f_k(X) \propto g\left(Z_k^{1:s} \mid X\right) \int \Phi_{k|k-1}\left(X \mid X'\right) f\left(X'\right) \delta X', \tag{3.20}$$

which avoids discarding the significant components. They proposed an efficient implementation based on Gibbs sampling for truncating the $\delta$-GLMB filtering density, leading to quadratic complexity in the number of hypothesized objects and linear complexity in the total number of measurements from all sensors.

The above mentioned measurement partitioning and Gibbs-sampling approaches have been utilized for multisensor PMB filtering design (with slight adjustment) [55]. Nevertheless, the measurement partitioning still suffers from an intractable combinatorial complexity. This problem, however, does not exist in the T-FoT framework—as addressed in Section 3.2.3—which from the beginning handles the multisensor measurements in a sequential fusion mode [30,31]. That is, all sensor measurements are used to update the T-FoT individually.

Finally, we note a common misunderstanding of the communication cost incurred by radar data, which has been argued to be much higher than MPD or likelihood-parameter communication. In fact, density fusion requires the communication of a large number of Gaussian component (GCs) with weights, means, and covariances or even particles (states and weights). They are actually more costly in communication than raw point measurements [56]; see a case study [36]. This is because each measurement generally corresponds to at least one GC, and the dimension of the state is usually higher than that of the measurement. This, however, may not hold for measurements like videos or images which are indeed communicatively costly [57].

## 3.4 MULTISENSOR SUBOPTIMAL MPD FUSION

Rather than the above data-level multitarget measurement fusion that performs fusion before filtering, another strategy is to fuse the filtering results obtained by each sensor. That is, the sensors first perform filtering calculations independently, and the filter results are communicated and fused at each filtering iteration. There are two basic types of results to be fused: point-estimates and posteriors (or their moments). In the former, the fusion may resort to a heuristic-data-association algorithm [58, Sec. 5], [59], and the fused result may not feedback to the filter (i.e., the fusion may yield better estimates but does not help to improve the filter). What is more commonly used is the latter, which improves the posterior/moment through fusion. In what follows, we focus on the latter and do not particularly distinguish PHD from MPD.

Notably, optimal posterior fusion requires either independence or prior knowledge of the cross-correlation between the fusing posteriors. However, the common information between fusion MPDs is typically unknown, especially when a large wireless sensor network (WSN) is involved, thus preventing Bayesian optimal fusion. Two cutting-edge suboptimal average approaches, namely arithmetic average (AA) and geometric average (GA), with a demonstrated ability to eschew double-counting the common information of fusing sources [60] are reviewed in this section. Simply, as long as the fused data can be split into two parts depending on whether they are shared by all the fusing parties: non-common part and common part, it is obvious that the both average fusion approaches avoid double-counting the common data as long as the fusing weights are normalized. This is the key to deal with common a-priori information and data incest [61] when a sensor network is employed where some information can easily be replicated and repeatedly transmitted to the local sensors. Meanwhile, non-common information will be counted less than a unit [62], which implies an over-conservativeness. In other words, the average fusion can always avoid information double-counting [60]. We refer to this property loosely as *robustness* and so both AA and GA fusion can be deemed a robust fusion approach. Both of them are characterized as conservative fusion [63,64]. More statistic and information-theoretic results of both average fusion approaches can be found in [65,66]. In parallel to the publication of this chapter, we notice further study/extension of both average fusion approaches for which the readers are kindly referred to Refs. [67, 68].

The benefit of being conservative is obvious in practice, especially when there are unknown inputs and model mismatching. However, MPD average fusion is boosted by the connection with the popular consensus algorithm in the context of distributed WSN. After the first WSN was realized in the 1990s, deployment of WSNs has been rapid, widespread, and fundamental. While its combination with the RFS approach has only recently become a focus, an upsurge in research has been aroused within the network consensus framework.

### 3.4.1 Properties of AA/GA Fusion

For the MPDs $f_i(X)$ from sensor $i \in \mathcal{I}$ where $\mathcal{I} = \{1,2,\ldots,I\}$ denotes the set of fusing sensors, their AA and GA fusions have the respective forms:

$$f_{\mathrm{AA}}(X) = \sum_{i \in \mathcal{I}} w_i f_i(X), \tag{3.21}$$

$$f_{\mathrm{GA}}(X) = C^{-1} \prod_{i \in \mathcal{I}} \left[ f_i(X) \right]^{w_i} \tag{3.22}$$

where $C = \int \prod_{i \in \mathcal{I}} \left[ f_i(X) \right]^{w_i} \delta X$, $w_i$ is the fusion weight assigned to sensor $i$, and $\sum_{i \in \mathcal{I}} w_i = 1$.

We note that, the idea of using the mixture of the Gaussian distributions to be fused as the fused result, which aiding by normalization is equivalent to the AA fusion, to rejuvenate particles appeared in Eq. (29) of Ref [35] which can be viewed as the embryonic form of the AA-Gaussian fusion. The GA fusion is also popularly known as exponential mixture density [69] and generalized covariance intersection [70,71]. It is often viewed as a generalization of the approach of covariance intersection [63,72], which was originally developed for the fusion of Gaussian probability density functions despite the design of the fusion weights. It is necessary to note that the covariance intersection (CI) and its generalization by Mahler [70,71] are a max-min optimization which does not only perform GA fusion but also optimizes the fusing weights via minimizing the fused covariance (such as in CI) or maximize the peaks of the fused distribution (such as proposed in Mahler's original generalized CI (GCI)). However, in most GCI-RFS implementations as to be reviewed next, there is usually nothing performed for optimizing the fusing weights in the sense of minimizing the fused covariance or maximizing the posterior peak but instead the fusing weights that ensure fast network consensus convergence such as the metropolis weights are used.

It was recently pointed out that both AA- and GA-fusion approaches are essentially Fréchet means [62] characterizing the central tendency of distributions in arbitrary metric spaces. For a metric space $(\mathbb{F}, d(\cdot,\cdot))$, the discrete Fréchet $p$-mean of the densities $f_1(\cdot),\ldots,f_n(\cdot) \in \mathbb{F}$ with non-negative weights $w_1,\ldots,w_n$ is defined by

$$\mu^p(f) = \arg\min_{g \in \mathbb{F}} \sum_{i \in \mathcal{I}} w_i d^p\big(g(\cdot), f_i(\cdot)\big), \tag{3.23}$$

where $d^p\big(g(\cdot), f(\cdot)\big)$ is a given metric on the distance between $f(\cdot)$ and $g(\cdot)$ at power $p$. $\sum_{i=1}^{n} w_i d^p\big(g(\cdot), f_i(\cdot)\big)$ is called the Fréchet functional. Refs. [62,73] proved that AA and GA fusions are Fréchet means using different Fréchet functionals as follows:

$$f_{\mathrm{AA}}(X) = \arg\min_{g \in \mathbb{F}} \sum_{i \in \mathcal{I}} w_i \left\| f_i - g \right\|^2, \tag{3.24}$$

$$f_{\mathrm{GA}}(X) = \arg\min_{\substack{g \in \mathbb{F}: \\ \int g(X)\delta X = 1}} \sum_{i \in \mathcal{I}} w_i \left\| \log f_i - \log g \right\|^2, \tag{3.25}$$

where $\|f\|^2 = \int [f(X)]^2 \delta X$.

Another property is that AA/GA fusion can be computed by minimizing the sum of the weighted Kullback-Leibler divergence (KLD) between fusing and fused MPDs [74,75], which are also referred to as I-projection and M-projection [76], respectively; that is,

$$f_{\mathrm{AA}}(X) = \arg\min_{g \in \mathbb{F}} \sum_{i \in \mathcal{I}} w_i D_{\mathrm{KL}}\left(f_i \| g\right), \tag{3.26}$$

$$f_{\mathrm{GA}}(X) = \arg\min_{\substack{g \in \mathbb{F}: \\ \int g(X)\delta X = 1}} \sum_{i \in \mathcal{I}} w_i D_{\mathrm{KL}}\left(g \| f_i\right), \tag{3.27}$$

where $D_{\mathrm{KL}}\left(f \| g\right) = \int f(X) \log\left(\frac{f(X)}{g(X)}\right) \delta X.$

The motivation behind these formulations (as shown in (3.24)–(3.27)) is that both averaging-fusion approaches provide a best fit of the mixture of the fusing distributions; see Ref. [77]. This *best-fit* property of both averaging-fusion rules has long been recognized [65,75,78–80], and has been elaborately/over emphasized using new terminologies such as Cauchy-Schwarz divergence fusion [82] or minimum information loss/gain fusion [83]. It is, however, important to note that:

1. This *best fit* is suboptimal and what is fitted is a *mixture of the fusing local posteriors*, not the true multisensor posterior [77]. The former can be viewed as a rule-of-the-thumb substitute for the latter. Through this substitution, the average fusion is connected with the optimal fusion. Therefore, we must emphasize that the fit objective is a mixture, not the optimal posterior in order to avoid overstating the sub-optimality of both fusion approaches.
2. Cauchy-Schwarz or KL divergence has never been really calculated in both AA and GA fusion operations in the existing implementations; exception will be shown in (3.30) and (3.31). Renaming the known AA and GA fusion using these specific divergences is quite misleading and makes things unnecessarily complicated and also limited (since both averages can be derived by (3.24) and (3.25), respectively). In contrast, both AA and GA are mathematically accurate in terminology which shows clearly their linear and log-linear fusion essence.

More convincingly, the AA is better (in the sense of having a smaller divergence relative to the true density) than all sub-densities on average. That is, the KLD of the true density $g(X)$ relative to the AA is given as (for which the proof is given in Ref. [66])

$$D_{\mathrm{KL}}\left(f_{\mathrm{AA}} \| g\right) = \sum_{i \in \mathcal{I}} w_i D_{\mathrm{KL}}\left(f_i \| g\right) - \sum_{i \in \mathcal{I}} w_i D_{\mathrm{KL}}\left(f_i \| f_{\mathrm{AA}}\right) \tag{3.28}$$

$$\leq \sum_{i \in \mathcal{I}} w_i D_{\mathrm{KL}}\left(f_i \| g\right) \tag{3.29}$$

where the equation holds only when all sub-densities $f_i, i \in \mathcal{I}$ are identical.

The above result provides an information-theoretic justification for the mixture. In fact, the AA is often better than the best fusing distribution if the fusion weights are properly designed; see the analysis and illustration given in Ref. [66]. More theoretic properties of the AA fusion approach and its relationship with the known *covariance union* and *covariance intersection* are available in Ref. [66], as well as the fault-tolerance, mode-preservation and mean square error (MSE) properties of the AA. Consequently, a suboptimal, namely diversity preference, weighting approach has been proposed therein based on the following suboptimal maximization problem—this for the first time determines the fusing weights from the fusion perspective (for better fusion performance) rather than from the network consensus perspective (for fast convergence). This has been utilized for fusing Gaussian [66] and Student's t [135] distributions, respectively, both of which exhibit promising results.

$$f_{\mathrm{AA}}\left(\mathbf{w}_{\mathrm{subopt}}\right) = \arg\max_{\mathbf{w}} \min_{g \in \mathbb{F}} \sum_{i \in \mathcal{I}} w_i D_{\mathrm{KL}}\left(f_i \| g\right), \tag{3.30}$$

$$f_{\mathrm{GA}}\left(\mathbf{w}_{\mathrm{subopt}}\right) = \arg\max_{\mathbf{w}} \min_{g \in \mathbb{F}} \sum_{i \in \mathcal{I}} w_i D_{\mathrm{KL}}\left(g \| f_i\right). \tag{3.31}$$

Hereafter, $\mathbf{w} := \{w_1, w_2, \ldots, w_I\}$, where $w_i > 0, \mathbf{w}^T 1_I = 1$.

### 3.4.2 Pros and Cons of AA- and GA-MPD Fusion

First, the targets that are mis-detected by any fusing sensor will not be contained in the GA. That is, misdetection constitutes a challenge to GA fusion. At the same time, the false alarms that occur in any fusing sensor can be entirely suppressed by GA fusion. By contrast, AA fusion preserves all target estimates from all sensors, including false alarms. This leads to a deferred decision-making framework, such that the measurement id is highly uncertain at present can be identified at a later time after further observation has been received [62]. Comparably speaking, GA fusion offers a better false-alarm-suppression capability, but is also more vulnerable to misdetection, whereas AA fusion can better avoid miss-detections but is weaker in rejecting false alarms [65]. Hence, GA fusion may deteriorate significantly with the increase in the number of sensors involved and with the decrease in the target-detection probability [62]. More fusing sensors and a lower target-detection probability imply more frequent misdetection. By contrast, AA fusion performs better as the number of sensors increases: more sensors can better suppress the impact of false alarms.

Second, [64] pointed out that the GA fusion of Bernoulli/Poisson/IIDC MPDs is prone to underestimating the number of targets, while the AA fusion maintains a proper cardinality unbiasedness in fusion [65,84]. Indeed, the AA of two unbiased variables remains unbiased while their GA is generally smaller than their AA. It is relevant that GA fusion has been found to suffer from a certain delay in detecting the newborn targets [64,85–87]. Even more importantly, the false mixand will not affect the good mixand of the information so much in the AA as it does in the GA, yielding a mode-preservation feature [66]. Arguably speaking, when component merging is adaptively applied with the AA fusion and the state estimates are extracted in the MAP mode, they perform equivalently to the estimator that adaptively switches between EAP and MAP. This can be illustrated in Figure 3.2.

Third, GA fusion is more complicated than AA fusion for implementation and computation. In the GM implementation, the power of GM in GA fusion is usually difficult to compute and needs to be approximated by [74,88]

$$\sum [N(\cdot)]^{w_i} \approx \sum [N(\cdot)]^{w_i}. \tag{3.32}$$

This approximation is only valid when Gaussian components are well separated. The particle implementation [89] of GA fusion requires a point-wise multiplication, which means that one MPD cannot be directly fused with others when the supports of particles differ from one another. Therefore, parametric-estimation approaches [90,91] for converting discrete particles into a continuous approximation are often resorted to. By contrast, the GM/particle implementation of AA fusion requires only union and re-weighting operations, which can be computed much faster than GA fusion. Moreover, AA fusion is easy to be paralleled to be performed with respect to those components/particles. This decomposable and uncoupled feature helps to develop parallel/distributed implementations which is still an open research direction of significant practical values for distributed-filtering algorithm design.

### 3.4.3 Formulations of AA- and GA-MPD Fusion

#### *3.4.3.1 Fusion for Unlabeled RFSs*

The following theoretical results play a core role in the AA/GA fusion of various RFS filters. Most of these results were first presented in Ref. [92] for the GA fusion and in Ref. [75] for the AA fusion; note that [75] was submitted for review on Feb. 28, 2019.

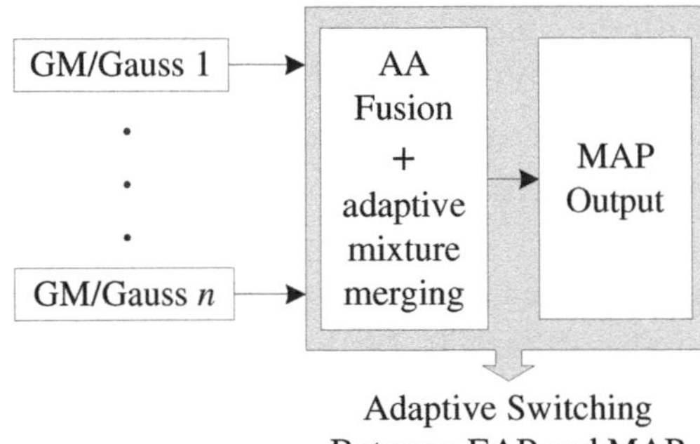


FIGURE 3.2 AA fusion with adaptive mixture merging and MAP performs equivalently as an estimator that adaptively switches between EAP and MAP.

**Theorem 3.1 (AA-Bernoulli)**

*Given the Bernoulli MPD $f_i^{\mathrm{b}}(X)$ from sensor $i \in \mathcal{I}$, the AA of these MPDs is still Bernoulli and has the form*

$$f_w^{\mathrm{b}}(X)=\begin{cases} 1-r_w, & X=\varnothing, \\ r_w s_w(x), & X=\{x\}, \end{cases} \tag{3.33}$$

*where* $r_w = \sum_{i\in\mathcal{I}} w_i r_i$ *and* $s_w(x)=\frac{1}{r_w}\sum_{i\in\mathcal{I}} w_i r_i s_i(x)$.

**Proof 3.1**

*The proof of this theorem can be found in* [75, *Theorem 2*] *and* [93, *Lemma 1*].

**Theorem 3.2 (GA-Bernoulli)**

*Given the Bernoulli MPD $f_i^{\mathrm{b}}(X)$ from sensor $i \in \mathcal{I}$, the GA of these MPDs is still Bernoulli and has the form*

$$f_w^{\mathrm{b}}(X)=\begin{cases} 1-r_w, & X=\varnothing \\ r_w s_w(x), & X=\{x\}, \end{cases} \tag{3.34}$$

*where*

$$r_w = 1-\frac{1}{K}\prod_{i\in\mathcal{I}}(1-r_i)^{w_i}, \tag{3.35}$$

$$s_w(x)=\frac{1}{r_w K}\prod_{i\in\mathcal{I}}[r_i s_i(x)]^{w_i}, \tag{3.36}$$

$$K=\prod_{i\in\mathcal{I}}(1-r_i)^{w_i}+\prod_{i\in\mathcal{I}} r_i^{w_i}\int\prod_{i\in\mathcal{I}}[s_i(x)]^{w_i}\,dx. \tag{3.37}$$

**Proof 3.2**

*The proof can be found in* [92, *Theorem in Sec.III.A*].

**Theorem 3.3 (AA-Poisson)**

*Given the Poisson MPD $f_i^{\mathrm{P}}(X)$ from sensor $i \in \mathcal{I}$, the AA of these MPDs has the form $f_w(X)=\sum_{i\in\mathcal{I}} w\_i\, f_i(X)$ with PHD*

$$D_w(x) = \sum_{i \in \mathcal{I}} w_i D_i(x). \tag{3.38}$$

**Proof 3.3**

*The proof can be found in* [75, *Theorem 3*] *and* [83, *Proposition 2*]. *We further note that the PHD filter does not maintain and update the Poisson MPD, but only its first-order moment—the PHD—over time. Therefore, one does not need to worry about the non-closure of Poisson-MPD-AA fusion. AA fusion is well motivated to be performed over the PHDs* [86,94].

**Theorem 3.4 (GA-Poisson)**

*Given the Poisson MPD $f_i^{\mathrm{P}}(X)$ from sensor $i \in \mathcal{I}$, the GA of these MPDs is still Poisson and has the form*

$$f_w^{\mathrm{P}}(X) = \mathrm{e}^{-\lambda_w} \prod_{x \in X} \lambda_w s_w(x), \tag{3.39}$$

*where* $\lambda_w = \prod_{i \in \mathcal{I}} \lambda_i^{w_i} \cdot \int \prod_{i \in \mathcal{I}} [s_i(x)]^{w_i} dx$, *and* $s_w(x) = \frac{\prod_{i \in \mathcal{I}} [s_i(x)]^{w_i}}{\int \prod_{i \in \mathcal{I}} [s_i(x)]^{w_i} dx}$.

**Proof 3.4**

*The proof can be found in* [92, *Theorem in Sec.III.B*].

**Theorem 3.5 (AA-IIDC)**

*Given the IIDC MPD $f_i^{\mathrm{iidc}}(X)$ from sensor $i \in \mathcal{I}$, the AA fusion of the MPDs has the form $f_{\mathrm{AA}}(X) = \sum_{i \in \mathcal{I}} w\_i f_i(X)$ with PHD $D_{\mathrm{AA}}(x) = \sum_{i \in \mathcal{I}} w_i D_i(x)$ and cardinality distribution $p_{\mathrm{AA}}(n) = \sum_{i \in \mathcal{I}} w_i p_i(n)$, which is unfortunately no longer an IIDC MPD 75, 83 (i.e., it does not have closure) and cannot be used in the next iteration directly. Instead, a closure IIDC-AA fusion with regard to each cardinality |X|=1,2,..., can be given by*

$$f_w^{\mathrm{iidc}}(X\_n) = n!\, p_{\mathrm{AA}}(n) f\_AA(x\_1,...,x\_n), \tag{3.40}$$

*where* $f_{\mathrm{AA}}(\mathrm{x}_1,...,\mathrm{x}_n) = \frac{1}{p_{\mathrm{AA}}(n)} \sum_{i \in \mathcal{I}} w_i p_i(n) f_i(\mathrm{x}_1,...,\mathrm{x}_n)$.

**Proof 3.5**

Combining (3.3) and (3.21) leads to

$$n!p_{\mathrm{AA}}(n)f_{\mathrm{AA}}(\mathrm{x}_1,\ldots,\mathrm{x}_n)=\sum_{i\in\mathcal{I}}w_i n!p_i(n)f_i(\mathrm{x}_1,\ldots,\mathrm{x}_n)$$

which result in the above expression for $f_{\mathrm{AA}}(\mathrm{x}_1,\ldots,\mathrm{x}_n)$ straightforwardly. Further on, one has for the cardinality fusion

$$\begin{aligned}p_{\mathrm{AA}}(n)&=\int_{|X|=n}f_{\mathrm{AA}}(X)\delta X\\&=\sum_{i\in\mathcal{I}}w_i p_i(n)\end{aligned}$$

**Theorem 3.6 (GA-IIDC)**

*Given the IIDC MPD $f_i^{\mathrm{iidc}}(X)$ from sensor $i\in\mathcal{I}$, the GA of these MPDs remains IIDC and has the form*

$$f_w^{\mathrm{iidc}}(X)=n!\,p_w(n)\prod_{x\in X}s_w(x), \tag{3.41}$$

*where* $p_w(n)=\dfrac{\prod_{i\in\mathcal{I}}[p_i(n)]^{w_i}C^n}{\sum_{m=0}^{\infty}\prod_{i\in\mathcal{I}}[p_i(m)]^{w_i}C^m}$, *and* $s_w(x)=\prod_{i\in\mathcal{I}}[s_i(x)]^{w_i}/C$, *and* $C=\int\prod_{i\in\mathcal{I}}[s_i(x)]^{w_i}dx$.

**Proof 3.6**

*The proof can be found in* [92, *Theorem in Sec.III.C*].

**Theorem 3.7 (AA-MBM)**

*Given the MBM MPD $f_i^{\mathrm{iidc}}(X)$ from sensor $i\in\mathcal{I}$, the AA of these MBMs remains an MBM.*

**Proof 3.7**

*The proof can be found in* [77, *Lemma 1*].

TABLE 3.1 Closure Property of GA and AA Fusion for Different MPDs

| | Bernoulli | Poisson | IIDC | MB | GLMB | MBM |
|---|---|---|---|---|---|---|
| AA | Yes | No | No | No | unknown | Yes |
| GA | Yes | Yes | Yes | No | No | No |

TABLE 3.2 Closure Property of GA and AA Fusion in Different Implementations

| | GM | Particles (Dealt Dirac Mixture) |
|---|---|---|
| AA | Yes | Yes |
| GA | No | Depends |

The closure properties[1] of GA and AA fusion for various MPDs are summarized in Table 3.1. However, it remains unclear whether GLMB-AA fusion admits closure. Furthermore, in view of the implementation based on GM or particles, the closure of AA and GA fusion is given in Table 3.2. The closure of the particle-GA fusion depends upon whether the fusing distribution is represented by particles of the same state (and different weights). If so, then GA fusion preserves the same particle states with the GA-fused weights. If, however, the particle states and weights are both different which is more common, then GA fusion must be approximated and no exact closure is guaranteed.

The non-closure of the AA fusion of Poisson and IIDC can be addressed through fusion of PHD and with regard to each cardinality as shown in Theorems 3.3 and 3.5, respectively. The non-closure of the AA/GA fusion of MB MPDs can be addressed in two ways [75]. The first is *fusing after association*, i.e., the BCs in one MB are associated with those in the others, enabling the closure fusion of the Bernoulli processes. This is referred to as the *target-wise fusion rule* [62], where two types of BC-association methods are presented based on either clustering or 2-D assignment. The second solution is to find a reasonable MB approximation to the AA/GA of MBs (e.g., with minor KLD). For example, [96] developed a two-step approximation method to realize GA fusion of MBs. They first approximated the GAs of two MBs to an unlabeled version of the $\delta$-GLMB distribution, which was then approximated to a new MB that matches the PHD of the former.

### 3.4.3.2 Fusion for Labeled RFSs

The fusion of labeled RFSs is more challenging due to the discrete labels. Under the ideal assumption that the labels of all fused sensors are unrealistically matched, the marginalized $\delta$-GLMB and labeled MB MPDs are closed under GA fusion [97], while only the $\delta$-GLMB MPDs are closed under AA fusion [98]. However, they suffer from a severe performance degradation when perfect label consistencies are violated [99–101]. It is unrealistic to have perfect matching between labels from distinct sensors in practice, considering the different geographical coverage and independent birth/clutter/noise processes. As such, some heuristic methods may be employed to deal with the label inconsistency problem.

[1] Closure property implies that the fused MPD still belongs to the same family of fusing MPDs.

TABLE 3.3 Mainstream Multisensor RFS Filters

| | Heuristic | Principled Approximation | |
|---|---|---|---|
| | | *Bayesian parallel combination* | *Multisensor optimal RFS-based filters* |
| Multisensor measurement/ likelihood fusion | IC multisensor filter [38]<br>Measurement clustering [37] | PCAM filter [42,43] | Multisensor Bernoulli filter [45,46]<br>Multisensor PHD filter [38,48]<br>Multisensor CPHD filter [47]<br>Multisensor MeMber filter [52]<br>Multisensor GLMB filter [27,53,54] |
| Multisensor MPD/ moments fusion | Track-to-track-fusion-based<br>PHD filter without feedback [58]<br>LMB filter without feedback [59] | **AA: best fit of mixture**<br>AA-Bernoulli [75,93]<br>AA-PHD [81,82,84,86,94,103]<br>AA-CPHD [73,83,95]<br>AA-MeMber [62]<br>AA-GLMB/LMB [98]<br>AA-PMBM [77]<br>AA-Gaussian [66]<br>AA-Student's t [135] | **GA: GCI/EMD**<br>GA-Bernoulli [92,104]<br>GA-PHD [89,92]<br>GA-CPHD [74,92]<br>GA-MeMber [96]<br>GA-GLMB/LMB [97,100–102] |

One possible solution is *fusing after association* [98,101]. Another method [100,102] is to marginalize the labels of each MPD to form unlabeled ones. Then, the fusion procedure is the same as their unlabeled counterparts, followed by a label-reconstruction step.

Existing labeled RFS-fusion methods essentially only fuse the target-state posterior but not the labels, which serve as no more than an indicator in the fusion. Hence, the target state posterior but not the labels at local sensors are fused. Properly fusing labels remains an open question, which is also an important part of labeled filter estimation. This is mainly due to the fact that the labels are estimates containing historical information (of uncertain length) and any fusion of them will involve backward changing the historical trajectories. Mainstream multisensor RFS filters are summarized in Table 3.3.

## 3.5 REMAINING CHALLENGES AND BEYOND

### 3.5.1 Fusing Weight and Communication Mode

The performance-versus-complexity trade-off is particularly acute in sensor networks since collaboration between sensors comes at the cost of exchanging information between them [105]. Various strategies have been employed to seek the best fusion performance while reducing the cost in both communication and fusion computation.

#### *3.5.1.1 Fusion Weight*

In general, the optimal fusion weights $\mathbf{w}=[w_1,w_2,\ldots]^T$ in GA/AA fusion need to be specified by minimizing a properly designed cost function [66]. In the point estimate viewpoint, the cost function can be the trace or determinant of the fused estimator. In the Bayesian viewpoint, the cost function can be set as Fréchet functions in (3.24)–(3.25), the sum of KLDs in (3.30)–(3.31), the Rényi divergence [106], etc. It is still a quite open problem.

In practice, however, it may turn out to be too complicated or impossible to compute the optimal/suboptimal weights in real time. Instead, the fusion weights $\mathbf{w}$ are often specified directly through prior knowledge. For example, to ensure fast convergence for

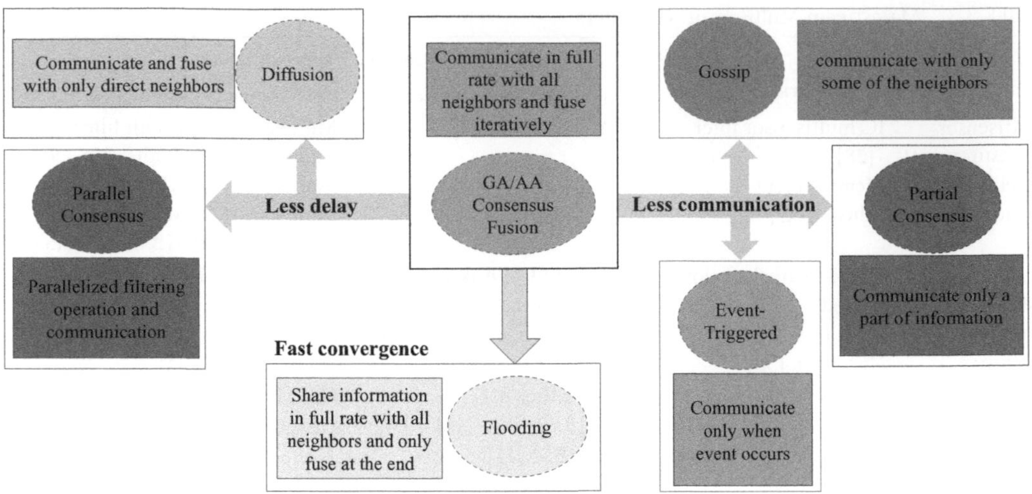


FIGURE 3.3 Various network-communication modes of distinct strengths.

consensus when the network is strongly connected, a popular choice is given by the so-called Metropolis weights [107]. More advanced fusion-weight designs should take into account more network issues, such as dynamics and controllability [45,108]. While in simple linear Gaussian case, the optimized fusing weights have proven useful [66,135], it was found that the KLD optimization as shown in (3.30) and (3.31) is intractable for most RFS distributions, for which sophisticated yet fast approximation methods is needed.

### *3.5.1.2 Communication Mode*

The majority of existing WSN-communication schemes can be classified as shown in Figure 3.3. Lying at the heart of these communication approaches is the distributed-consensus approach, which usually iterates for many rounds and communicates with all neighbors to achieve network convergence [107,109]. However, this protocol has not taken into account the communication constraints and convergence needs in real-time filtering applications. In order to reduce the communication cost, three types of fusion strategies may be employed.

- Under the Gossip algorithm [95,110,111], each sensor only communicates with one/some of its neighboring sensors at each iteration, for which the communication is reduced at the price of slower network convergence.
- In an event-triggered strategy [112,113], communication happens only when an event occurs, with the triggering conditions being a significant difference between the current information and that of the latest broadcast or when the targets of interest are found.
- The partial-consensus method [86,94] only transmits partial information, rather than all information, between sensors in each iteration, while the insignificant residues are retained locally. The significant information (in terms of GCs typically) is

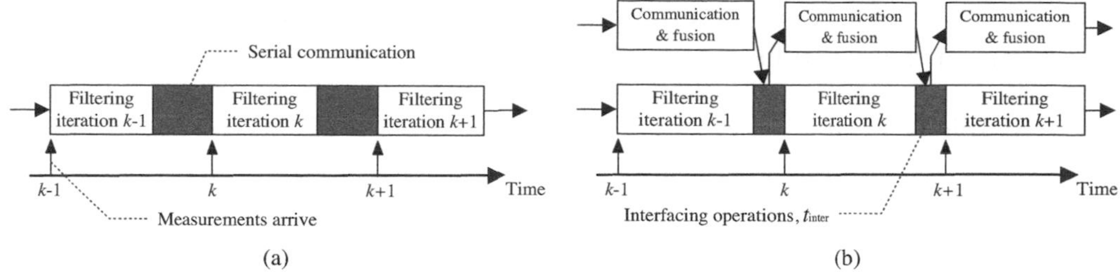


FIGURE 3.4 (a and b) Comparison between serial and parallel filtering-communication modes.

distinguished based on its probabilities/weights, or its locations in the field of view. Of sufficient similarity, a lower costly communication can be achieved by reducing the dimensionality of the state that involved in the communication [114].

To accelerate convergence, the distributed flooding scheme [35] can be applied in which each sensor collects the relevant information via iterative neighborhood communication and performs fusion only at the end of all communication iterations. The distributed flooding algorithm has the best deterministic convergence efficiency in all peer-to-peer protocols and will reach full convergence if the number of iterations is greater than the network's diameter. Furthermore, to reduce the communication delay, the following two network-communication modes can be considered:

- The diffusion strategy [103,115] only requires the sensor to communicate with (all or some of) its neighbor sensors once at each filtering iteration, no matter the convergence.
- The parallel-filtering-communication mode [94,116] aims to carry out filtering and communication/fusion in parallel, thereby preventing any delay caused to the filter due to communication; more detail is given in the following section.

#### 3.5.1.3 *Parallel-Filtering-Communication Mode*

The majority of existing distributed-filtering algorithms rely on a serial filtering-communication mode, which can be illustrated in Figure 3.4a. As shown, the internode communication is performed iteratively with the filtering calculation. To avoid sensor-data delay, the sum of the calculation times for communication and for each filtering iteration must be smaller than the sensing iteration. This mode is, however, restrictive in practice [94,116]. By contrast, the more practical and useful yet difficult mode is the parallel-filtering-communication mode, as shown in Figure 3.4b and originally demonstrated in Ref. [94,116]. Here, communication can be performed in parallel with filtering operations and only a small amount of time (the green block) needs to be reserved for the interfacing operation.

Thanks to the importance-sampling approach, the particle implementation [94] renders a higher level of parallel filtering communication than does the GM implementation [116]. Recently, we further demonstrate the possibility of carrying out the fusion in parallel with the flooding communication [117], where fusion and communication are parallelized largely. Further investigation is certainly significant in parallelizing the filtering and fusion/communication in other distributed filters.

### 3.5.2 Dynamic or Restrictive WSN

#### *3.5.2.1 Sensor Registration*

The majority of existing multisensor tracking algorithms presume that the relative location of each sensor is known to others, which is the basis for relevant information fusion. This, however, is not satisfied when direct measurements of the sensors' location and orientation are unreliable. Thus, there must be more or less bias in matching the coordinates of different sensors.

Assuming that the sensors are independent and their measurements are conditionally independent, we have

$$f_k\left(\mathbf{b},X \mid Z_k^{1:s}\right) \propto \prod_{i=1}^{s} g_k^i\left(Z_k^i \mid X,b^i\right) f_{k|k-1}\left(\mathbf{b},X \mid Z_{k-1}^{1:s}\right), \tag{3.42}$$

where $\mathbf{b}=\left(b^1 \ldots b^s\right)^T$ is the joint-sensor-bias vector. Joint multisensor registration and MTT serves to estimate $\mathbf{b}$ and $X$ simultaneously. A natural idea for this purpose is using the common information concerning the targets that the sensors observe. Representative examples are the Bayesian unified registration and tracking filter [118,119], distributed estimation based on pairwise Markov random-field models [120,121], and the Wasserstein-like distance between state estimates [122]. The above approaches only consider translational biases, but ignore the rotational biases of the coordinate systems. Ref. [123] estimates the rotational biases by minimizing the KLD between the two posterior densities. However, it is necessary to note that the KLD between densities is not only due to the posterior difference in the coordinate systems but is also affected by many factors, such as false alarms and missed detections which can be very different among sensors at any specific time-instants.

More dynamic network issues include time-varying typology and node addition/removal which all affect the fusion-weight design and the algorithm convergence. There have also been fruitful works on sensor control. MTT and sensor control can be unified as a partially observed Markov-decision process, where the control command is usually given by solving an optimization problem, e.g., increasing the observability or reducing the error bound of the system.

#### *3.5.2.2 Limited Field of View*

The fields of view (FoVs) of the sensors are usually limited and point in different directions [124], presenting a great challenge for multisensor fusion. One solution is to extend the prior FoVs of all sensors to an identical and larger FoV before fusion [125,126], but this may underestimate the existence probabilities of targets in non-public areas. Another widely used method is to split the information according to the FoVs' intersection [73,103,127], where the first attempt of the AA fusion under FoVs is given in Ref.

[103] which carried out the fusion only in the common FoVs. Fusion will be executed in each splitting area separately, followed by a reunion operation for all areas. But this method needs to accurately know the FoVs' information of all sensors. The clustering-based method [59,128–130] clusters and fuses the estimates (from different sensors) referring to the same target, which is a way of solving the problem of unknown FoVs. Relevantly, a significant line of research is to constrain the directions and ranges of FoVs through sensor management [131–133] to achieve the balance between the surveillance range and the tracking performance.

#### *3.5.2.3 More Open Issues*

MTT fusion is clearly not a simple extension of single-target information fusion. Simply put, it does not make any sense to fuse between two different targets or between a target and a false alarm. Thus, data association is explicitly or implicitly needed in MTT fusion, for which achievements in traditional/non-RFS-based (multisensor) MTT approaches might be useful. As one of the most focused information-fusion realms, multisensor RFS fusion deserves further research and still faces many challenges. To name a few:

1. *Fusion beyond state:* Fusion of labels when the trajectories of targets are considered; fusion of extension properties when extended targets are considered; fusion between different types of MPDs including model information; fusion of continuous-time T-FoT/trajectories—as addressed in Section 3.2.3—defined via parametric curves like polynomials, non-parametric Gaussian process or even a "black-box" neural network.
2. *Fusion of low cost:* Communication and calculation saving strategies as addressed in Section 3.5.1.2; fusion in parallel to the filtering calculation—as addressed in Section 3.5.1.3—to accommodate different types of filters without networking delay–real-time performance is a need that will never be overemphasized in reality.
3. *Fusion under abnormal/non-standard WSN conditions:* Sensor failure, malicious nodes, time-varying topology and number of nodes, heterogeneous sensor configurations [134], network attack, and so on.

## ACKNOWLEDGMENT

This work was partially supported by National Natural Science Foundation of China (62071389), JWKJW Foundation (2021-JCJQ-JJ-0897, 2020-JCJQ-ZD-150-12), and Key Laboratory Foundation of National Defence Technology (JKWATR-210504).

## REFERENCES

1. Ronald P.S. Mahler. *Advances in Statistical Multisource-Multitarget Information Fusion*, Norwood, MA, USA: Artech House, 2014.
2. Ba-Ngu Vo, Mahendra Mallick, Yaakov Bar-Shalom, Stefano Coraluppi, Richard Richard, Ronald Mahler, and Ba-Tuong Vo. Multitarget Tracking. *Wiley Encyclopedia*, pages 1–15, New York, NY, USA: 2015.

3. S. Hamed Javadi and Alfonso Farina. Radar networks: A review of features and challenges. *Inf. Fusion*, 61: 48–55, 2020.
4. Shaoming He, Hyo-Sang Shin, Shuoyuan Xu, and Antonios Tsourdos. Distributed estimation over a low-cost sensor network: A review of state-of-the-art. *Inf. Fusion*, 54: 21–43, 2020.
5. Kai Da, Tiancheng Li, Yongfeng Zhu, Hongqi Fan, and Qiang Fu. Recent advances in multisensor multitarget tracking using random finite set. *Front. Inf. Technol. Electron. Eng.*, 22(1): 5–24, 2021.
6. Ronald P.S. Mahler. Multitarget Bayes filtering via first-order multitarget moments. *IEEE Trans. Aerosp. Electron. Syst.*, 39(4): 1152–1178, 2003.
7. Ba-Ngu Vo, Sumeetpal Singh, and Arnaud Doucet. Sequential Monte Carlo methods for multitarget filtering with random finite sets. *IEEE Trans. Aerosp. Electron. Syst.*, 41(4): 1224–1245, 2005.
8. Ba-Ngu Vo and Wing-Kin Ma. The Gaussian mixture probability hypothesis density filter. *IEEE Trans. Signal Process.*, 54(11): 4091, 2006.
9. Ronald P.S. Mahler. PHD filters of higher order in target number. *IEEE Trans. Aerosp. Electron. Syst.*, 43(4): 1523–1543, 2007.
10. Ba-Tuong Vo, Ba-Ngu Vo, and Antonio Cantoni. Analytic implementations of the cardinalized probability hypothesis density filter. *IEEE Trans. Signal Process.*, 55(7): 3553–3567, 2007.
11. Ba-Tuong Vo. Random finite sets in multi-object filtering. Ph.D. Dissertation, School of Electrical, Electronic and Computer Engineering, The University of Western Australia, 2008.
12. Branko Ristic, Ba-Tuong Vo, Ba-Ngu Vo, and Alfonso Farina. A tutorial on Bernoulli filters: theory, implementation and applications. *IEEE Trans. Signal Process.*, 61(13): 3406–3430, 2013.
13. Ba Tuong Vo, Ba-Ngu Vo, and Antonio Cantoni. The cardinality balanced multi-target multi-Bernoulli filter and its implementations. *IEEE Trans. Signal Process.*, 57(2): 409–423, 2009.
14. Jason L. Williams. Marginal multi-Bernoulli filters: RFS derivation of MHT, JIPDA, and association-based MeMBer. *IEEE Trans. Aerosp. Electron. Syst.*, 51(3): 1664–1687, 2015.
15. Karl Granstrĺom, Peter Willett, and Yaakov Bar-Shalom. Approximate multi-hypothesis multi-Bernoulli multi-object filtering made multi-easy. *IEEE Trans. Signal Process.*, 64(7): 1784–1797, 2015.
16. Ángel F. García-Fernández, Jason L. Williams, Karl Granstrĺom, and Lennart Svensson. Poisson multi-Bernoulli mixture filter: direct derivation and implementation. *IEEE Trans. Aerosp. Electron. Syst.*, 54(4): 1883–1901, 2018.
17. Yuxuan Xia, Karl Granstrcom, Lennart Svensson, and Ángel F. García-Fernández. Performance evaluation of multi-Bernoulli conjugate priors for multi-target filtering. In *Proc. Int. Conf. Inf. Fusion*, Xi'an, China, pages 1–8, 2017.
18. Julian Smith, Florian Particke, Markus Hiller, and Jĺorn Thielecke. Systematic analysis of the PMBM, PHD, JPDA and GNN multi-target tracking filters. In *Proc. Int. Conf. Inf. Fusion*, Ottawa, ON, Canada, pages 1–8, 2019.
19. Ángel F. García-Fernández, Yuxuan Xia, Karl Granstrĺom, Lennart Svensson, and Jason L Williams. Gaussian implementation of the multi-Bernoulli mixture filter. In *Proc. Int. Conf. Inf. Fusion*, Ottawa, ON, Canada, pages 1–8, 2019.
20. Jason L. Williams. An efficient, variational approximation of the best fitting multi-Bernoulli filter. *IEEE Trans. Signal Process.*, 63(1): 258–273, 2014.
21. Ba-Tuong Vo and Ba-Ngu Vo. Labeled random finite sets and multi-object conjugate priors. *IEEE Trans. Signal Process.*, 61(13): 3460–3475, 2013.
22. Ba-Ngu Vo, Ba-Tuong Vo, and Dinh Phung. Labeled random finite sets and the Bayes multi-target tracking filter. *IEEE Trans. Signal Process.*, 62(24): 6554–6567, 2014.

23. Ba-Ngu Vo and Ba-Tuong Vo. A multi-scan labeled random finite set model for multi-object state estimation. *IEEE Trans. Signal Process.*, 67(19): 4948–4963, 2019.
24. Ba-Ngu Vo, Ba-Tuong Vo, and Hung Gia Hoang. An efficient implementation of the generalized labeled multi-bernoulli filter. *IEEE Trans. Signal Process.*, 65(8): 1975–1987, 2016.
25. Bin Yang, Jun Wang, and Wenguang Wang. An efficient approximate implementation for labeled random finite set filtering. *Signal Process.*, 150: 215–227, 2018.
26. Stephan Reuter, Ba-Tuong Vo, Ba-Ngu Vo, and Klaus Dietmayer. The labeled Multi-Bernoulli filter. *IEEE Trans. Signal Process.*, 62(12): 3246–3260, 2014.
27. Claudio Fantacci and Francesco Papi. Scalable multisensor multitarget tracking using the marginalized $\delta$-GLMB density. *IEEE Signal Process. Lett.*, 23(6): 863–867, 2016.
28. Lingji Chen. From labels to tracks: it's complicated. In *Proc. SPIE Conf. Signal Process. Sensor Fusion Target Recognit. XXVII*, Orlando, Florida, USA, volume 10646, 2018.
29. Angel F. García-Fernández, Lennart Svensson, and Mark R. Morelande. Multiple target tracking based on sets of trajectories. *IEEE Trans. Aerosp. Electron. Syst.*, 56(3): 1685–1707, 2020.
30. Tiancheng Li, Huimin Chen, Shudong Sun, and Juan M. Corchado. Joint smoothing and tracking based on continuous-time target trajectory function fitting. *IEEE Trans. Autom. Sci. Eng.*, 16(3): 1476–1483, 2019.
31. Tiancheng Li. Single-road-constrained positioning based on deterministic trajectory geometry. *IEEE Comm. Lett.*, 23(1): 80–83, 2019.
32. Jinyang Zhou, Tiancheng Li, Xiaoxu Wang, and Litao Zheng. Target tracking with equality/inequality constraints based on trajectory function of time. *IEEE Signal Process. Lett.*, 28: 1330–1334, 2021.
33. Tiancheng Li, Xiaoxu Wang, Yan Liang, Junkun Yan, and Hongqi Fan. A track-oriented approach to target tracking with random finite set observations. In *Proc. Int. Conf. Control Autom. Inf. Sci.*, Chengdu, China, pages 1–6, 2019.
34. Tiancheng Li and Hongqi Fan. From target tracking to targeting track: A data-driven approach to non-cooperative target detection and tracking. 2021. arXiv:2104.11122.
35. Tiancheng Li, Juan M. Corchado, and Javier Prieto. Convergence of distributed flooding and its application for distributed Bayesian filtering. *IEEE Trans. Signal Inf. Process. Netw.*, 3(3): 580–591, 2017.
36. Tiancheng Li, Juan M. Corchado, and Huimin Chen. Distributed flooding-then-clustering: A lazy networking approach for distributed multiple target tracking. In *Proc. Int. Conf. Inf. Fusion*, Cambridge, UK, pages 2415–2422, 2018.
37. Tiancheng Li, Javier Prieto, Hongqi Fan, and Juan M. Corchado. A robust multi-sensor PHD filter based on multi-sensor measurement clustering. *IEEE Commun. Lett.*, 22(10): 2064–2067, 2018.
38. Ronald P.S. Mahler. The multisensor PHD filter: I. general solution via multitarget calculus. In *Proc. SPIE Conf. Signal Process. Sensor Fusion Target Recognit. XVIII*, Orlando, Florida, USA, volume 7336, 2009.
39. Nam Trung Pham, Weimin Huang, and Sim Heng Ong. Multiple sensor multiple object tracking with GMPHD filter. In *Proc. Int. Conf. Inf. Fusion*, Quebec, QC, Canada, pages 1–7, 2007.
40. Sharad Nagappa and Daniel E. Clark. On the ordering of the sensors in the iterated-corrector probability hypothesis density (PHD) filter. In *Proc. SPIE Conf. Signal Process. Sensor Fusion Target Recognit. XVIII*, Orlando, Florida, USA, volume 8050, 2011.
41. Long Liu, Hongbing Ji, and Zhenhua Fan. Improved iterated-corrector PHD with Gaussian mixture implementation. *Signal Process.*, 114:89–99, 2015.
42. Ronald P.S. Mahler. Approximate multisensor CPHD and PHD filters. In *Proc. Int. Conf. Inf. Fusion*, Edinburgh, UK, pages 1–8, 2010.
43. Cheng Ouyang and Hongbing Ji. Scale unbalance problem in product multi-sensor PHD filter. *Electron Lett.*, 47(22): 1247–1249, 2011.

44. Long Liu, Hongbing Ji, and Zhenhua Fan. A cardinality modified product multi-sensor PHD. *Inf. Fusion*, 31: 87–99, 2016.
45. Giuseppe Papa, Rene Repp, Florian Meyer, Paolo Braca, and Franz Hlawatsch. Distributed Bernoulli filtering using likelihood consensus. *IEEE Trans. Signal Inf. Process. Netw.*, 5(2): 218–233, 2018.
46. Ondrej Hlinka, Ondrej Sluciak, Franz Hlawatsch, Petar M. Djuric, and Markus Rupp. Likelihood consensus and its application to distributed particle filtering. *IEEE Trans. Signal Process.*, 60(8): 4334–4349, 2012.
47. Santosh Nannuru, Stephane Blouin, Mark Coates, and Michael Rabbat. Multisensor CPHD filter. *IEEE Trans. Aerosp. Electron. Syst.*, 52(4): 1834–1854, 2016.
48. Emmanuel Delande, Emmanuel Duflos, Philippe Vanheeghe, and Dominique Heurguier. Multi-sensor PHD: Construction and implementation by space partitioning. In *Proc. Int. Conf. Control Autom. Inf. Sci.*, Prague, Czech Republic, pages 3632–3635, 2011.
49. Ronald P.S. Mahler. PHD filters for nonstandard targets, I: Extended targets. In *Proc. Int. Conf. Inf. Fusion*, Seattle, WA, USA, pages 915–921, 2009.
50. Umut Orguner, Christian Lundquist, and Karl Granstrĺom. Extended target tracking with a cardinalized probability hypothesis density filter. In *Proc. Int. Conf. Inf. Fusion*, Chicago, IL, USA, pages 1–8, 2011.
51. Karl Granstrĺom, Christian Lundquist, and Umut Orguner. A Gaussian mixture PHD filter for extended target tracking. In *Proc. Int. Conf. Inf. Fusion*, Edinburgh, UK, pages 1–8, 2010.
52. Augustin-Alexandru Saucan, Mark J. Coates, and Michael Rabbat. A multisensor multi-Bernoulli filter. *IEEE Trans. Signal Process.*, 65(20): 5495–5509, 2017.
53. Baishen Wei, Brett Nener, Weifeng Liu, and Liang Ma. Centralized multi-sensor multi-target tracking with labeled random finite sets. In *Proc. IEEE Int. Conf. Control, Auto. Inf. Sci.*, Ansan, South Korea, pages 82–87, 2016.
54. Ba-Ngu Vo, Ba-Tuong Vo, and Michael Beard. Multi-sensor multi-object tracking with the generalized labeled multi-Bernoulli filter. *IEEE Trans. Signal Process.*, 67(23): 5952–5967, 2019.
55. Weijian Si, Hongfan Zhu, and Zhiyu Qu. Multi-sensor Poisson multi-Bernoulli filter based on partitioned measurements. *IET Radar, Sonar Navig.*, 14(6): 860–869, 2020.
56. Augustus Buonviri, Matthew York, Keith LeGrand, and James Meub. Survey of challenges in labeled random finite set distributed multi-sensor multi-object tracking. In *Proc. IEEE Aerosp. Conf.*, Big Sky, MT, USA, pages 1–12, 2019.
57. Du Yong Kim, Ba-Ngu Vo, Ba-Tuong Vo, and Moongu Jeon. A labeled random finite set online multi-object tracker for video data. *Pattern Recognit.*, 90: 377–389, 2019.
58. Tiancheng Li, Juan M. Corchado, Shudong Sun, and Javier Bajo. Clustering for filtering: Multi-object detection and estimation using multiple/massive sensors. *Inf. Sci.*, 388–389: 172–190, 2017.
59. Hoa Van Nguyen, Hamid Rezatofighi, Ba-Ngu Vo, and Damith C. Ranasinghe. Distributed multi-object tracking under limited field of view sensors. *IEEE Trans. Signal Process.*, 69: 5329–5344, 2021.
60. Tim Bailey, Simon Julier, and Gabriel Agamennoni. On conservative fusion of information with unknown non-Gaussian dependence. In *Proc. Int. Conf. Inf. Fusion*, Singapore, pages 1876–1883, 2012.
61. Bahador Khaleghi, Alaa Khamis, Fakhreddine O. Karray, and Saiedeh N. Razavi. Multisensor data fusion: A review of the state-of-the-art. *Inf. Fusion*, 14(1): 28–44, 2013.
62. Tiancheng Li, Xiaoxu Wang, Yan Liang, and Quan Pan. On arithmetic average fusion and its application for distributed multi-Bernoulli multitarget tracking. *IEEE Trans. Signal Process.*, 68: 2883–2896, 2020.

63. Simon J. Julier, Tim Bailey, and Jeffrey K. Uhlmann. Using exponential mixture models for suboptimal distributed data fusion. In *Proc. IEEE Nonlinear Statist. Signal Process. Workshop*, Cambridge, UK, pages 160–163, 2006.
64. Murat Ī. Uney, Jérémie Houssineau, Emmanuel Delande, Simon J Julier, and Daniel E Clark. Fusion of finite-set distributions: Pointwise consistency and global cardinality. *IEEE Trans. Aerosp. Electron. Syst.*, 55(6): 2759–2773, 2019.
65. Tiancheng Li, Hongqi Fan, Jesús García, and Juan M. Corchado. Second-order statistics analysis and comparison between arithmetic and geometric average fusion: Application to multisensor target tracking. *Inf. Fusion*, 51:233–243, 2019.
66. Tiancheng Li, Yue Xin, Yan Song, Enbin Song, and Hongqi Fan. Some statistic and information-theoretic results on arithmetic average density fusion. 2021. arXiv:2110.01440.
67. Gunther Koliander, Yousef El-Laham, Petar M. Djuric, and Franz Hlawatsch. Fusion of probability density functions. Proceedings of the IEEE, 110(4):404-453, 2022.
68. Mert Kayaalp, Yunus Inan, Emre Telatar, and Ali. H Sayed. On the arithmetic and geometric fusion of beliefs for distributed inference. 2022. arXiv:2204.13741.
69. Murat Ī. Uney, Daniel E. Clark, and Simon J. Julier. Information measures in distributed multitarget tracking. In *Proc. IEEE Int. Conf. Control, Auto. Inf. Sci.*, Chicago, IL, USA, pages 1–8, 2011.
70. Ronald P.S. Mahler. Optimal/robust distributed data fusion: a unified approach. In *Proc. Signal Process., Sensor Fusion, Target Recognit. IX*, Orlando, FL, USA, volume 4052, pages 128–138, 2000.
71. Ronald P.S Mahler. Toward a theoretical foundation for distributed fusion. In David Hall, Chee-Yee Chong, James Llinas, Martin Liggins II, *Distributed Data Fusion for Network-Centric Operations*, pages 199–224. CRC Press, Boca Raton, FL, 2012.
72. Jeffrey K. Uhlmann. General data fusion for estimates with unknown cross covariances. In *Proc. SPIE Conf. Signal Process. Sensor Fusion Target Recognit. V*, Orlando, FL, USA, volume 2755, 1996.
73. Kai Da, Tiancheng Li, Yongfeng Zhu, and Qiang Fu. Gaussian mixture particle jump-Markov-CPHD fusion for multitarget tracking using sensors with limited views. *IEEE Trans. Signal Inf. Process. Netw*, 6: 605–616, 2020.
74. Giorgio Battistelli, Luigi Chisci, Claudio Fantacci, Alfonso Farina, and Antonio Graziano. Consensus CPHD filter for distributed multitarget tracking. *IEEE J. Sel. Top. Sign. Process*, 7(3): 508–520, 2013.
75. Kai Da, Tiancheng Li, Yongfeng Zhu, Hongqi Fan, and Qiang Fu. Kullback-Leibler averaging for multitarget density fusion. In *Proc. Symp. Dsitributed Comput. Artif. Intell.*, Ávila, Spain, pages 253–261, 2019.
76. Daphne Koller and Nir Friedman. *Probabilistic Graphical Models: Principles and Techniques*. MIT Press, Cambridge, MA, 2009.
77. Taincheng Li, Kai Da, Zhunga Liu, Xiaoxu Wang, and Yan Liang. Best fit of mixture for computationally efficient Poisson multi-Bernoulli mixture filtering. Signal Processing DOI: 10.36227/techrxiv.12351710.
78. Tom Heskes. Selecting weighting factors in logarithmic opinion pools. In *Advances in Neural Information Processing Systems*, pages 266–272. MIT Press, Cambridge, MA, 1998.
79. Michael B. Hurley. An information theoretic justification for covariance intersection and its generalization. In *Proc. Int. Conf. Inf. Fusion*, Annapolis, MD, USA, pages 505–511, 2002.
80. Ali E. Abbas. A Kullback-Leibler view of linear and log-linear pools. *Decis. Anal.*, 6(1): 25–37, 2009.
81. Tiancheng Li, Juan M. Corchado, and Shudong Sun. On generalized covariance intersection for distributed PHD filtering and a simple but better alternative. In *Proc. Int. Conf. Inf. Fusion*, Xi'an, China, pages 1–8, 2017.
82. Amirali K. Gostar, Reza Hoseinnezhad, and Alireza Bab-Hadiashar. Cauchy-Schwarz divergence-based distributed fusion with Poisson random finite sets. In *Proc. Int. Conf. Control Autom. Inf. Sci.*, Chiang Mai, Thailand, pages 112–116, 2017.

83. Lin Gao, Giorgio Battistelli, and Luigi Chisci. Multiobject fusion with minimum information loss. *IEEE Signal Process. Lett.*, 27: 201–205, 2020.
84. Tiancheng Li, Franz Hlawatsch, and Petar M. Djurić. Cardinality-consensus-based PHD filtering for distributed multitarget tracking. *IEEE Signal Process. Lett.*, 26(1): 49–53, 2019.
85. Melih Gunay, Umut Orguner, and Mubeccel Demirekler. Chernoff fusion of Gaussian mixtures based on sigma-point approximation. *IEEE Trans. Aerosp. Electron. Syst.*, 52(6): 2732–2746, 2016.
86. Tiancheng Li, Juan M Corchado, and Shudong Sun. Partial consensus and conservative fusion of Gaussian mixtures for distributed PHD fusion. *IEEE Trans. Aerosp. Electron. Syst.*, 55(5): 2150–2163, 2019.
87. Giorgio Battistelli, Luigi Chisci, Claudio Fantacci, Nicola Forti, Alfonso Farina, and Antonio Graziano. Distributed peer-to-peer multitarget tracking with association-based track fusion. In *Proc. Int. Conf. Inf. Fusion*, Salamanca, Spain, pages 1–7, 2014.
88. Simon J. Julier. An empirical study into the use of chernoff information for robust, distributed fusion of Gaussian mixture models. In *Proc. Int. Conf. Inf. Fusion*, Florence, Italy, pages 1–8, 2006.
89. Murat Ĭ. Uney, Daniel E. Clark, and Simon J. Julier. Distributed fusion of PHD filters via exponential mixture densities. *IEEE J. Sel. Top. Signal Process.*, 7(3): 521–531, 2013.
90. Erik B. Sudderth, Alexander T. Ihler, Michael Isard, William T. Freeman, and Alan S. Willsky. Nonparametric belief propagation. *Commun. ACM*, 53(10): 95–103, 2010.
91. Alexandre B. Tsybakov. *Introduction to Nonparametric Estimation*. Springer Science & Business Media, Berlin/Heidelberg, Germany, 2008.
92. Daniel Clark, Simon Julier, Ronald Mahler, and Branko Ristic. Robust multi-object sensor fusion with unknown correlations. In *Proc. IEEE Sens. Signal Process. Defence*, London, UK, pages 1–5, 2010.
93. Tiancheng Li, Zhunga Liu, and Quan Pan. Distributed Bernoulli filtering for target detection and tracking based on arithmetic average fusion. *IEEE Signal Process. Lett.*, 26(12): 1812–1816, 2019.
94. Tiancheng Li and Franz Hlawatsch. A distributed particle-PHD filter using arithmetic-average fusion of Gaussian mixture parameters. *Inf. Fusion*, 73: 111–124, 2021.
95. Jun Ye Yu, Mark Coates, and Michael Rabbat. Distributed multi-sensor CPHD filter using pairwise gossiping. In *Proc. Int. Conf. Control Autom. Inf. Sci.*, Shanghai, China, pages 3176–3180, 2016.
96. Bailu Wang, Wei Yi, Reza Hoseinnezhad, Suqi Li, Lingjiang Kong, and Xiaobo Yang. Distributed fusion with multi-Bernoulli filter based on generalized covariance intersection. *IEEE Trans. Signal Process.*, 65(1): 242–255, 2016.
97. Claudio Fantacci, Ba-Ngu Vo, Ba-Tuong Vo, Giorgio Battistelli, and Luigi Chisci. Robust fusion for multisensor multiobject tracking. *IEEE Signal Process. Lett.*, 25(5): 640–644, 2018.
98. Lin Gao, Giorgio Battistelli, and Luigi Chisci. Fusion of labeled RFS densities with minimum information loss. *IEEE Trans. Signal Process.*, 68: 5855–5868, 2020.
99. Bailu Wang, Wei Yi, Suqi Li, Mark. R Morelande, Lingjiang Kong, and Xiaobo Yang. Distributed multi-target tracking via generalized multi-Bernoulli random finite sets. In *Proc. Int. Conf. Inf. Fusion*, Washington, DC, USA, pages 253–261, 2015.
100. Suqi Li, Wei Yi, Reza Hoseinnezhad, Giorgio Battistelli, Bailu Wang, and Lingjiang Kong. Robust distributed fusion with labeled random finite sets. *IEEE Trans. Signal Process.*, 66(2): 278–293, 2017.
101. Suqi Li, Giorgio Battistelli, Luigi Chisci, Wei Yi, Bailu Wang, and Lingjiang Kong. Computationally efficient multi-agent multi-object tracking with labeled random finite sets. *IEEE Trans. Signal Process.*, 67(1): 260–275, 2019.

102. Meng Jiang, Wei Yi, Reza Hoseinnezhad, and Lingjiang Kong. Distributed multi-sensor fusion using generalized multi-Bernoulli densities. In *Proc. Int. Conf. Inf. Fusion*, Heidelberg, Germany, pages 1332–1339, 2016.
103. Tiancheng Li, Víctor Elvira, Hongqi Fan, and Juan M. Corchado. Local-diffusion-based distributed SMC-PHD filtering using sensors with limited sensing range. *IEEE Sens. J.*, 19(4): 1580–1589, 2019.
104. Mehmet B. Guldogan. Consensus Bernoulli filter for distributed detection and tracking using multi-static Doppler shifts. *IEEE Signal Process. Lett.*, 21(6): 672–676, 2014.
105. Dan Li, Kerry D. Wong, Yu Hen Hu, and Akbar M. Sayeed. Detection, classification, and tracking of targets. *IEEE Signal Process. Mag.*, 19(2): 17–29, 2002.
106. Branko Ristic, Ba-Ngu Vo, and Daniel Clark. A note on the reward function for PHD filters with sensor control. *IEEE Trans. Aerosp. Electron. Syst.*, 47(2): 1521–1529, 2011.
107. Lin Xiao and Stephen Boyd. Fast linear iterations for distributed averaging. *Syst. Control Lett.*, 53(1): 65–78, 2004.
108. Linying Xiang, Fei Chen, Wei Ren, and Guanrong Chen. Advances in network controllability. *IEEE Circuits Syst. Mag.*, 19: 8–21, 2019.
109. Reza Olfati-Saber, J. Alex Fax, and Richard M. Murray. Consensus and cooperation in networked multi-agent systems. *Proc. IEEE*, 95(1): 215–233, 2007.
110. Stephen Boyd, Arpita Ghosh, Balaji Prabhakar, and Devavrat Shah. Randomized gossip algorithms. *IEEE Trans. Inf. Theory*, 52(6): 2508–2530, 2006.
111. Stiven S. Dias and Marcelo G.S. Bruno. Distributed Bernoulli filters for joint detection and tracking in sensor networks. *IEEE Trans. Signal Inf. Process. Netw.*, 2(3): 260–275, 2016.
112. Dawei Shi, Tongwen Chen, and Ling Shi. An event-triggered approach to state estimation with multiple point-and set-valued measurements. *Automatica*, 50(6): 1641–1648, 2014.
113. Lin Gao, Giorgio Battistelli, and Luigi Chisci. Event-triggered distributed multitarget tracking. *IEEE Trans. Signal Inf. Process. Netw.*, 5(3): 570–584, 2019.
114. Bo Chen, Daniel W. C. Ho, Wen-An Zhang, and Li Yu. Distributed dimensionality reduction fusion estimation for cyber-physical systems under dos attacks. *IEEE Trans. Syst., Man, Cybern., Syst.*, 49(2):455–468, 2019.
115. Ali H. Sayed, Petar M. Djurić, and Franz Hlawatsch. *Chapter 6- Distributed Kalman and Particle Filtering* in Proc. Cooperative Graph Signal Process. pages 169–207. New York, NY, USA: Academic Press, 2018.
116. Tiancheng Li, Mahendra Mallick, and Quan Pan. A parallel filtering-communication based cardinality consensus approach for real-time distributed PHD filtering. *IEEE Sens. J.*, 20(22): 13824–13832, 2020.
117. Feng Yang, Litao Zheng, Tiancheng Li, and Lihong Shi. A computationally efficient distributed Bayesian filter with random finite set observations. *Signal Process.*, 194: 108454, 2022.
118. Feng Lian, Chongzhao Han, and Weifeng Liu. Joint spatial registration and multi-target tracking using an extended probability hypothesis density filter. *IET Radar Sonar Nav.*, 5(4): 441–448, 2011.
119. Branko Ristic, Daniel E. Clark, and Neil Gordon. Calibration of multi-target tracking algorithms using non-cooperative targets. *IEEE J. Sel. Top. Signal Process.*, 7(3): 390–398, 2013.
120. Murat Í. Uney, Bernard Mulgrew, and Daniel E Clark. A cooperative approach to sensor localisation in distributed fusion networks. *IEEE Trans. Signal Process.*, 64(5): 1187–1199, 2015.
121. Murat Í. Uney, Bernard Mulgrew, and Daniel E. Clark. Latent parameter estimation in fusion networks using separable likelihoods. *IEEE Trans. Signal Inf. Process. Netw.*, 4(4): 752–768, 2018.
122. Kai Da, Tiancheng Li, Yongfeng Zhu, and Qiang Fu. A computationally efficient approach for distributed sensor localization and multitarget tracking. *IEEE Commun. Lett.*, 24(2): 335–338, 2020.

123. Lin Gao, Giorgio Battistelli, Luigi Chisci, and Ping Wei. Distributed joint sensor registration and multitarget tracking via sensor network. *IEEE Trans. Aerosp. Electron. Syst.*, 56(2): 1301–1317, 2020.
124. Weifeng Liu, Yimei Chen, Hailong Cui, and Quanbo Ge. Multi-sensor tracking with non-overlapping field for the GLMB filter. In *Proc. IEEE Int. Conf. Control, Auto. Inf. Sci.*, Chiang Mai, Thailand, pages 197–202, 2017.
125. Suqi Li, Giorgio Battistelli, Luigi Chisci, Wei Yi, Bailu Wang, and Lingjiang Kong. Multi-sensor multi-object tracking with different fields-of-view using the LMB filter. In *Proc. 21th Int. Conf. Inf. Fusion*, Cambridge, UK, pages 1201–1208, 2018.
126. Giorgio Battistelli, Luigi Chisci, and Arturo Laurenzi. Random set approach to distributed multivehicle SLAM. *IFAC-PapersOnLine*, 50(1):2457–2464, 2017.
127. Jonathan Gan, Milos Vasic, and Alcherio Martinoli. Cooperative multiple dynamic object tracking on moving vehicles based on sequential Monte Carlo probability hypothesis density filter. In *Proc. Intell. Transp. Syst. Conf.*, Rio de Janeiro, Brazil, pages 2163–2170, 2016.
128. Milos Vasic, David Mansolino, and Alcherio Martinoli. A system implementation and evaluation of a cooperative fusion and tracking algorithm based on a Gaussian mixture PHD filter. In *Proc. Int. Conf. Robot. Intell. Syst.*, Daejeon, South Korea, pages 4172–4179, 2016.
129. Wei Yi, Guchong Li, and Giorgio Battistelli. Distributed multi-sensor fusion of PHD filters with different sensor fields of view. *IEEE Trans. Signal Process.*, 68: 5204–5218, 2020.
130. Guchong Li, Giorgio Battistelli, Luigi Chisci, Wei Yi, and Lingjiang Kong. Distributed multi-view multi-target tracking based on CPHD filtering. *Signal Process.*, 188: 108210, 2021.
131. Ronald P.S. Mahler. Unified sensor management using CPHD filters. In *Proc. 10th Int. Conf. Inf. Fusion*, Quebec, QC, Canada, pages 1–7, 2007.
132. Feng Lian, Liming Hou, Jing Liu, and Chongzhao Han. Constrained multi-sensor control using a multi-target MSE bound and a δ-GLMB filter. *Sensors*, 18(7): 2308, 2018.
133. Y. Chen, Q. Zhang, Y. Luo, and K. Li. Multi-target radar imaging based on phased-MIMO technique-Part II: Adaptive resource allocation. *IEEE Sens. J.*, 17(19): 6198–6209, 2017.
134. Benru Yu, Tiancheng Li, Shaojia Ge, and Hong Gu. Robust CPHD fusion for distributed multitarget tracking using asynchronous sensors. *IEEE Sens. J.*, 22(1): 1030–1040, 2022.
135. Tiancheng Li, Zheng Hu, Zhunga Liu, and Xiaoxu Wang. Multi-sensor suboptimal fusion Student's t filter. IEEE Trans. Aerosp. Electron. Syst. arXiv:2204.11098.

CHAPTER 4

# Next-Generation Connected Traffic Using UAVs/Drones

Kashif Naseer Qureshi

*Bahria University*

Ajay Sikandar

*GL Bajaj Institute of Technology and Management*

Piyush Dhawankar

*University of York*

## CONTENTS



## LIST OF ABBREVIATION

| | |
|---|---|
| **AODV** | Ad hoc on-demand distance vector |
| **DAG** | Directed acyclic graph |
| **DDoS** | Distributed denial of service |
| **DOLSR** | Directional optimized link state routing |
| **DSR** | Dynamic source routing |
| **DVRP** | Distance vector routing protocol |
| **ECM** | Electronic countermeasure |
| **EMM-DSR** | Extended max-min dynamic source, routing protocol |
| **GCs** | ground control stations |

DOI: 10.1201/b22998-4

| | |
|---|---|
| **GPS** | Global positioning system |
| **GPSR** | Greedy perimeter stateless routing |
| **HAE** | High altitude endurance |
| **HRC** | Hybrid routing based on clustering |
| **HRP** | Hybrid routing protocol |
| **LCAD** | Load-carry-and-deliver |
| **LSRP** | Link state routing protocol |
| **MAC** | Medium access control |
| **ML-OLSR** | Mobility and load aware OLSR |
| **MPCA** | Mobility prediction clustering algorithm |
| **MPR** | Multi-point relay |
| **MRU** | MAVLink utility relay |
| **OLSR** | Optimized link state routing |
| **POLSR** | Predictive-OLSR |
| **QoE** | Quality of experience |
| **RC5** | Rivest cipher |
| **RGR** | Reactive-greedy-reactive |
| **RSGFF** | Recovery strategy greedy forwarding failure |
| **TBRPF** | Topology broadcast based on reverse-path forwarding |
| **TORA** | Temporally ordered routing algorithm |
| **UANET** | UAV ad hoc network |
| **UAVs** | Unmanned ariel vehicles |
| **UDP** | User datagram protocol |
| **UVAR** | UAV-assisted routing |
| **UVAR-G** | Ground to UAV communication |
| **UVAR-S** | UAV to UAV communication |

## 4.1 UNMANNED ARIEL VEHICLES

The most advanced and popular real-time technology is Unmanned Ariel Vehicles (UAVs). Recently UAVs, commonly known as "drones," have attracted the attention of this world due to their flexible and advanced communication processes [1]. The UAV system is a group of various elements like payloads, control elements, data links, support elements, and system users. The payload is the capacity of an aircraft usually measured in terms of weight. Control elements are the GC and monitoring stations equipped with the latest and modern communication systems. Data links are initiated using radio frequencies for receiving or transmitting information among UAVs with or without prior infrastructure. The routing protocols are used to find the UAV's location information and determine the distance for routing decisions. The most well-known usage of UAVs in the military and defense field is to monitor and sense the hostile environment [2]. It is observed that UAVs are used in other fields as well as for real-time monitoring and sensing fields such as intelligent transportation systems, agriculture systems, weather, and healthcare services [3]. This technology is also useful for security, wireless coverage, and logistics. The UAV is a prominent part of the whole system that is necessary to fly the aircraft even if no pilot is present in the aircraft, but it doesn't mean that it can fly itself [4]. Figure 4.1 shows the UAV networks architecture.

The UAVs are controlled by Ground Control (GC) stations by using wireless and communication technologies. The GC stations need reliable methods for fast and accurate data delivery. These stations require permission from local air traffic authorities for high-altitude flying. These networks are shifted toward user-centric, mobility pattern aware, and Quality of Experience. Then came CRAN [5]. For data forwarding, different routing protocols have been adopted for UAVs like Extended Max-Min Dynamic Source, Routing Protocol, distance vector routing protocol, and flow augmentation routing protocols. The UAV node contains a drone controller, GPS, battery, motors, transmitter, and receiver. Figure 4.2 shows the basic components of a UAV.

The UAVs are operated remotely and designed based on network requirements. The UAV-based network architecture has GCs, payload, and communication systems. Nowadays, the usability of UAVs has been acknowledged in many domains. Normally on the civilian level, the UAVs are used for video/photography, pipeline inspection, agriculture, coastguard,

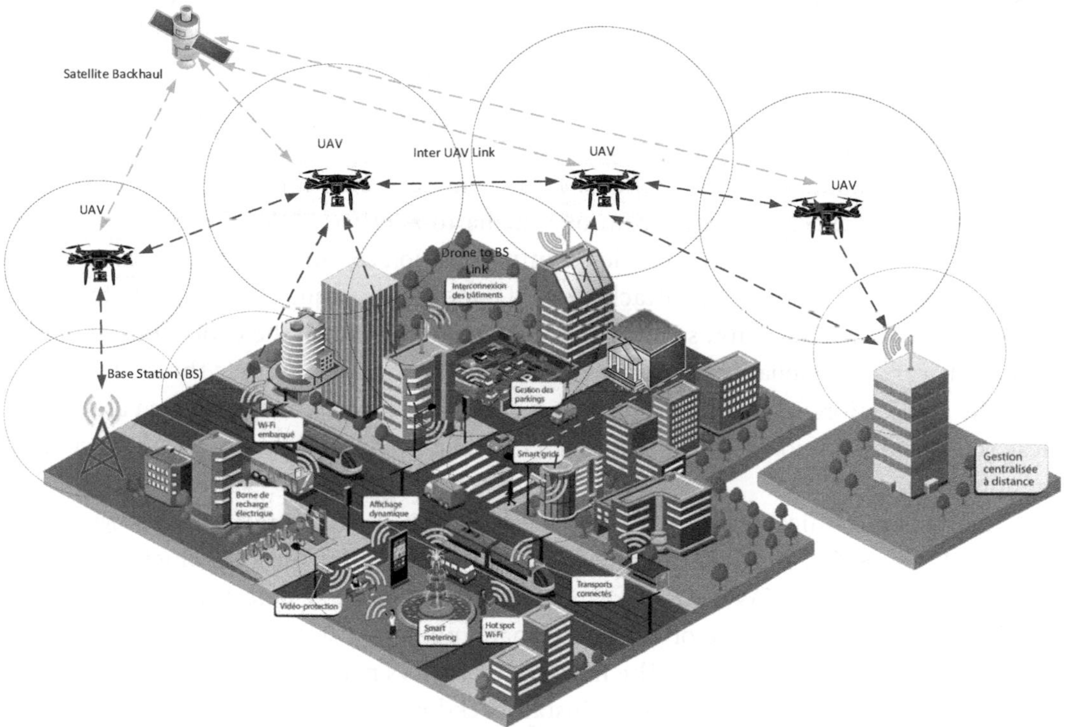


FIGURE 4.1 UAV networks architecture.

FIGURE 4.2 UAV basic structure and components.

marine life, forestry, police authority, and news reporting. Whereas, in the military, the UAVs are playing a major role as combat machines. They are being used for drone attacks, surveillance, border monitoring, and package inspection.

The classification of UAVs can be based on their payload and characteristics. Based on UAV usage these are categorized into tactical UAV, operative UAV, and strategic High Altitude Endurance. The common classification of UAVs consists of domestic (home use), commercial or business, and military. However, not every type has been made secure equally by the manufacturers. For toys or normal domestic use, UAVs are not strong in terms of security mechanisms, while in business or commercial UAVs the limited extent of security mechanisms is used. For military UAVs, the cyber security factor is taken into consideration on many levels. Depending on the role and task of UAVs, different communication links are established by using communication technologies and standards. Different types of network and routing techniques like deterministic, stochastic routing, epidemic routing, coding-based approaches, static, proactive, reactive, and hybrid protocols are used to establish smooth communication in UAVs networks.

The research community is trying to figure out secure communication and routing protocol, and seamless handover from one media to another. UAVs can pose important security and privacy threats. The wireless communication channels attract several types of remote attacks. For example, an attacker could endeavor to acquire critical data by eavesdropping on the wireless link, sending malevolent commands to the drone, or even altering its software functionality. If a drone is hacked, it can pose a serious threat depending upon the application for which it has been used. UAVs are highly sensor-driven assets, and these sensors are mission-critical. The sensors are deployed inside or close to the inspected phenomenon.

Wireless communication channels are used in Unnamed Aircraft Systems which are highly vulnerable. The main concern for the security of UAV is a burglar, which can be able to take the control of a UAV and perform adversary actions, like defeat the mission, deactivate or switch off the engine of UAV, terminate mission, distract UAV from waypoints, and direct to an unknown point. The illegal use of UAVs may cause destruction, monitor sensor status, penetrate the file system, install backdoors, steal information, and can perform many other unwanted activities. The security threats to UAV communication can be categorized into two main types: one is linked to humans while the other is independent of humans. The threats that are independent of human include natural phenomenon which affects the quality of transmission. Environmental ranges like components shelf life, temperature pressure, vibration in-flight, shock, turbulence, power variation, and lightning can change the behavior of electronic components, which UAV can withstand that causes loss of data link, which may lead to the destruction of UAVs. The human-related threats are those that are intentionally posed by humans like Jamming, GPS Spoofing, Eavesdropping, Loss of data confidentiality, and man in the middle. Some of these potential threats are discussed below.

### 4.1.1 GPS Spoofing

The location of the UAV can be faked (it seems to be on a location where it is not) if the latitude and longitude coordinates are sent to the navigation tracker after compromising them. This type of cyber-attack would cause the UAV to divert from its waypoints. Watermarking would make GPS spoofing hard but not impossible.

### 4.1.2 Jamming

Jamming is used to disturb the communication of the target network deliberately causing meddling, saturation, and collision at the receiver end. Frequency hopping, channel surfing, and spatial retreats are methods to avoid jamming.

### 4.1.3 Parameter-Based System Attack

Where autopilot framework intended to be reusable over various arrangements of the airplane and bolster various missions, the flight control can parameterize. Parameters like fuel limit, battery status, and mode state, and pitch, yaw, turning range, waypoints, home area, suitable elevation, and flight plan can be changed by an enemy utilizing Trojan pony that upsets the capacity of reconnaissance and assault and adjusts flight plan furthermore, waypoints. A few potential assault vectors are found against Gimbal System [6]. The UAV camera utilizes GPS area to spare recordings and pictures, which can be modified. The administrator doesn't think of it as an assault rather glitches by any segment.

### 4.1.4 Hardware Security

Hardware equipment security against assembling and configuration is significant. Malevolent equipment usefulness can cause secondary passage [7], information decoding utilizing power utilization examination, and convention identification.

### 4.1.5 Link Integrity/ECM Robustness

Connection respectability or Electronic Countermeasure (ECM) power in a UAV is most significant on the ground where every single principal work like weapons control and flight mode change relies upon information interface framework control and observing. Besides, ECM condition is imagined to be exceptionally serious, as UAVs are wanted to be utilized likewise in past the line-of-sight missions. Connection trustworthiness and ECM strength prerequisites among UAVs and GCs are very challenging.

### 4.1.6 Link Availability and Reliability

Generally, high unwavering quality is required for the information connection. Climate conditions like barometrical blurring, multi-way misfortunes, receiving wires misalignment, and so forth can impact connect accessibility and unwavering quality. The UAV's activities need the information to connect accessibility higher than 99% for controlling and ceaseless observing of sensors yield from UAV.

### 4.1.7 Beyond Line-of-Sight UAV Control

It is associated with the execution of the military UAV missions at a scope of several kilometers.

The information interface framework is required to guarantee high band and low inactivity associations with permit persistent and solid UAV control [8]. In Ref. [9], another calculation is proposed to approve between the UAV and the GC station by including new equipment like NTP server and Raspberry Pi-3 that may cause handling overhead in UAVs for business applications. In Ref. [10], various leveled interruption location and reaction conspire has been proposed yet they didn't talk about the countermeasures of the recognized risk. In Ref. [29], the authors proposed an arranging calculation, which includes well-being highlights if there should be an occurrence of contradictions or blurring system association and the MAVLink Utility Relay. It is a specially crafted programming segment that uses the MAVLink correspondence convention to circulate messages also, directions to numerous areas over sequential or User Datagram protocol. Verifying MAVLink convention is proposed in Ref. [11], which can actualize just when the mission is in progress; however, that may cause handling overhead, more utilization of battery, and system idleness during the mission. Fluffing the MAVLink convention acted in Ref. [12], utilizing altered GCs, and result recognize mistake imperfections of the MAVLink programming execution. HIL cloned GCs are utilized by [13], to abuse the MAVLink convention and propelled extraordinary sorts of cybersecurity assaults on UAV effectively.

### 4.1.8 UAV Threat Model

The threat model is dependent on fundamental data security administrations, classification, trustworthiness, and accessibility. There are pernicious assaults and unintended or normal occasions under the area of three security administrations. Pernicious assaults on UAV correspondence classification are an infection, malware, trogons, Key-lumberjacks, hacking, seizing, convention-based, cross-layer, character satirizing, multiprotocol handover, and listening in. While unintended occasions that may cause loss of classification are social building, covetousness, and life danger. Loss of uprightness can be brought about by noxious assaults on bargaining the connection, increase sign to commotion proportion, subroutine abuses, and tapping, catching feed, and retransmitting. An interloper utilizing DoS can play out the assault on the accessibility or Distributed Denial of Service (DDoS), misrepresenting control signal order.

Further dangers are arranged in pernicious assaults and unattended/normal occasions. AR drones can affect by an infection copter during flight [14]. Skyjack automaton can identify distinctive UAVs during flight and can hack them. Aggressor ramble takes ace control of injured individual automaton. MalDrone can be introduced on the automaton as indirect access and take over full control of UAV [34]. A typical correspondence convention between the business application automaton and GCS is the MAVLink convention. MAVLink convention is attempting to turn into an overall standard for UAV gadgets correspondence yet it does not have any security system actualized. A technique to evaluate the expense of verifying the MAVLink convention through the estimation of system inertness, power utilization, and adventure achievement is proposed. Verifying MAVLink convention is achieved by utilizing cryptographic arrangements yet it includes greater overhead [15]. RC5 encryption calculation as a countermeasure to digital assaults was proposed yet huge computational overhead forestall the UAV working appropriately. By verifying the remote channel isn't sufficient yet secure usage is moreover required.

## 4.2 ROUTING PROTOCOLS FOR UAVs

Drones are used in military and private departments. UAV networks are stabilized with the help of mobile and ad hoc networks. Due to the high mobility and density of traffic in urban areas, the UAV's protocols and algorithms do not work directly with the UAV network. Drones are small in size and have the efficient ability to fly modules that work as a data communication link. To enable effective and useful data packets transmission, a group of UAV's data communicates and cooperates self-organizing into a network is called the UAV ad hoc network (UANET). Compared with the single UAV system the multi UAV systems are better to achieve good results; multi UAV systems have some unique qualities such as fast multitasking capacity, long life network, high performance, and higher scalability. Due to high mobility and high scale implementation, some challenges and problems come in the UAVs network; one of the most common and basic problems is data communication between the UAVs. The UAVs or drone-assisted routing protocols are categories into five classifications including single-hop routing, proactive routing, reactive routing, hybrid routing, and position-based routing [16].

### 4.2.1 Single-Hop Routing Protocols

Single-hop routing is a static routing scheme working without updating the routing table. Drones are used in single-hop routing as a carrier for data transmission from source to destination in UAVs networks, and these protocols use fixed topologies. Single-hop routing protocols are not suitable for dynamic environments. Load-carry-and-deliver (LCAD) routing protocol is a static protocol [17]. The flying UAVs have carried the data packet and delivered it to its destination. The LCAD routing protocol has main three phases including loading packets from the source, carrying the packet when ready to fly, and delivery of the packet to its destination node. LCAD can be used in delay-tolerant networks. However, with the increased distance between source and destination, the packet delivery latency increases. Figure 4.3 shows the single-hop routing example.

### 4.2.2 Multi-Hop Routing Protocols

In multi-hop routing data packets are transferred hop-to-hop. In multi-hop, there are two main classes of routing protocols: position base protocols and topology-based protocols. The topology base is further divided into three main categories including proactive routing

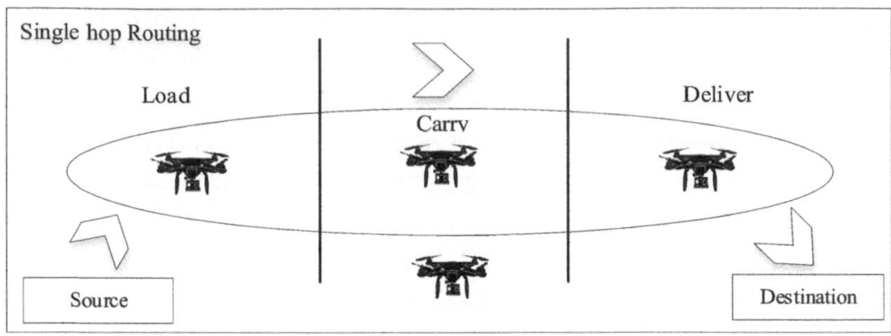


FIGURE 4.3 Single-hop routing example.

protocols, reactive routing protocols, and hybrid routing protocols. In proactive routing protocols [16], tables are predefined and the source to the destination address is stored on the UAV in advance. These protocols are also called active routing protocols. In these types of routing, protocols routing path is selected and transmitted data packet without any wait or query. The disadvantage of proactive routing protocol is when a huge amount of packets are transmitted the control packets need to create the path for data communication but due to high mobility these links cannot be established and the link breaks due to high mobility. Proactive routing protocols are not good for UAV's communication for high mobility networks. However, the disadvantages are also obvious; for example, a large number of control packets are needed for route establishment, which introduces higher communication overhead. Besides, proactive routing protocols are not suitable for high-mobility networks.

Topology Broadcast based on Reverse-Path Forwarding (TBRPF) protocol [18] is the first UANET protocol with physical deployment. In this protocol, the minimum path cost path is selected. Anyhow the minimum cast of hop path is not every time-optimal, especially in the dynamic network along with high-speed UAV networks. In this protocol, minimum hop cast means a high-performance communication link that is used to communicate data packets. Optimized Link State Routing (OLSR) protocol is also used in Ad Hoc Network as compared to other routing protocols (e.g., flooding routing protocols). OLSR reduces the message overhead by using multi-point relay (MPR) into effects. In OLSR MPR selection is very important. Based on OLSR, the new protocol Directional Optimized Link State Routing (DOLSR) [19] is proposed for better performance [4]. In DOLSR, an additional directional is used for data communication. It reduced the overhead of traffic.

First, DOLSR calculates the total distance from the source to destination if the source to destination distance is much longer than $D_{max}/2$ ($D_{max}$ is the maximum distance that can be achieved by a directional antenna), the DOLSR protocol is selected for data transmission otherwise OLSR is used to transmit the data. Compared to OLSR, DOLSR uses less MPR to achieve better performance in terms of lower communication overhead and shorter end-to-end delay [4]. Predictive-OLSR protocol [20] is based on the speed of data transmission, in this protocol, GPS plays a very important role to communicate between two nodes and provide information about the speed of the data packet. UAVs with higher link quality will be selected for data transformation between nodes.

The other protocol based on OLSR is the Mobility and Load aware OLSR (ML-OLSR) protocol. ML-OLSR [21] is similar to Predictive-OLSR proactive routing protocol, which is position base and speed based. The ML-OLSR routing protocol is used to abstain from choosing a rapid node as MPR between neighbor nodes. In ML-OLSR adjacent node is used that is used to avoid packet conflicts, and moreover, the data packet is loaded in every UAV took an account to discover the stable route path without any obstruction.

Reactive routing protocols [1] are on-demand routing protocols. These protocols are also called passive routing protocols. The working of these protocols is when we want to send the data packet to destination these protocols find the path on demand when needed. In reactive protocols control message overhead is efficient but end-to-end delay increases. Dynamic source routing (DSR) [22] is a proactive routing protocol that is effective and suitable for high mobility in the UAV network and is a proactive routing protocol that is better as compared to other protocols concerning performance. DSR maintains the route path when needed, and this is the cause of dynamic link failure. Ad Hoc On-Demand Distance Vector (AODV) is also a reactive routing protocol that appoints committed channels for data transferring and controlling traffic jams to increase the performance of the network. In the AODV routing protocol, there are two stages to process the building routing path when the next hop is not reachable and to improve the data packet delivery ratio.

Based on the AODV routing protocol a new Reactive-Greedy-Reactive (RGR) [23] protocol is planned to increase the performance of the UAV network. In RGR, some hops are performed greatly and, the shortest path does not mean the best or most reliable one. After the RGR protocol, the Modified RGR [24] routing protocol is proposed to improve the route selecting, scalability, and reliability of the data link. In the Modified RGR routing protocol, the high link is used for data packets forwarding. This protocol is based on GPS information because it calculates the distance between two nodes. UAV-assisted routing (UVAR) protocol [25] is proposed for the UAV network to enhance the performance of the network. There are two main phases in the UVAR protocol including UVAR-G (for the ground to UAV communication) and UVAR-S (for UAV to UAV communication). In the UVAR-G phase, the communication mode is ground to UAVs which use multi-hop paths to send the data packet from the ground to UAVs. However, in UVAR-S dynamic routing is used for data packet forwarding. In UAV to UAV communication, the UAV will select the nearest hop for data packet forwarding, and if the UAV created the alternate path and sends the data packet to the destination, the intelligent system will maintain the alternative path otherwise reestablish the new path. UVAR-S is efficient as a reactive routing protocol because it is the demand routing protocol. A comparison is provided in Table 4.1.

The hybrid routing protocols [1] are the combo of reactive routing protocol and proactive routing protocol. In hybrid routing protocols, we find the quality of reactive routing protocol and proactive routing protocol. In a hybrid routing protocol, when the need for proactive routing protocol it uses this protocol otherwise reactive protocol. Hybrid Routing Protocol is a network routing protocol that combines the Distance Vector Routing Protocol and Link State Routing Protocol features. Hybrid Routing Protocol is used to determine optimal network destination routes and report network topology data modifications. Hybrid Routing based on Clustering (HRC) [26] is a proactive routing protocol

TABLE 4.1 Comparison of Proactive and Reactive Routing

| Parameters | Proactive Routing | Reactive Routing |
|---|---|---|
| Provided information on the routing table | Source to destination complete information available | Only accessible when needed |
| Time by time update route | Requisite | Not requisite |
| Cope with motility | Convey the message to other nodes for routing table update | Create a route when needed |
| Dense traffic of control packet | Expeditiously node mobility increasing | Longer than proactive routing protocols |
| Packet communication delay | Directly data packet can be transmitted | Indirectly data packet transmitted |
| Overhead of protocol | High | Low |

that is used to intra-communication (inside the cluster communication it may be nodes communication or it may be with cluster head communication) in a cluster, and while reactive routing protocol is used in inter-communication that means it will communicate with in group or prefix within two groups. Anyhow, the main purpose of HRC is to create an environment for UAVs network and cluster for drone communication in a swarm. In HRC, the ground stations calculate initial 3-D coordination informative data and location information of drone. The node which has the highest energy is selected as a cluster head because of a long communication period, and at the end, clusters are upgraded on the base of the charismatic structure of UAVs. In HRC, routing protocol clusters cannot be updated on time [11].

Mobility Prediction Clustering Algorithm (MPCA) protocol [27] is proposed for the UAV network. In MPCA, link expiration time is calculated by using GPS between the two drones. LET creates the tree structure (tri-structure) and selects the node which has the highest power energy efficiency in the cluster as a cluster head established the connection for UAV communication. Link expiration time enhances the performance of the UAV network. LET sets the limits of clusters and the improvement of network scalable. The disadvantage of MPCA is energy efficiency for a large-scale aggressive network with a huge speed of UAVs. When UAVs move fast, the cluster is required to update fast but the cluster cannot update frequently, and overhead increase.

Temporally Ordered Routing Algorithm (TORA) routing protocol there are different routes established which depend on directed acyclic graph (DAG) graphs. These graphs are used to transfer the data packet from source to destination. When any link breaks due to any reason then the data packets move on an alternative path to covey the data packet to the destination. In TORA, multiple links are created for data packets communication as a backup route. Sometimes multiple paths strategy is not grateful for data communication to avoid link failure in a sparse network. In this situation, a huge amount of control packets is broadcast to every node and overhead increases in the network and delays occur. Due to delay re-establishment of the route take more time and performance goes down. Rapid-reestablish Temporally Ordered Routing Algorithm (RTORA) routing protocol [28] is introduced for UAVs network, and it is the advanced version of the TORA. RTORA routing protocol increased the performance of the UAV network. In the RTORA, routing protocol overhead is reduced as compared to the TORA routing protocol. In TORA link failure is less and better performance in different parameters. This protocol has controlled the useless control packets from flooding that were used in the TORA routing protocol. Simulation results show that the RTORA routing protocol is better than the TORA routing protocol by using different parameters like lower control overhead and better end-to-end delay performance.

Position-Based Routing Protocols are the type of Hybrid Routing Protocols; these protocols are known as Geographic Routing Protocols. Geographic Routing Protocols are GPS enabled to discover the routes from source to destination. In high speed drones routing table maintenance is not easy and possible in proactive routing protocols in an energetic environment. Reactive routing protocols are expensive to maintain the routes again and again for each data packet transmission. Due to reduced cost and increased performance, the researchers introduced Position Base Routing Protocols that provides the location information using GPS.

Greedy Perimeter Stateless Routing (GPSR) [29] is the first geographic routing that is beginning routing protocol proposed for the geographical location in the UAV network. In GPSR, routing protocol greedy forwarding packet scheme is used, and this protocol is implemented in different wireless sensor networks, Rooftop networks, Vehicular networks, Ad-hoc networks, and UAVs networks. GPSR performance as compared to OLSR and AODV routing protocol is not better because GPSR data packet delivery delay is more than OLSR and AODV routing protocol. GPSR depends on greedy search or greedy data forwarding techniques that fail data packet received at the node and there is no neighbor node near to destination node, in this case, the data packet will not send any more to another neighbor node closer to the destination [14].

The other position-based routing protocol was the proposed Recovery Strategy Greedy Forwarding Failure (RSGFF) [30] for better performance in the UAV network. RSGFF has three main plans of action designed to achieve a long-term or overall aim. When the greedy search algorithm fails and data packets are lost or dropped, RSGFF waits and retries. The second RSGFF sends data packets to its near node for data transmission. In the end, the data packet is sent to the fast-moving node which is near the destination. The third choice is best for average data packet delivery transmission.

## 4.3 UAVs BASED OTHER SOLUTIONS

In Multi UAVs, systems communicate by using a wireless connection in an ad hoc manner. UAV network has new features which are different from mobile ad hoc networks. The different routing protocols like distance-based greedy protocols are worked to reduce route setup time in UAVs network. Multi-path non-delay-tolerant network routing protocol is based on geographical information of nodes in the network [31]. In UAVs networks, the high mobility of nodes is a challenging task to handle with limited energy resources in UAVs

[32]. According to the speedy UAVs, the network topologies are continuously changed. The connectivity time is very short and the energy of nodes is limited. To address this problem, the proposed solution is a Multidimensional Perception and Energy Awareness (MPEAOLSR) routing protocol [33]. The main concern is communication between UAVs with ground systems so that ground systems can assign a task to UAVs. Moreover, how different communication standards of UAVs and ground systems can be made.

UAVs have limited energy resources and it is not possible to process all the collected data by UAVs. Another challenge is how communication is possible between Drone and Drone. As different technologies are introduced to design an architecture for UAVs by using different technologies for data communication. Wireless technologies like Wi-Fi, WiMAX, 3G, 4G, and LTE have been adopted for UAVs to UAVs. To overcome the issue of processing data and limited energy of UAVs, the Computation Offloading (CO) technique is used to conserve the energy resources and increase the capacity of processing.

Multiagent UAVs problem raised when multiple UAVs are getting the information from the same operation area. As Each UAV is assigned the same task, they cannot rely on a suit communication surface to interact with each other. We can define the problem as it's like a game where every UAV is the player and has different routes. The above problem can be divided into two parts which are dependent on the environment and application. The formulation of all UAVs is necessary to cooperate or coordinate directly. In this method, UAVs can decide the path on their own [34]. The UAV controls are very difficult sometimes specifically when UAVs are gone beyond the road. To handle this kind of problem, to find the location and share the data solutions have been proposed. In this method, the 920 MHz band is used to cover a large distance area. To address this issue, the authors in [35] proposed the machine learning technique which analyzes the data for communication. This technique reduces the cost of establishing a model and getting more information about UAV communication.

The Extended Max-Min DSR Protocol is another protocol that maximizes energy efficiency through finding the short path which is based on energy. The performance of this protocol is to maintain the end-to-end delay and data throughput. Flow augmentation routing (FAR) [36] is the power optimization protocol. It assumes a static network that finds the optimal routing path from source to destination. It finds the path which minimizes the cost. Routing in UAV is a serious issue because of the sudden change in direction and high speed. Routing protocol periodically sends a hello message to check the establishment and maintenance of the routes. Sometimes the hello message increases the bandwidth. Sending of hello message has fixed interval which increases the bandwidth. Therefore, due to increasing the bandwidth UAV consumes energy. A new method is proposed in Ref. [37], called EE-Hello. The reason for the implementation of this method over the existing protocol is an exchange of messages without change.

There is a static routing protocol that has a routing table computed and loaded during the operation or when the job is started but it can't update during the operation [38]. Another problem of packet forwarding of UAVs network is the unstable link of connection.

To reduce this problem authors in Ref. [39], a proposed solution called location aided delay-tolerant routing protocol which used in the post-disaster operation area.

Another routing protocol is the AODV protocol. This protocol is used when the node has the requirements of communication. The node sends the request to establish a communication link and the node does not need to maintain the path every time and does not need to maintain the path. In this case, the AODV dynamic algorithm provides high-speed moving nodes and creates the routes which will find the destination [36]. Some communities of researchers are getting more interest in autonomous UAVs with the increased availability of wireless communication networks. These networks are very helpful and cooperative to UAVs to achieve their specific task in an ad hoc manner. With the help of these networks, UAVs navigate autonomously and are linked with other nodes via radio links. This connectivity is maintained strongly when drones are in mobile mode. Because through this transmission UAVs continuously achieve data routing from source to destination [40].

To tackle path failure and loss of packets, authors [41] proposed OLSR (enhance Optimize Link State Routing protocol) based on a prediction of mobility and delay prediction abbreviated as in UAV networks. To find the stable neighbor nodes in terms of MPR this mobility prediction model is using the Kalman filter algorithm to improve the routing protocol stability. To achieve the balancing traffic and end-to-end delay a cross-layer delay prediction model is used as a basic requirement in UAVs networks. As nodes are very dynamic and due to this mostly partitioning of networks occurs. To overcome this problem a solution was proposed with a relay node that can connect partitioned networks and provide temporarily alternative backup [42]. Relay node could be anything like a vehicle, personal solider and UAV is very suitable for quick action and coverage. There are still challenges in UAVs network while applying relay node like congestion and in terms of load balancing. To overcome these challenges, the proposed solution uses a relay function routing protocol in an ad hoc network environment. This protocol can control paths according to communication types like pair of destination, source, and services. By using this protocol, the packet delivery ratio is higher and end-to-end delay is very short as compared to the geographical routing protocol

UAVs are very efficient to complete the task in the very crucial environment (search operation, border surveillance, and monitoring of disaster) due to versatility, small operating expenses, flexibility, and ease of installation. However, due to high mobility, changing of topology the existing routing protocols and algorithms in ad hoc networks cannot be directly applied due to some limitations. To solve this problem, the proposed solution [43] called LTA-OLSR (link quality and traffic load aware optimized link-state routing protocol). The link quality scheme can differentiate the link qualities between nodes and its neighbor by RSSI (received signal strength indication) of received packets. A traffic load scheme is introduced to indicate the light load path of Medium Access control (MAC) layer channel information and measure how many packets are stored in the buffer. A comparison is proved in Table 4.2.

TABLE 4.2 Comparison of UAV-Based Solutions

| Protocols | Technique | Advantage | Limitations |
|---|---|---|---|
| UAV-based architecture [44] | Computation offloading (CO) technique | Cloudlets conserve the energy resources and increase the capacity of processing | Path planning, data collection, communication, and networking, data delivery, power consumption, security and privacy |
| UVAR routing protocol [25] | UVAR-G (air-to-ground) and UVAR-S(air-to-air) | Very effective in urban vehicular environments to be aware of traffic issues | Only for urban areas<br>Less security<br>Not for high way or rural areas |
| Genetic algorithm [45] | Iterative genetic algorithm with a rival algorithm | GA, they improve the flight range | Very difficult while applying in a group of UAVs |
| BR-AODV [40] | Flocking-based routing algorithm | Can easily work for moderate size UAVs | Shows inefficiency when applying to large size UAVs |
| OLSR (enhance optimized link state routing protocol) [41] | Kalman filter algorithm | Reduce end-to-end delay and packet loss | Lack of security |
| UAV relay function routing protocol [42] | UAV relayed tactical MANET | Can control path according to communication types (destination, source, and services) | MANET formation and collaboration with ground MANET is just proposed but not implemented |
| LTA-OLSR routing protocols [43] | Link quality scheme | Perform well as compared to OLSR and DSR | Radio propagation and channel dynamics are not covered |
| Optimized link-state routing protocol [46] | Optimized link state routing protocol (OLSR) | Provide the communication without interruption, used GPS to calculate the link node expiration time and residual energy | Just made for the ocean environment. doesn't fit in rainy or wind environment and heightened place |

(Continued)

TABLE 4.2 (*Continued*) Comparison of UAV-Based Solutions

| Protocols | Technique | Advantage | Limitations |
|---|---|---|---|
| AODV, OLSR, and DSDV routing protocols [47] | Comparison between AODV, OLSR, and DSDV | OLSR has the best performance as compared to AODV and DSDV | The experiment was based on speed and box size but data delivery and the end-to-end delay were not covered |
| AODV and OLSR routing protocols [48] | Comparison between AODV and OLSR | AODV performs better in FANETs as compared to OLSR | In this evaluation simulations and the real system, related issues are not resolved |
| Machine learning technique [35] | Data analysis | Reduces the cost establishing model, information about UAVs | Very time-consuming and packet loss is not covered in it |
| GeoUAVs routing protocol [49] | Comparison of GeoUAVs with AntHocNet and BeeAdHoc protocols | Main features achieved by this protocol (nodes mobility, 3D movement with changing, maximum reliability, end-to-end delay) | Just proposed and Real-time calculation is not achieved |
| SIL and clustering schemes [32] | 3D SIL. SIL is worked in UAV | For emergency communication | Very time-consuming architecture and is only suited to a 3D environment. As in the case of UAVs, there can be any environment |
| Game-theoretic framework [34] | Game theory analysis with tight price of anarchy bounds | The first is for formulation all UAVs cooperate and coordinates are directly or some other path or both and the second one is also based on formulation. in this method, they decide the path its own | Deciding the path is very time-consuming and end-to-end delay and packet loss is not covered |
| UAV-assisted routing protocols [50] | Flooding changing in topology | Scheme of different assisting routing protocols to achieve better output | Very time-consuming difficult to suggest which protocol can be used as assistance |

## REFERENCES

1. K. Kumar, S. Kumar, O. Kaiwartya, A. Sikandar, R. Kharel, and J. L. Mauri, "Internet of unmanned aerial vehicles: QoS provisioning in aerial ad-hoc networks," *Sensors*, vol. 20, no. 11, p. 3160, 2020.
2. X. Cai, J. Rodríguez-Piñeiro, X. Yin, N. Wang, B. Ai, G. F. Pedersen, and A. P. Yuste, "An empirical air-to-ground channel model based on passive measurements in LTE," *IEEE Transactions on Vehicular Technology*, vol. 68, no. 2, pp. 1140–1154, 2018.
3. K. Kumar, S. Kumar, O. Kaiwartya, P. K. Kashyap, J. Lloret, and H. Song, "Drone assisted flying ad-hoc networks: Mobility and service oriented modeling using neuro-fuzzy," *Ad Hoc Networks*, vol. 106, p. 102242, 2020.
4. M. Mozaffari, W. Saad, M. Bennis, and M. Debbah, "Efficient deployment of multiple unmanned aerial vehicles for optimal wireless coverage," *IEEE Communications Letters*, vol. 20, no. 8, pp. 1647–1650, 2016.
5. M. Chen, M. Mozaffari, W. Saad, C. Yin, M. Debbah, and C. S. Hong, "Caching in the sky: Proactive deployment of cache-enabled unmanned aerial vehicles for optimized quality-of-experience," *IEEE Journal on Selected Areas in Communications*, vol. 35, no. 5, pp. 1046–1061, 2017.
6. T.D. Tran, J.-M. Thiriet, N. Marchand, A. El Mrabti, and G. Luculli, "Methodology for risk management related to cyber-security of Unmanned Aircraft Systems," in *2019 24th IEEE International Conference on Emerging Technologies and Factory Automation (ETFA)*, Zaragoza, Spain, 2019, pp. 695–702: IEEE.
7. X. Kan, J. Thomas, H. Teng, H. G. Tanner, V. Kumar, and K. Karydis, "Analysis of ground effect for small-scale UAVs in forward flight," *IEEE Robotics and Automation Letters*, vol. 4, no. 4, pp. 3860–3867, 2019.
8. M. Petkovic and M. Narandzic, "Overview of UAV based free-space optical communication systems," in *International Conference on Interactive Collaborative Robotics*, St. Petersburg, Russia, 2019, pp. 270–277, Springer.
9. M. Rodrigues, J. Amaro, F. S. Osório, and B. K. RLJC, "Authentication methods for UAV communication," in *2019 IEEE Symposium on Computers and Communications (ISCC)*, Barcelona, Spain, 2019, pp. 1210–1215: IEEE.
10. H. Sedjelmaci, A. Boudguiga, I. B. Jemaa, and S. M. Senouci, "An efficient cyber defense framework for UAV-Edge computing network," *Ad Hoc Networks*, vol. 94, p. 101970, 2019.
11. B. J. Stinson, A. M. Duncan, B. Vacaliuc, A. Harter, I. Roberts, and T. Thompson, *"MAVNet: Design of a Reliable Beyond Visual Line of Sight Communication System for Unmanned Vehicles,"* Oak Ridge National Lab (ORNL), Oak Ridge, TN, 2019.
12. O. M. Alhawi, M. A. Mustafa, and L. C. Cordeiro, "Finding security vulnerabilities in unmanned aerial vehicles using software verification," *arXiv preprint arXiv*:1906.11488, 2019.
13. A. Allouch, O. Cheikhrouhou, A. Koubâa, M. Khalgui, and T. Abbes, "MAVSec: Securing the MAVLink protocol for ardupilot/PX4 unmanned aerial systems," in *2019 15th International Wireless Communications & Mobile Computing Conference (IWCMC)*, Tangier, Morocco, 2019, pp. 621–628: IEEE.
14. L. Vessels, K. Heffner, and D. Johnson, "Cybersecurity risk assessment for space systems," in *2019 IEEE Space Computing Conference (SCC)*, Pasadena, CA, USA, 2019, pp. 11–19: IEEE.
15. W. W. Weinstein, J. M. Zagami, and J. B. Weader, "Cryptographic system for secure command and control of remotely controlled devices," ed: Google Patents, 2019.
16. J. Jiang and G. Han, "Routing protocols for unmanned aerial vehicles," *IEEE Communications Magazine*, vol. 56, no. 1, pp. 58–63, 2018.
17. C.-M. Cheng, P.-H. Hsiao, H. Kung, and D. Vlah, "Maximizing throughput of UAV-relaying networks with the load-carry-and-deliver paradigm," in *2007 IEEE Wireless Communications and Networking Conference*, Hong Kong, China, 2007, pp. 4417–4424: IEEE.

18. B. Bellur, M. Lewis, and F. Templin, "An ad-hoc network for teams of autonomous vehicles," in *Proceedings of the First Annual Symposium on Autonomous Intelligence Networks and Systems*, 2002, pp. 1–6.
19. A. I. Alshabtat, L. Dong, J. Li, and F. Yang, "Low latency routing algorithm for unmanned aerial vehicles ad-hoc networks," *International Journal of Electrical and Computer Engineering*, vol. 6, no. 1, pp. 48–54, 2010.
20. S. Rosati, K. Krużelecki, L. Traynard, and B. R. Mobile, "Speed-aware routing for UAV ad-hoc networks," in *2013 IEEE Globecom Workshops (GC Wkshps)*, Atlanta, GA, USA, 2013, pp. 1367–1373: IEEE.
21. Y. Zheng, Y. Wang, Z. Li, L. Dong, Y. Jiang, and H. Zhang, "A mobility and load aware OLSR routing protocol for UAV mobile ad-hoc networks," 2014.
22. V. R. Khare, F. Z. Wang, S. Wu, Y. Deng, and C. Thompson, "Ad-hoc network of unmanned aerial vehicle swarms for search & destroy tasks," in *2008 4th International IEEE Conference Intelligent Systems*, Varna, Bulgaria, 2008, vol. 1, pp. 6-65–6-72: IEEE.
23. J. H. Forsmann, R. E. Hiromoto, and J. Svoboda, "A time-slotted on-demand routing protocol for mobile ad hoc unmanned vehicle systems," in *Unmanned Systems Technology IX*, 2007, vol. 6561, p. 65611P: International Society for Optics and Photonics.
24. J.-D. M. M. Biomo, T. Kunz, and M. St-Hilaire, "Routing in unmanned aerial ad hoc networks: Introducing a route reliability criterion," in *2014 7th IFIP wireless and mobile networking conference (WMNC)*, Vilamoura, Portugal, 2014, pp. 1–7: IEEE.
25. O. S. Oubbati, A. Lakas, F. Zhou, M. Güneş, N. Lagraa, and M. B. Yagoubi, "Intelligent UAV-assisted routing protocol for urban VANETs," *Computer Communications*, vol. 107, pp. 93–111, 2017.
26. K. Liu, J. Zhang, and T. Zhang, "The clustering algorithm of UAV networking in near-space," in *2008 8th International Symposium on Antennas, Propagation and EM Theory*, 2008, pp. 1550–1553: IEEE.
27. C. Zang and S. Zang, "Mobility prediction clustering algorithm for UAV networking," in *2011 IEEE GLOBECOM Workshops (GC Wkshps)*, Houston, TX, USA, 2011, pp. 1158–1161: IEEE.
28. Z. Zhai, J. Du, and Y. Ren, "The application and improvement of temporally ordered routing algorithm in swarm network with unmanned aerial vehicle nodes," in *Proceedings of IEEE ICWMC*, Niagara Falls, ON, Canada, 2013, pp. 7–12.
29. M. Hyland, B. E. Mullins, R. O. Baldwin, and M. A. Temple, "Simulation-based performance evaluation of mobile ad hoc routing protocols in a swarm of unmanned aerial vehicles," in *21st International Conference on Advanced Information Networking and Applications Workshops (AINAW'07)*, Niagara Falls, ON, Canada, 2007, vol. 2, pp. 249–256: IEEE.
30. J.-D. M. M. Biomo, T. Kunz, and M. St-Hilaire, "Routing in unmanned aerial ad hoc networks: A recovery strategy for greedy geographic forwarding failure," in *2014 IEEE Wireless Communications and Networking Conference (WCNC)*, Istanbul, Turkey, 2014, pp. 2236–2241: IEEE.
31. M. Y. Arafat and S. Moh, "Routing protocols for unmanned aerial vehicle networks: A survey," *IEEE Access*, vol. 7, pp. 99694–99720, 2019.
32. M. Y. Arafat and S. Moh, "Localization and clustering based on swarm intelligence in UAV networks for emergency communications," *IEEE Internet of Things Journal*, vol. 6, no. 5, pp. 8958–8976, 2019.
33. S. Y. Dong, "Optimization of OLSR routing protocol in UAV ad hoc network," in *2016 13th International Computer Conference on Wavelet Active Media Technology and Information Processing (ICCWAMTIP)*, Chengdu, China, 2016, pp. 90–94: IEEE
34. O. Thakoor, J. Garg, and R. Nagi, "Multiagent UAV routing: A game theory analysis with tight price of anarchy bounds," *IEEE Transactions on Automation Science and Engineering*, vol. 17, no. 1, pp. 100–116, 2019.

35. L. Shan, R. Miura, T. Kagawa, F. Ono, H.-B. Li, and F. Kojima, "Machine learning-based field data analysis and modeling for drone communications," *IEEE Access*, vol. 7, pp. 79127–79135, 2019.
36. H. R. Hussen, S.-C. Choi, J.-h. Park, and J. Kim, "Performance analysis of MANET routing protocols for UAV communications," in *2018 Tenth International Conference on Ubiquitous and Future Networks (ICUFN)*, Prague, Czech Republic, 2018, pp. 70–72: IEEE.
37. I. Mahmud and Y.-Z. Cho, "Adaptive hello interval in FANET routing protocols for green UAVs," *IEEE Access*, vol. 7, pp. 63004–63015, 2019.
38. L. Gupta, R. Jain, and G. Vaszkun, "Survey of important issues in UAV communication networks," *IEEE Communications Surveys & Tutorials*, vol. 18, no. 2, pp. 1123–1152, 2015.
39. M. Y. Arafat and S. Moh, "Location-aided delay tolerant routing protocol in UAV networks for post-disaster operation," *IEEE Access*, vol. 6, pp. 59891–59906, 2018.
40. N. E. H. Bahloul, S. Boudjit, M. Abdennebi, and D. E. Boubiche, "A flocking-based on demand routing protocol for unmanned aerial vehicles," *Journal of Computer Science and Technology*, vol. 33, no. 2, pp. 263–276, 2018.
41. M. Song, J. Liu, and S. Yang, "A mobility prediction and delay prediction routing protocol for UAV networks," in *2018 10th International Conference on Wireless Communications and Signal Processing (WCSP)*, Hangzhou, China, 2018, pp. 1–6: IEEE.
42. B.-S. Kim, K.-I. Kim, B. Roh, and H. Choi, "A new routing protocol for UAV relayed tactical mobile ad hoc networks," in *2018 Wireless Telecommunications Symposium (WTS)*, Phoenix, AZ, USA, 2018, pp. 1–4: IEEE.
43. C. Pu, "Link-quality and traffic-load aware routing for UAV ad hoc networks," in *2018 IEEE 4th International Conference on Collaboration and Internet Computing (CIC)*, Philadelphia, PA, USA, 2018, pp. 71–79: IEEE.
44. N. H. Motlagh, T. Taleb, and O. Arouk, "Low-altitude unmanned aerial vehicles-based internet of things services: Comprehensive survey and future perspectives," *IEEE Internet of Things Journal*, vol. 3, no. 6, pp. 899–922, 2016.
45. M. Karakaya and E. Sevinç, "An efficient genetic algorithm for routing multiple UAVs under flight range and service time window constraints," *Bilişim Teknolojileri Dergisi*, vol. 10, no. 1, p. 113, 2017.
46. P. Xie, "An enhanced OLSR routing protocol based on node link expiration time and residual energy in ocean FANETS," in *2018 24th Asia-Pacific Conference on Communications (APCC)*, Ningbo, China, 2018, pp. 598–603: IEEE.
47. A. C. Zucchi and R. M. Silveira, "Performance analysis of routing protocol for ad hoc UAV network," in *Proceedings of the 10th Latin America Networking Conference*, São Paulo Brazil, 2018, pp. 73–80.
48. A. V. Leonov and G. A. Litvinov, "Applying AODV and OLSR routing protocols to air-to-air scenario in flying ad hoc networks formed by mini-UAVs," in *2018 Systems of Signals Generating and Processing in the Field of on Board Communications*, Moscow, Russia, 2018, pp. 1–10: IEEE.
49. F. Z. Bousbaa, C. A. Kerrache, Z. Mahi, A. E. K. Tahari, N. Lagraa, and M. B. Yagoubi, "GeoUAVs: A new geocast routing protocol for fleet of UAVs," *Computer Communications*, vol. 149, pp. 259–269, 2020.
50. O. Sami Oubbati, N. Chaib, A. Lakas, S. Bitam, and P. Lorenz, "U2RV: UAV-assisted reactive routing protocol for VANETs," *International Journal of Communication Systems*, vol. 33, no. 10, p. e4104, 2020.

CHAPTER 5

# Machine Learning for UAV Communication-Assisted Computing Networks

Quoc-Viet Pham

*Pusan National University*

Zhaohui Yang

*University College London*

Chongwen Huang and Qianqian Yang

*Zhejiang University*

Tianhao Guo

*Shanxi University*

Yongjun Xu

*Chongqing University of Posts and Telecommunications*

Zhaoyang Zhang

*Zhejiang University*

CONTENTS



DOI: 10.1201/b22998-5

## LIST OF ABBREVIATIONS

| | |
|---|---|
| **AANs** | Aerial access networks |
| **AI** | Artificial intelligence |
| **DL** | Deep learning |
| **DRL** | Deep reinforcement learning |
| **5G** | Fifth-generation |
| **FL** | Federated learning |
| **HAPs** | High-altitude platforms |
| **IoT** | Internet-of-Things |
| **KPIs** | Key performance indicators |
| **LoS** | Line-of-sight |
| **LAPs** | Low-altitude platforms |
| **LEO** | Low-Earth-orbit |
| **MEC** | Mobile edge computing |
| **ML** | Machine learning |
| **NFV** | Network function virtualization |
| **QoS** | Quality of service |
| **RANs** | Radio access networks |
| **RL** | Reinforcement learning |
| **6G** | Sixth-generation |
| **SDN** | Software-defined networking |
| **3D** | Three-dimensional |
| **UAVs** | Unmanned aerial vehicles |
| **VMs** | Virtual machines |
| **VR** | Virtual reality |

## 5.1 INTRODUCTION

Fifth-generation (5G) systems are not yet commercially deployed worldwide. However, research communities have started to define roadmaps for future sixth-generation (6G) wireless networks [1,2]. As various novel Internet-of-Things (IoT) applications are set to emerge in 6G, additional data need to be collected and transmitted. However, IoT devices are constrained by factors such as batteries, transmission power, and processing capacity. To facilitate data transmission from IoT devices worldwide, several studies have investigated aerial access networks (AANs), which can offer line-of-sight communication links, favorable channels, and improved coverage. An AAN generally consists of three primary components: low-altitude platforms, high-altitude platforms (HAPs), and low-Earth-orbit (LEO) satellite systems [3]. These platforms fully complement the conventional terrestrial access network to create a future access network in 6G that can provide wireless services with global coverage and diverse quality of service (QoS). Moreover, to better support emerging compute-intensive applications, such as fully autonomous vehicles, holographic communications, and virtual reality, edge computing (e.g., fog computing and mobile edge computing (MEC)) is a promising concept owing to the availability of powerful computing and storage resources at the edge of the network [4,5]. The amalgamation of AANs and edge computing introduces a novel concept referred to as aerial computing. Aerial computing is expected to provide advanced services such as communication, computing, caching, sensing, navigation, and control at a global scale.

Aerial computing is considered a promising paradigm that enables local data analysis and real-time service provisioning in air. Facilitated by advantages such as high mobility, fast deployment, global availability, scalability, and flexibility, aerial computing complements conventional computing paradigms (e.g., cloud computing, fog computing, and MEC) and is thus considered a pillar of the comprehensive computing infrastructure in future 6G wireless systems. For example, in Ref. [6], IoT data over small areas could be collected and processed cooperatively by a set of unmanned aerial vehicles (UAVs) at low altitudes, whereas IoT data over large-scale areas could be executed by HAPs at high altitudes. In Ref. [7], UAVs were leveraged as intermediate points between vehicles and the network edge to enable data collection and task pre-processing and thus facilitate data transfer in smart vertical domain environments. Meanwhile, in the race toward global Internet connectivity among tech giants, such as SpaceX, Google, Amazon, and Facebook, several LEO satellite systems have been deployed to provide Internet services to users, particularly in rural and hard-to-reach areas, with a performance that is comparable to that of the current terrestrial mobile network [8]. However, these systems have not yet been integrated into current computing systems as key components.

The aforementioned observations emphasize the importance of a comprehensive computing architecture for future 6G wireless systems. As a native constituent, aerial computing is composed of low-altitude and high-altitude computing platforms as well as satellite computing platforms. It can carry its full complement of conventional terrestrial computing paradigms. Conceptually, aerial computing can facilitate various smart industrial applications, such as smart cities, smart vehicles, smart factories, and smart grids. Moreover, it is expected to provide a reference 6G computing architecture for future studies.

In recent years, several surveys on edge computing and mobile networks have been conducted. For example, enabling MEC technologies, including virtual machines, network function virtualization, software-defined networking, and network slicing were reviewed in Ref. [9]. As MEC and radio access networks can complement each other well, the synergy between MEC and radio access network concepts has been reviewed in several studies, such as Refs. [10] and [11]. Further, the use of MEC for numerous IoT applications, such as critical/massive IoT communications, wearable IoT, smart energy, and IoT automotive, has been discussed in Ref. [12]. Computation offloading and resource management perspectives, which are key components that enable many edge MEC services and applications, have been discussed in Refs. [13–15]. The amalgamation of MEC and Artificial intelligence (AI), referred to as edge intelligence and intelligent edge, has been reviewed in several studies [16–18]. In addition, certain review articles have focused on the integration of edge computing with emerging technologies such as blockchain [19] and 5G and beyond networks [20]. More recently, privacy/security issues and solutions of MEC services were reviewed [21].

MEC is regarded as a promising technology and a key enabler of many wireless services offering better QoS. Moreover, MEC has the potential to be closely integrated with emerging technologies and systems [12,18,19]. In addition, these studies elaborate on the various challenges of MEC that need to be examined in beyond-5G networks, such as resource optimization, security and privacy, and real-time implementation. With the expected extraordinary popularization of the next-generation Internet and the emergence of killer applications such as VR, space tourism, fully autonomous vehicles, holographic communications, and deep-sea sight, the current MEC infrastructure needs to be further expanded both vertically and horizontally to complement existing edge computing systems. However, the aforementioned reviews [9–21] primarily focus on 5G network scenarios, technologies, and applications, but the new requirements for a comprehensive 6G computing infrastructure have not been reviewed yet.

Future 6G wireless systems have been envisioned in recent studies [22]. For example, the authors of Ref. [10] discussed their vision for use cases, enabling technologies, and overcoming the challenges of 6G networks. In Ref. [23], a set of key performance indicators were defined for 6G, such as peak data rate >100 Gbps, traffic density >100 Tb/s/km$^2$, network density >10 million connections per km$^2$, mobility >1000 km/h, centimeter-level positioning accuracy, reliability >99.999%, receiver sensitivity <–130 dBm, 3-times spectral efficiency, and 10-times energy efficiency. Further, six technical trends were discussed in Ref. [23], namely, coverage expansion, new resources and bandwidth spectrum, new modulation techniques, enhancing the system capacity, adding computing and intelligence capabilities, and three-dimensional (3D) network architecture. Regarding 3D networks, several studies have reviewed UAV communications [24], HAP networks [25], airborne communications [26], satellite communications [27,28], and terrestrial-satellite integrated networks [29]. Recently, the primary aspects of aerial radio access networks, such as architectures, network design, enabling technologies, and applications, were reviewed in the context of 6G [30]. Owing to excellent features and numerous applications, many researchers believe that AI techniques are a key component of many 6G systems. For example, the alliance of machine learning (ML) and privacy in 6G was presented in Ref. [31]. Particularly, ML is considered a double-edged sword for privacy, that is, ML can be leveraged to enhance privacy, which can be violated if the learning model is attacked by intentional parties. Moreover, AI has become an integral part of future network systems such as sensing AI at receivers, on-device AI, access AI (i.e., transmission at PHY, MAC, and network layers), and data-provenance AI [32].

## 5.2 PROJECTS FOR AERIAL COMPUTING IN 6G

Several 6G international projects and standards development organizations are presently working on a global level. In this section, we summarize few research projects and standardization efforts toward a comprehensive 6G computing infrastructure.

The 6G Flagship project focuses on the development of wireless communication techniques, implementation of 5G, and formulation of 6G standards. Flagship conducts large-scale pilot projects through the test network to support the industry in impelling the 5G standard to the commercialization stage. In addition, it develops the basic technical components required for future 6G, including wireless access and collaborative intelligent computing, as well as new applications in these areas. Instead of communicating with people, 6G Flagship discusses the communication between objectives and devices, including low altitude computing (LAC), high altitude computing (HAC), and satellite computing platforms, from the perspective of intelligence and aerial computing.

The Korea Master of Science in Information Technology (MSIT) assumes that 6G technology will realize preemptive development. The data transmission speed based on low-orbit satellites exceeds 10 km, resulting in technological evolution beyond 5G, such as remote surgery, fully automated driving, and flying cars, and reducing the delay time to one-tenth of that of 5G services. The goal is to design core technology for the first commercialization of 6G services in the world and thus prepare for a leading role in the 6G global market. MSIT predicts that the initial 6G network will be deployed in 2028, and the commercialization of this technology will be realized tentatively in 2030. According to the South Korea MSIT 6G research program, the architectural requirements of future wireless networks include the limited computation capability of mobile devices, and ML starts from the initial phase of technology development. This reiterates the potential of aerial computing in solving the limited computation capacity of wireless networks and the importance of integrating ML and aerial computing.

The National Institute of Information and Communication Technology (NICT) released the Beyond 5G/6G and Quantum Network white papers to accelerate R&D cooperation with other research institutions. The white paper summarizes the social visions and use cases that are expected to be realized through the advancement of Beyond 5G/6G and quantum network technology, as well as the key technologies and R&D roadmaps needed to realize them. Moreover, in the white paper, edge computing is listed as an important application for implementing 6G, and an omid-cloud gateway is proposed, which can provide computing, communication, and power resources. A drone that stays close to the served user can be used as a gateway to provide an aerial computing service.

The leading telecommunications industry has launched a number of over-the-air computing and Industry IoT projects. Several Industry IoT standards have been proposed, which are still evolving. One important Industry IoT scenario is the machine-to-machine (M2M) network, which provides a service layer that can be embedded in hardware and software to connect devices. Contiki, a lightweight Unix-like operating system aimed at WSNs, is an open-source operating system (OS) for low-cost, low-power IoT devices. Owing to the low energy budget and the multitude of devices, the combination of aerial computing and over-the-air transmission is a promising solution.

An industry organization supported by the Chinese government has released a 6G technology white paper, focusing on the application of next-generation wireless technology and potential key technologies, which may be ten times faster than the maximum rate of 5G wireless networks. Analysts have stated that the paper will provide clear guidance for industry participants and provide a good start for the 6G R&D plan, which is expected to be implemented in commercial use in approximately 2030 as the competition between China and the United States is intensifying. The China 6G white paper provides the overall vision while expounding right business application use cases, ten emerging key technologies, and views on the development of the industry. The China Academy of Information and Communications Technology IMT-2030 6G Promotion Group released a white paper, which pointed out that 6G applications will present three potential development directions (immersion, intelligence, and generalization), as well as eight business applications, including holographic communication, immersive XR, digital twins, intelligent interactive communication, perception interconnection, and global interconnection. Further, in this white paper, the computing resources include cloud computing (the same as aerial computing), fog computing, and edge computing, which are crucial for intelligent deployment. In addition, for aerial computing, both distributed and centralized deployments have been involved. Moreover, the trend is to implement a multi-tier aerial computing network, including LAC, HAC, and satellite computing platforms.

In Europe, the first batch of 6G projects worth 60 million euros was launched under the 5G-PPP. Among them, the Hexa-X flagship developed the first 6G system concept, supplemented by eight research projects on 6G-specific technologies. These technologies will form the foundation of the people-centered Next-Generation Internet and achieve the Sustainable Development Goals. The committee has passed a legislative proposal for the upcoming European Smart Networks and Services partnership for 6G. According to Europe's vision for the 6G ecosystem, 6G will usher in a new era wherein billions of devices, people, UAVs, vehicles, and robots will produce a massive amount of digital information. In addition, 6G will manage more challenging use cases, such as holographic telepresence and immersive stringent communication requirements. The 2030s are expected to be the beginning of the era of the widespread use of personal mobile robots. Therefore, 6G is envisioned as one of the foundations of human society. To achieve the sustainable progress of society and satisfy the requirements of the United Nations sustainable development, 6G must provide new functions while effectively addressing urgent social needs. This evolution must adhere to the main social values of Europe, such as inclusiveness, transparency, security, and privacy. Further, digital technology is also becoming an indispensable and important means of protecting national sovereignty. Thus, to meet this goal, the Europe 6G program has introduced plans for implementing UAVs and satellites to assist 6G communications, including UAV and satellite-assisted MEC, that is, aerial computing.

The National Science Foundation (NSF) has targeted 6G and launched its largest single public–private partnership, recruiting nine cloud, technology, and telecommunications giants to help academia develop technologies that will define the next generation of networks. Newly launched Resilient and Intelligent Next Generation Systems program partners include Apple, Ericsson, Google, IBM, Intel, Microsoft, Nokia, Qualcomm, and VMware. The US Department of Defense and the National Institute of Standards and Technology are also involved in this program. Through this plan, the NSF will allocate a funding of approximately 40 million USD to promote academic research projects focused on next-generation connectivity. Moreover, the participants in selected projects of the program can collaborate with the aforementioned industry and government partners as well as use the NSF's four wireless test platform facilities to test their ideas. In efforts to build a network beyond 5G, the Cloud Enhanced Open Software-Defined Mobile Wireless Testbed for City-Scale Deployment platform was deployed.

## 5.3 ML FOR AERIAL COMPUTING NETWORKS

AI technology aims to train machines to perform human tasks. It has been applied to various areas, such as image recognition [33], robotic vehicles [34], machine translation [35], and game AI [36]. Further, ML, which is a promising subset of the AI technique to learn from the data and impart intelligence to existing systems, has been employed to render wireless communication and networks highly efficient and adaptable. The advantage of applying ML in wireless network operation is that it enables the network to monitor, learn, and predict various communication-related parameters, such as wireless channels, traffic patterns, user context, and device locations [37]. ML algorithms include supervised learning, unsupervised learning, reinforcement learning (RL), deep learning (DL), deep reinforcement learning (DRL), and federated learning (FL) [38].

1. *Supervised learning/unsupervised learning:* In supervised learning algorithms, both the input and desired outputs of the used datasets are available. However, supervised learning algorithms can only be employed in scenarios with sufficient labeled data, for example, classification. In contrast, in unsupervised learning, the dataset used for training does not include labeled output or target values. The purpose of unsupervised learning is to extract key features of the data for better prediction. Further, unsupervised learning algorithms can be applied to scenarios such as clustering based on the available data.

2. *Deep learning:* DL algorithms have been developed to deal with complex input–output mappings. DL consists of multiple layers for feature extraction and transformation. Learning may be supervised and/or unsupervised. DL can be used for in-depth analysis in a complex scenario with massive data and to realize different control schemes for different protocol layers. For example, in Ref. [39], DL was used to classify cloud images captured by satellite clouds in aerial computing systems.

3. *Reinforcement learning:* The main idea of RL is to train the agent to generate actions according to the current environment. In RL, the problems are solved by employing a sequence of actions that use the trial-and-error rule [37]. RL algorithms have been extensively used in wireless network optimization to obtain the optimal policy, for example, user grouping decisions or actions. DRL is a DL framework developed based on RL; it relies on updated samples in practice instead of the ideal transition probability in theory. DRL involves learning from the feedback that evaluates the actions taken rather than learning from the correct actions. DRL algorithms have been applied to wireless networks for multiple aspects, including mobile networking, resource allocation, schedular design, and routing [40]. Furthermore, it can promptly

make a decision under dynamically changing network conditions, such as channel state information [41].

4. *Federated learning:* FL is a distributed collaborative ML approach that enables users to collaboratively learn a shared model with their data maintained on their own devices [42]. In contrast to a standard ML algorithm that requires a centralized training dataset in the data center, FL allows devices to train a learning model locally and transmit the training parameters instead. Thus, it addresses concerns such as user privacy and limited data transmission resources (e.g., bandwidth) [42,43]. A potential application of aerial computing for FL-enabled wireless networks is aerial aggregation, that is, a computing server (e.g., LAC, HAC, and satellite computing) in aerial computing acts as the aerial server and performs model aggregation in the air, as shown in Figure 5.1. Furthermore, as shown in Figure 5.1, aerial computing can extend the learning coverage of terrestrial FL-enabled networks, particularly when certain obstacles prevent model updating and broadcasting between FL users and the aggregation server.

5. *Big data:* The features of big data have been commonly described by five Vs: volume, variety, velocity, veracity, and value. Certain attempts have been undertaken to apply big data analytics to wireless communications [44]. The increasing complexity of networks and complicated wireless traffic patterns make big data analytics appealing. ML algorithms coupled with computing techniques, specifically edge computing, can be used to process big data in 6G [45]. MEC has become the primary computing method for big data analysis in heterogeneous 6G.

   AI and big data, combined with other enabling technologies mentioned in this section, have been widely introduced to aerial computing in 6G owing to the advantages of intelligent management and automatic self-improvement. In particular, AI/ML can be used in intelligent UAV/drone trajectory design, virtual resource management, data processing, task computation, and channel estimation. A detailed survey on the application of AI/ML tools for UAV-based networks is presented in

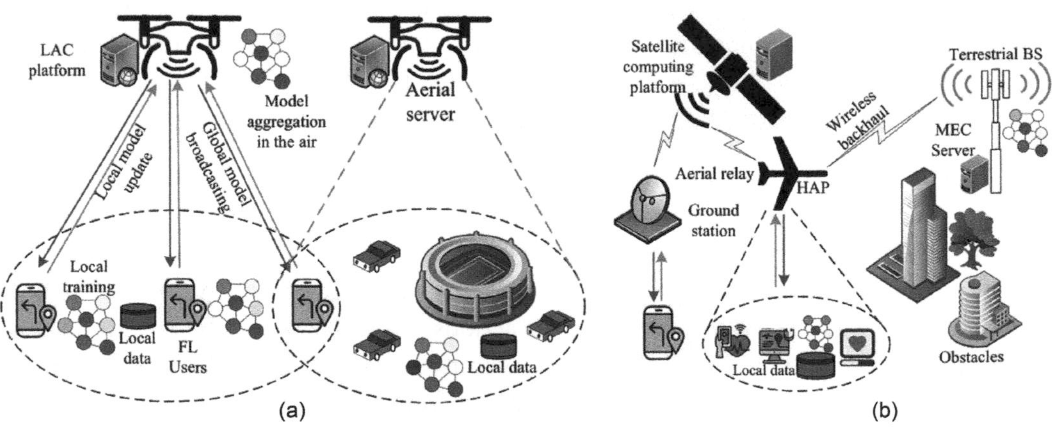


FIGURE 5.1 Applications of aerial computing for FL-enabled wireless systems. (a) FL model aggregation in the air, (b) FL model relaying in the air.

Ref. [37]. In addition, AI and big data technology can be used in network function virtualization and software-defined networking-based networks to achieve intelligent network management and optimization [46]. Studies have also been conducted to use RL approaches to improve resource utilization and reduce service provision [47–49]. Certain challenges need to be addressed in the future to realize aerial computing in 6G, including data security/privacy, learning efficiency, communication cost, and the tradeoff between learning accuracy and convergence. AI/ML is recognized as the most promising solution for providing intelligent wireless communications and signal processing computation task control for increasingly complex and heterogeneous networks.

6. *Summary and discussion:* In this section, we have presented a set of enabling technologies of aerial computing. These technologies can be flexibly implemented in future 6G networks to efficiently provide fast computing services, better mobility, and higher scalability and availability. Further, these technologies can facilitate the process of addressing challenges such as intelligent communication and computation resource allocation design, joint optimization of UAV trajectory and placement design, user association and grouping design, secure computing and communications. However, certain challenges need to be addressed, as highlighted below.

   - *Intelligent Control:* The cooperation of enabling technologies should allow for intelligence in aerial computing. The joint intelligent optimization of computing resources (LAC, HAC, and satellite computing platforms), communication link design with different multi-access techniques, and network control/scheduling can be investigated in future studies.
   - *Secure Control:* For aerial computing, the user data need to be transmitted and shared with the aerial server, which leads to security concerns and data transmission overhead. Therefore, FL and blockchain can be exploited to protect data privacy and improve security by transmitting the training parameter to the server for global aggregation [50,51].
   - *Aerial Intelligence:* AC, HAC, and satellite computing platforms can enable aerial intelligence by collecting surrounding data and then executing onboard intelligent algorithms. The aims of intelligent algorithms include autonomous collision avoidance, adaptive flight gesture adjustment, and trajectory optimization for data collection. Therefore, the design of intelligent algorithms can efficiently support aerial computing in terms of energy savings, large coverage, delay minimization, etc.

## 5.4 CHALLENGES AND FUTURE DIRECTIONS

In this section, we identify interesting research challenges and highlight possible future directions for aerial computing.

### 5.4.1 Energy Efficiency

In aerial computing systems, achieving sustainable energy management at LAC and HAC platforms, as well as energy-efficient satellite computing, is a major concern. Data communications and service delivery, e.g., data transmission for packet delivery in satellite environments and space travel with airplanes, require large energy resources to ensure network operations. In addition, each base station in AANs typically consumes 2.5–4 kW [52]. Thus, the deployment of large-scale aerial networks with thousands of stations at LAC and HAC platforms results in enormous energy consumption, which also increases carbon emissions. Thus, designing energy-efficient communication protocols via optimization and learning techniques is desirable to realize sustainable and green aerial computing systems. For example, the authors of Ref. [53] suggested a space-air-ground architecture with a focus on maximizing the system energy efficiency enabled by the joint optimization of uplink transmission power control, sub-channel selection, and deployment of aerial relays such as UAVs. This can be achieved by dividing the original problem into two sub-problems, optimal sub-channel selection and power control policy, which are obtained by available aerial relay deployment. Further, the policy of aerial relay deployment is then acquired using the solved solution in the first sub-problem. In addition, energy refilling techniques to exploit renewable energy resources can be useful for building sustainable aerial computing systems. Aerial devices and satellites can harvest power from ambient environments, such as wind, solar, vibration, and thermal power, at the LAC platform to support their operations, e.g., data communications over aerial links [54].

### 5.4.2 Resource Management

Aerial computing accommodates data collection and computation tasks to serve end IoT devices via different computing platforms (e.g., LAC and HAC) and support edge services in future 6G networks. Compared to terrestrial computing infrastructures such as cloud data centers, computing platforms in aerial computing possess limited storage resources and battery capabilities, which would hinder the deployment of aerial computing services such as airplane-based civil monitoring for terrestrial IoT environments [55]. Thus, resource management strategies are urgently required to satisfy seamless service delivery in space. Moreover, in the future 6G era, intelligent aerial computing is expected to be a dominant research area, where AI functions can be integrated into aerial devices at the LAC platform to enable self-controlled and autonomous aerial systems. In this context, training DL models on these aerial devices with massive datasets may be infeasible because of the high demands of computation and memory resources, particularly when training with large-scale audio and image data [56]. Therefore, optimizing on-device AI/DL models is of paramount importance for solving the computational burden posed on aerial devices in air. Consequently, several solutions have been proposed to facilitate resource management in aerial computing. For example, the authors of Ref. [57] considered a resource management solution for aerial computing systems wherein a joint optimization of task offloading, resource allocation mechanism, and the trajectory of aerial devices in the air was derived with respect to terrestrial users' latency requirements. Thus, the energy of the aerial devices was minimized, while a longer flight duration was achieved. Moreover, it is also necessary to develop on-device learning solutions to support self-learning aerial computing systems.

For example, improved network architecture, training optimization, and hardware design were used to accelerate on-device data training [58]. A streamlined slimming framework was developed and combined with a consecutive tensor layer to improve the training rates. Simulation results show that the proposed method can enhance the training rate by up to 30% compared to traditional approaches without compromising learning accuracy. Thus, this approach provides opportunities for designing intelligent flying devices in aerial computing environments.

### 5.4.3 Network Stability

The topology of aerial computing fluctuates with the number of nodes, battery levels, and varying space communication conditions. The trajectories and speeds of aerial components and flying devices also vary owing to the terrestrial application requirements and environmental dynamics, e.g., different altitudes of buildings in smart cities in the LAC platform and different orbital altitudes of multiple LEO satellites [59]. Further, the operational complexity and unpredictability of aerial components in space can make the involved aerial computing system unstable. In fact, aerial nodes can join and leave unpredictably, or their flight can be stationary, slow, or fast. Thus, achieving network sustainability in aerial systems is critical. As a promising method, an optimal tracking policy was constructed in Ref. [60] for each aerial device in an aerial-based network to mitigate the varying network topology issues in aerial computing operations, e.g., flying trajectories of UAVs. The key focus was on achieving an adaptive surrounding network configuration for varying channel quality and communication bandwidth resources. Subsequently, a particle swarm optimization algorithm was developed to optimally schedule the energy allocation among a set of aerial nodes, while the prediction error of the surrounding node locations was minimized. Another study in Ref. [61] built an observer, which could monitor surrounding unmanned aerial nodes in an aerial environment using Kalman filtering with respect to the maximum number of parallel targets, measurement time, measurement success rate, and measurement noise. Thus, the optimal measurement policy was obtained for network topology monitoring, aiming to achieve an accurate prediction of trajectory topology in autonomous aerial computing systems.

### 5.4.4 Large-Scale Network Optimization

In aerial computing systems, aerial devices operate on a large scale, and cooperative optimization is needed to utilize the advantages from multiple and distributed network datasets, such as diverse channel features and environment properties, while improving the network quality. For example, performing an optimal trajectory control policy for all manned and unmanned aerial components is a challenge on the HAC platform if only the characteristics of an aerial device are obtained [62]. More importantly, the datasets of future intelligent aerial networks are distributed over large-scale networks rather than being centrally located. Therefore, there is an urgent need for distributed and large-scale optimization approaches to enable scalable and intelligent aerial computing applications. A large-scale trajectory optimization solution was proposed in Ref. [63] for the Internet of aerial devices, for example, UAVs and balloons, through the use of a multi-agent DRL algorithm. This enables aerial devices to collaboratively develop a distributed sense-and-send protocol in an aerial computing setting, facilitating large-scale sensing and data task transfer in cellular networks. Another cooperative optimization method was considered in Ref. [64], wherein UAVs participated in a federated classification process based on the FL concept. Each aerial device can train and classify images of the covered geographic area, captured using onboard cameras, and then transmit the learned parameters to a central server for aggregation.

### 5.4.5 Security, Privacy, and Trust

Although 6G-based aerial computing can offer global coverage and diverse QoS provision to industrial applications, critical issues related to security, privacy, and trust must be solved [65]. Adversaries may attack aerial communication channels and deploy data breaches in the flying BSs on the HAC platform, as the management of lower altitude-based servers is limited owing to the physical distance [66]. AI techniques are extensively used to enable intelligent aerial computing, but they often require centralized data collection for training, which raises potential privacy issues owing to the exposure of sensitive information in the air. Moreover, the deployment of satellite-terrestrial communications in space can face critical challenges caused by untrusted environments, as third parties and attackers can compromise the data exchange among aerial devices, BSs, and terrestrial IoT users. Therefore, blockchain is a promising solution [67] for building trust and establishing secure decentralized satellite-ground communications for aerial computing systems. In this context, aerial components and network providers can communicate securely via immutable block ledgers without the need for central authorities. In addition, smart contracts can be used as self-executing software running on blockchain technology for providing automatic authentication and verification to ensure reliable aerial communications [67]. This technique attains further relevance in the 6G era because aerial computing systems tend to be decentralized and deployed on a large scale, which can be realized using the decentralization feature of blockchain. To preserve data privacy in aerial computing, perturbation techniques such as differential privacy and dummy operations are particularly helpful in protecting data leakage against external threats during data exchange in aerial networks. For example, differential privacy was adopted in Ref. [68] to realize privacy-enhanced intelligent aerial communications by introducing artificial noise into the trained gradients at each aerial device to protect user privacy during data learning. This solution helps hide sensitive information while guaranteeing convergence, such that adversaries cannot retrieve useful data samples.

## 5.5 CONCLUSIONS

Edge computing has become an indispensable component of the present network infrastructure, but it also has various limitations owing to the emergence of many new services and applications and the expansion of the network. We highlighted the fact that the current computing infrastructure does not meet new demands and requirements. In this regard, the concept of aerial computing was introduced and comprehensively reviewed in this study. First, we presented an overview of aerial computing with ML. Finally, we discussed the research challenges and promising future directions pertaining to aerial computing.

As the development of aerial computing is still in a preliminary stage and there are many unexplored issues, we believe that this chapter has revealed certain important lessons and key ideas that will drive further research and unlock the full potential of a comprehensive 6G computing infrastructure in the future.

## REFERENCES


1. H. Tataria, M. Shafi, A. F. Molisch, M. Dohler, H. Sjöland, and F. Tufvesson, "6G wireless systems: Vision, requirements, challenges, insights, and opportunities," *Proceedings of the IEEE*, vol. 109, no. 7, pp. 1166–1199, 2021.
2. C. de Alwis, A. Kalla, Q.-V. Pham, P. Kumar, K. Dev, W.-J. Hwang, and M. Liyanage, "Survey on 6G frontiers: Trends, applications, requirements, technologies and future research," *IEEE Open Journal of the Communications Society*, vol. 2, pp. 836–886, 2021.
3. E. Yaacoub and M.-S. Alouini, "A key 6G challenge and opportunity connecting the base of the pyramid: A survey on rural connectivity," *Proceedings of the IEEE*, vol. 108, no. 4, pp. 533–582, 2020.
4. F. Jiang, K. Wang, L. Dong, C. Pan, W. Xu, and K. Yang, "AI driven heterogeneous MEC system with UAV assistance for dynamic environment: Challenges and solutions," *IEEE Network*, vol. 35, no. 1, pp. 400–408, 2021.
5. Y. Siriwardhana, P. Porambage, M. Liyanage, and M. Ylianttila, "A survey on mobile augmented reality with 5G mobile edge computing: Architectures, applications, and technical aspects," *IEEE Communications Surveys & Tutorials*, vol. 23, no. 2, pp. 1160–1192, 2021.
6. Z. Yu, Y. Gong, S. Gong, and Y. Guo, "Joint task offloading and resource allocation in UAV-enabled mobile edge computing," *IEEE Internet of Things Journal*, vol. 7, no. 4, pp. 3147–3159, 2020.
7. S. Garg, A. Singh, S. Batra, N. Kumar, and L. T. Yang, "UAV-empowered edge computing environment for cyber-threat detection in smart vehicles," *IEEE Network*, vol. 32, no. 3, pp. 42–51, 2018.
8. A. Borrett, "Satellite broadband is the future of the $1trn space economy," Last Accessed 2 Mar. 2021. [Online]. Available: https://techmonitor.ai/internet-of-things/connectivity/satellite-broadband-future-1trn-space-economy.
9. T. Taleb, K. Samdanis, B. Mada, H. Flinck, S. Dutta, and D. Sabella, "On multi-access edge computing: A survey of the emerging 5G network edge cloud architecture and orchestration," *IEEE Communications Surveys & Tutorials*, vol. 19, no. 3, pp. 1657–1681, 2017.
10. M. Peng, S. Yan, K. Zhang, and C. Wang, "Fog-computing-based radio access networks: Issues and challenges," *IEEE Network*, vol. 30, no. 4, pp. 46–53, 2016.
11. Z. Yang, C. Pan, K. Wang and M. Shikh-Bahaei, "Energy Efficient Resource Allocation in UAV-Enabled Mobile Edge Computing Networks," *IEEE Transactions on Wireless Communications*, vol. 18, no. 9, pp. 4576–4589, Sept. 2019.
12. P. Porambage, J. Okwuibe, M. Liyanage, M. Ylianttila, and T. Taleb, "Survey on multi-access edge computing for internet of things realization," *IEEE Communications Surveys & Tutorials*, vol. 20, no. 4, pp. 2961–2991, 2018.
13. Y. Mao, C. You, J. Zhang, K. Huang, and K. B. Letaief, "A survey on mobile edge computing: The communication perspective," *IEEE Communications Surveys & Tutorials*, vol. 19, no. 4, pp. 2322–2358, 2017.
14. P. Mach and Z. Becvar, "Mobile edge computing: A survey on architecture and computation offloading," *IEEE Communications Surveys & Tutorials*, vol. 19, no. 3, pp. 1628–1656, 2017.
15. C.-H. Hong and B. Varghese, "Resource management in fog/edge computing: A survey on architectures, infrastructure, and algorithms," *ACM Computing Surveys*, vol. 52, no. 5, pp. 1–37, 2019.

16. Z. Zhou, X. Chen, E. Li, L. Zeng, K. Luo, and J. Zhang, "Edge intelligence: Paving the last mile of artificial intelligence with edge computing," *Proceedings of the IEEE*, vol. 107, no. 8, pp. 1738–1762, 2019.
17. S. Deng, H. Zhao, W. Fang, J. Yin, S. Dustdar, and A. Y. Zomaya, "Edge intelligence: The confluence of edge computing and artificial intelligence," *IEEE Internet of Things Journal*, vol. 7, no. 8, pp. 7457–7469, 2020.
18. X. Wang, Y. Han, V. C. Leung, D. Niyato, X. Yan, and X. Chen, "Convergence of edge computing and deep learning: A comprehensive survey," *IEEE Communications Surveys & Tutorials*, vol. 22, no. 2, pp. 869–904, 2020.
19. R. Yang, F. R. Yu, P. Si, Z. Yang, and Y. Zhang, "Integrated blockchain and edge computing systems: A survey, some research issues and challenges," *IEEE Communications Surveys & Tutorials*, vol. 21, no. 2, pp. 1508–1532, 2019.
20. Q.-V. Pham, F. Fang, V. N. Ha, M. J. Piran, M. Le, L. B. Le, W.-J. Hwang, and Z. Ding, "A survey of multi-access edge computing in 5G and beyond: Fundamentals, technology integration, and state-of-the-art," *IEEE Access*, vol. 8, pp. 116974–117017, 2020.
21. P. S. Ranaweera, A. D. Jurcut, and M. Liyanage, "Survey on multi-access edge computing security and privacy," *IEEE Communications Surveys & Tutorials Computing*, vol. 23, no. 2, pp. 1078–1124, 2021.
22. F. Tariq, M. R. Khandaker, K.-K. Wong, M. A. Imran, M. Bennis, and M. Debbah, "A speculative study on 6G," *IEEE Wireless Communications*, vol. 27, no. 4, pp. 118–125, 2020.
23. S. Chen, Y.-C. Liang, S. Sun, S. Kang, W. Cheng, and M. Peng, "Vision, requirements, and technology trend of 6G: How to tackle the challenges of system coverage, capacity, user datarate and movement speed," *IEEE Wireless Communications*, vol. 27, no. 2, pp. 218–228, 2020.
24. A. Fotouhi, H. Qiang, M. Ding, M. Hassan, L. G. Giordano, A. Garcia-Rodriguez, and J. Yuan, "Survey on UAV cellular communications: Practical aspects, standardization advancements, regulation, and security challenges," *IEEE Communications Surveys & Tutorials*, vol. 21, no. 4, pp. 3417–3442, 2019.
25. G. Karabulut Kurt, M. G. Khoshkholgh, S. Alfattani, A. Ibrahim, T. S. J. Darwish, M. S. Alam, H. Yanikomeroglu, and A. Yongacoglu, "A vision and framework for the high altitude platform station (HAPS) networks of the future," *IEEE Communications Surveys & Tutorials*, vol. 23, no. 2, pp. 729–779, 2021.
26. X. Cao, P. Yang, M. Alzenad, X. Xi, D. Wu, and H. Yanikomeroglu, "Airborne communication networks: A survey," *IEEE Journal on Selected Areas in Communications*, vol. 36, no. 9, pp. 1907–1926, 2018.
27. O. Kodheli, E. Lagunas, N. Maturo, S. K. Sharma, B. Shankar, J. F. M. Montoya, J. C. M. Duncan, D. Spano, S. Chatzinotas, S. Kisseleff, et al., "Satellite communications in the new space era: A survey and future challenges," *IEEE Communications Surveys & Tutorials*, vol. 23, no. 1, pp. 70–109, 2021.
28. N. Saeed, A. Elzanaty, H. Almorad, H. Dahrouj, T. Y. Al-Naffouri, and M.-S. Alouini, "CubeSat communications: Recent advances and future challenges," *IEEE Communications Surveys & Tutorials*, vol. 22, no. 3, pp. 1839–1862, 2020.
29. P. Wang, J. Zhang, X. Zhang, Z. Yan, B. G. Evans, and W. Wang, "Convergence of satellite and terrestrial networks: A comprehensive survey," *IEEE Access*, vol. 8, pp. 5550–5588, 2019.
30. N.-N. Dao, Q.-V. Pham, N. H. Tu, T. T. Thanh, V. N. Q. Bao, D. S. Lakew, and S. Cho, "Survey on aerial radio access networks: Toward a comprehensive 6G access infrastructure," *IEEE Communications Surveys & Tutorials Computing*, vol. 23, no. 2, pp. 1193–1225, 2021.
31. Y. Sun, J. Liu, J. Wang, Y. Cao, and N. Kato, "When machine learning meets privacy in 6G: A survey," *IEEE Communications Surveys & Tutorials*, vol. 22, no. 4, pp. 2694–2724, 2020.
32. D. C. Nguyen, P. Cheng, M. Ding, D. Lopez-Perez, P. N. Pathirana, J. Li, A. Seneviratne, Y. Li, and H. V. Poor, "Enabling AI in future wireless networks: A data life cycle perspective," *IEEE Communications Surveys & Tutorials*, vol. 23, no. 1, pp. 553–595, 2021.

33. J. Zhang, S. L. Li, and X. L. Zhou, "Application and analysis of image recognition technology based on artificial intelligence – machine learning algorithm as an example," in *2020 International Conference on Computer Vision, Image and Deep Learning (CVIDL)*, Nanchang, China, 2020, pp. 173–176.
34. L. Chen, S. Lin, X. Lu, D. Cao, H. Wu, C. Guo, C. Liu, and F.-Y. Wang, "Deep neural network based vehicle and pedestrian detection for autonomous driving: A survey," *IEEE Transactions on Intelligent Transportation Systems*, vol. 22, no. 6, pp. 3234–3246, 2021.
35. K. Chen, T. Zhao, M. Yang, L. Liu, A. Tamura, R. Wang, M. Utiyama, and E. Sumita, "A neural approach to source dependence based context model for statistical machine translation," *IEEE/ACM Transactions on Audio, Speech, and Language Processing*, vol. 26, no. 2, pp. 266–280, 2018.
36. G. N. Yannakakis and J. Togelius, "A panorama of artificial and computational intelligence in games," *IEEE Transactions on Computational Intelligence and AI in Games*, vol. 7, no. 4, pp. 317–335, 2015.
37. P. S. Bithas, E. T. Michailidis, N. Nomikos, D. Vouyioukas, and A. G. Kanatas, "A survey on machine-learning techniques for UAV-based communications," *Sensors*, vol. 19, no. 23, p. 5170, 2019.
38. N. C. Luong, D. T. Hoang, S. Gong, D. Niyato, P. Wang, Y.-C. Liang, and D. I. Kim, "Applications of deep reinforcement learning in communications and networking: A survey," *IEEE Communications Surveys & Tutorials*, vol. 21, no. 4, pp. 3133–3174, 2019.
39. C. Bai, M. Zhang, J. Zhang, J. Zheng, and S. Chen, "LSCIDMR: Large-scale satellite cloud image database for meteorological research," *IEEE Transactions on Cybernetics*, 2021, in press.
40. C. She, C. Sun, Z. Gu, Y. Li, C. Yang, H. V. Poor, and B. Vucetic, "A tutorial on ultrareliable and low-latency communications in 6G: Integrating domain knowledge into deep learning," *Proceedings of the IEEE*, vol. 109, no. 3, pp. 204–246, 2021.
41. L. Qian, Y. Wu, F. Jiang, N. Yu, W. Lu, and B. Lin, "NOMA assisted multi-task multi-access mobile edge computing via deep reinforcement learning for industrial internet of things," *IEEE Transactions on Industrial Informatics*, vol. 17, no. 8, pp. 5688–5698, 2021.
42. M. Chen, Z. Yang, W. Saad, C. Yin, H. V. Poor, and S. Cui, "A joint learning and communications framework for federated learning over wireless networks," *IEEE Transactions on Wireless Communications*, vol. 20, no. 1, pp. 269–283, 2021.
43. Q.-V. Pham, M. Zeng, R. Ruby, T. Huynh-The, and W.-J. Hwang, "UAV communications for sustainable federated learning," *IEEE Transactions on Vehicular Technology*, vol. 70, no. 4, pp. 3944–3948, 2021.
44. A. Imran, A. Zoha, and A. Abu-Dayya, "Challenges in 5G: How to empower SON with big data for enabling 5G," *IEEE Network*, vol. 28, no. 6, pp. 27–33, 2014.
45. Y. Liu, S. Bi, Z. Shi, and L. Hanzo, "When machine learning meets big data: A wireless communication perspective," *IEEE Vehicular Technology Magazine*, vol. 15, no. 1, pp. 63–72, 2020.
46. A. Guo, C. Yuan, G. He, and L. Xu, "Research on SDN/NFV network traffic management and optimization based on big data and artificial intelligence," in *2018 18th International Symposium on Communications and Information Technologies (ISCIT)*, Bangkok , Thailand, 2018, pp. 377–382.
47. S. I. Kim and H. S. Kim, "A research on dynamic service function chaining based on reinforcement learning using resource usage," in *2017 Ninth International Conference on Ubiquitous and Future Networks (ICUFN)*, Milan, Italy, 2017, pp. 582–586.
48. K. He, X. Zhang, S. Ren, and J. Sun, "Deep Residual Learning for Image Recognition," in *2016 IEEE Conference on Computer Vision and Pattern Recognition (CVPR)*, 2016, Las Vegas, NV, USA, pp. 770–778.
49. A. J. Davison, I. D. Reid, N. D. Molton, and O. Stasse, "MonoSLAM: Real-Time Single Camera SLAM," *IEEE Transactions on Pattern Analysis and Machine Intelligence*, vol. 29, no. 6, pp. 1052–1067, 2007.

50. W. Y. B. Lim, N. C. Luong, D. T. Hoang, Y. Jiao, Y.-C. Liang, Q. Yang, D. Niyato, and C. Miao, "Federated learning in mobile edge networks: A comprehensive survey," *IEEE Communications Surveys & Tutorials*, vol. 22, no. 3, pp. 2031–2063, 2020.
51. D. C. Nguyen, M. Ding, Q.-V. Pham, P. N. Pathirana, L. B. Le, A. Seneviratne, J. Li, D. Niyato, and H. V. Poor, "Federated learning meets blockchain in edge computing: Opportunities and challenges," *IEEE Internet of Things Journal*, 2021, in press.
52. S. Kandeepan, K. Gomez, L. Reynaud, and T. Rasheed, "Aerial-terrestrial communications: Terrestrial cooperation and energy-efficient transmissions to aerial base stations," *IEEE Transactions on Aerospace and Electronic Systems*, vol. 50, no. 4, pp. 2715–2735, 2014.
53. Z. Li, Y. Wang, M. Liu, R. Sun, Y. Chen, J. Yuan, and J. Li, "Energy efficient resource allocation for UAV-assisted space-air-ground internet of remote things networks," *IEEE Access*, vol. 7, pp. 145348–145362, 2019.
54. Z. Yang, W. Xu, and M. Shikh-Bahaei, "Energy efficient UAV communication with energy harvesting," *IEEE Transactions on Vehicular Technology*, vol. 69, no. 2, pp. 1913–1927, 2019.
55. R. Verdone and S. Mignardi, "Joint aerial-terrestrial resource management in UAV-aided mobile radio networks," *IEEE Network*, vol. 32, no. 5, pp. 70–75, 2018.
56. U. Challita, A. Ferdowsi, M. Chen, and W. Saad, "Machine learning for wireless connectivity and security of cellular-connected UAVs," *IEEE Wireless Communications*, vol. 26, no. 1, pp. 28–35, 2019.
57. Y. K. Tun, Y. M. Park, N. H. Tran, W. Saad, S. R. Pandey, and C. S. Hong, "Energy-efficient resource management in UAV-assisted mobile edge computing," *IEEE Communications Letters*, vol. 25, no. 1, pp. 249–253, 2020.
58. Q. Zhou, Z. Qu, S. Guo, B. Luo, J. Guo, Z. Xu, and R. Akerkar, "On-device learning systems for edge intelligence: A software and hardware synergy perspective," *IEEE Internet of Things Journal*, 2021, in press.
59. J. Peng, H. Gao, L. Liu, Y. Wu, and X. Xu, "FNTAR: A future network topology-aware routing protocol in UAV networks," in *2020 IEEE Wireless Communications and Networking Conference (WCNC)*, virtual, 2020, pp. 1–6.
60. A. Razi, F. Afghah, and J. Chakareski, "Optimal measurement policy for predicting UAV network topology," in *51st Asilomar Conference on Signals, Systems, and Computers*, Pacific Grove, CA, USA, 2017, pp. 1374–1378.
61. A. Razi, "Optimal measurement policy for linear measurement systems with applications to UAV network topology prediction," *IEEE Transactions on Vehicular Technology*, vol. 69, no. 2, pp. 1970–1981, 2019.
62. H. Qiu and H. Duan, "A multi-objective pigeon-inspired optimization approach to UAV distributed flocking among obstacles," *Information Sciences*, vol. 509, pp. 515–529, 2020.
63. J. Hu, H. Zhang, L. Song, R. Schober, and H. V. Poor, "Cooperative internet of UAVs: Distributed trajectory design by multi-agent deep reinforcement learning," *IEEE Transactions on Communications*, vol. 68, no. 11, pp. 6807–6821, 2020.
64. H. Zhang and L. Hanzo, "Federated learning assisted multi-UAV networks," *IEEE Transactions on Vehicular Technology*, vol. 69, no. 11, pp. 14104–14109, 2020.
65. S. Aggarwal, N. Kumar, and S. Tanwar, "Blockchain-envisioned UAV communication using 6G networks: Open issues, use cases, and future directions," *IEEE Internet of Things Journal*, vol. 8, no. 7, pp. 5416–5441, 2021.
66. W. Li, Z. Su, R. Li, K. Zhang, and Y. Wang, "Blockchain-based data security for artificial intelligence applications in 6G networks," *IEEE Network*, vol. 34, no. 6, pp. 31–37, 2020.
67. W. Sun, L. Wang, P. Wang, and Y. Zhang, "Collaborative blockchain for space-air-ground integrated networks," *IEEE Wireless Communications*, vol. 27, no. 6, pp. 82–89, 2020.
68. Y. Wang, Z. Su, N. Zhang, and A. Benslimane, "Learning in the air: Secure federated learning for UAV-assisted crowdsensing," *IEEE Transactions on Network Science and Engineering*, vol. 8, no. 2, pp. 1055–1069, 2021.

CHAPTER 6

# A Review of Intrusion Detection in FANETs

Xueru Du and Yue Cao

*Wuhan University*

Lei Wen

*National University of Defense Technology*

Zhaohui Yang

*University College London*

## CONTENTS



## LIST OF ABBREVIATIONS

| | |
|---|---|
| **2D** | Two-dimensional |
| **ALORID** | Airborne LiDAR |
| **APTs** | Advanced persistent threats |
| **BGM** | Bayesian game model |
| **CSS** | Cyber security system |
| **DDoS** | Distributed denial of service |

DOI: 10.1201/b22998-6

| | |
|---|---|
| **DIDS** | Decentralized intrusion detection system |
| **DLAIDS** | Deep learning-based and adaptive intrusion detection system |
| **DoS** | Denial of service |
| **FANET** | Flight ad hoc network |
| **FJUAV** | Unmanned aerial vehicle equipped with an air-to-ground friendly jammer |
| **HDRS** | Hierarchical detection and response system |
| **HIDS** | Hybrid intrusion detection system |
| **IDS** | Intrusion detection system |
| **IES** | Intrusion ejection system |
| **IPSR** | Intercept probability security region |
| **MANET** | Mobile ad hoc networks |
| **MTF** | Multi-tier framework |
| **ORIDA** | Obstacle recognition and intrusion detection algorithm |
| **SAF** | Statistical analysis framework |
| **SIDS** | Specification based intrusion detection system |
| **sUAS** | Small unmanned aerial system |
| **SVM** | Support vector machine |
| **UAVs** | Unmanned aerial vehicles |

## 6.1 INTRODUCTION

As improved technology leads to lots of available unmanned systems, more and more unmanned aerial vehicles (UAVs) are applied in various fields. Initially, UAVs were used for surveillance, combat, target tracking, reconnaissance, etc. in the military field. Nowadays, UAVs are also introduced to the civil field to explore whole new applications. For instance, UAVs have been utilized in aerial photography, observation of wildlife, emergency rescue, monitoring of infectious diseases, inspection of electricity, research, plant protection, transportation, and so on [1]. Figure 6.1 shows military and civil applications of UAVs.

To deal with complicated tasks, UAVs are usually organized as a collaborative group, namely, flight ad hoc network (FANETs). However, due to the nature of UAVs, it's necessary to specific features in FANETs, and Table 6.1 lists these features and corresponding requirements [2]. Generally speaking, the speed of UAVs in FANETs is much faster than that of nodes in other ad hoc networks. Thus, the high mobility of UAVs requires the network to pay more attention to maintain the stability. At the same time, it also leads to frequent changes in the topology of the network, so the network must ensure the connection of the nodes in FANETs. In addition, the limited time of communication needs to ensure the message can be transmitted successfully. Due to the small size and poor carrying capacity of UAVs, energy resources and computing resources are restricted. Therefore, UAVs must use resources reasonably when completing tasks to maximize resource utilization. Environmental factors such as noise, communication distance, and obstacles will reduce the quality of communication, so FANETs must reduce the impact of interference. Apart from these, some scenarios, such as disaster rescue, have high requirements for real-time performance, which requires UAVs to reduce latency of propagation in communications, so that messages could be transmitted back to the ground station as quickly as possible.

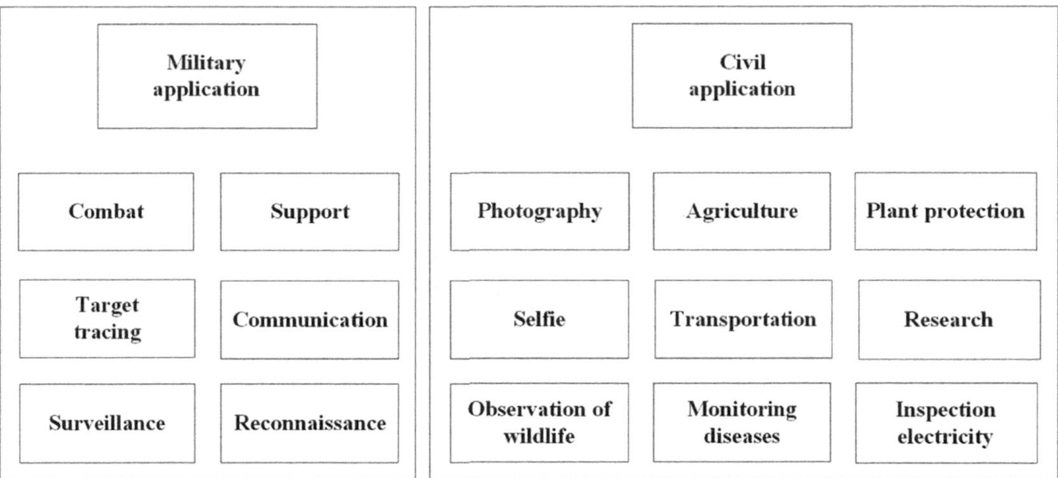


FIGURE 6.1 Applications of UAVs.

TABLE 6.1 Features and Corresponding Requirements of FANETs

| Feature | Requirement |
|---|---|
| High mobility | Maintain the stability |
| Frequent topological changes | Ensure nodes' connectivity |
| Connection time constraints | Ensure message transmission |
| Resource constraints | Make rational use of resources |
| Environmental interference | Resist interference |
| Real-time | Reduce latency |

Features of FANETs considerably increase the difficulties of the security defense and bring huge threats in aspects of the communications between the UAVs. There are active defense and passive defense in FANETs to deal with attacks. The active defense means that UAVs prevent potential dangers before they are attacked, for example, using encryption techniques to defend against potential eavesdropping attacks on information. By contrast, the passive defense refers that UAVs detect anomalies when they are attacked and then take protective measures to reduce losses.

Intrusion detection system (IDS) is a typical example of passive defense. IDS generally analyzes behaviors of nodes in the network to detect abnormal or illegal actions according to security guidelines of FANETs [3]. IDS usually collects configuration information of UAVs, transmission data, network traffic, and others to identify malicious nodes. When attacks are distinguished, IDS will take measures against attackers to reduce the impact of malicious attacks on the network. In addition, selfish nodes, which harm the overall performance for their own interests, can be found out as well.

The remainder of the chapter is organized as follows: Section 6.2 discusses the taxonomy of attacks from the perspective of information security and lists common attacks in detail. Section 6.3 introduces the IDS architecture and presents a variety of IDSs categorized by the type of targeted attacks. Section 6.4 compares the proposed IDSs from multiple views and summarizes the evaluation metrics of IDSs.

## 6.2 TAXONOMY OF CYBER ATTACKS

Different taxonomies are applied to categorize the attacks in FANETs. In Ref. [4], in accordance with the data flow in the autopilot, attacks are categorized into hardware attack (attacker accesses to the UAV autopilot components directly), wireless attack (attacker carries out the attacks through one of the wireless communication channels), and sensor spoofing (attacker passes false data through the on-board sensors of the UAV autopilot). In light of target layers, attacks fall into physical attack, data-link attack, network attack, transport attack, application attack, and multi-layer attack [5].

Here, we consider the perspective of information security, from which attacks could be classified into three groups, that is, confidentiality attack, integrity attack, and availability attack [6]:

- *Confidentiality attack:* Confidentiality is to protect data from unauthorized access, that is, information is only available to authorized users. The most common attack on confidentiality is intercepting information, and UAVs, ground stations, and communication links in the architecture of UAV systems are vulnerable to this kind of attack. It's hacking and eavesdropping that launch attacks on UAVs. Threats to ground stations are mostly software vulnerabilities, viruses, malware, Trojans, and key loggers. Moreover, security threats to the communication links between various components involve password cracking, identity spoofing, cross-layer attacks [7], and multi-protocol attacks [8].
- *Integrity attack:* Integrity consists of network integrity and data integrity. Data integrity aims to protect the integrity of data and integrity of data-related attributes, which refer to correctness, consistency, and compatibility of data. Integrity of a system can be compromised using two basic operations, which are modification of existing information and fabrication of new information. Modification aims to alter data during transit or while in storage. The common threats of modification have three major categories: jamming, compromising signal integrity, and capturing the signal. Next comes, fabricating or modifying information which includes the use of malicious code or existing subroutines of the system. Apart from the modification and fabrication, interference with wireless data links or sensors can also affect data integrity and even eavesdrop on transmitted data.
- *Availability attack:* Availability refers to the timely response of the system to legitimate access, that is, the system can correctly access required information during normal operation, and it can be quickly restored to use when the system is attacked. Attacks on availability are to make UAV unable to obtain the corresponding data or make the corresponding response according to the instructions. What's more, transmitting false commands or control signals can be a major threat to the availability of an UAV system as false signals can actually make the UAV land or attack somewhere else. Primary cyber attacks which might affect availability are black hole attacks, falsifying signals and Denial of Service (DoS) attacks [9].

According to the above taxonomy, common attacks are listed in Table 6.2, which considers eavesdropping, identity spoofing, GPS spoofing, jamming attacks, DoS and

TABLE 6.2 Common Attacks

| Security Attributes | Attacks | Outline |
|---|---|---|
| Confidentiality | Eavesdropping | Eavesdropping means attackers stealing information transmitted by normal nodes through the network. In passive eavesdropping, eavesdroppers monitor the wireless channel and obtain information from it. Meanwhile, it doesn't affect legitimate users to receive messages. Thus, it's difficult for legitimate users to detect passive eavesdropping. In addition, active eavesdroppers can move to the best eavesdropping location and use professional wireless equipment. |
| | Identity spoofing | Identity spoofing refers to an attack that uses a small number of nodes in the network to control multiple false identities, thereby using these identities to control or affect a large number of normal nodes in the network. |
| Integrity | GPS spoofing | GPS spoofing is to make GPS receivers receive false information by forging or replaying GPS signals, and then calculate the wrong location and time information. GPS spoofing can be sorted into forwarding GPS spoofing and generative GPS spoofing by the way of signal generation. Forwarding GPS spoofing records the real GPS signal and then transmits it to the GPS receiver. While generative GPS spoofing generates the fake GPS signal based on the specific location and time using a specific program, and then transmits it to the GPS receiver. Forwarding GPS spoofing is less difficult to implement, but it is less flexible. |
| | Jamming attacks | The jamming attack is an attack in which an attacker transfers interfering signals on wireless network intentionally. As a result, it decreases the signal-to-noise ratio at the receiver side and disrupts existing wireless communication. The jamming attack uses intentional radio interference and keeps the communicating medium busy. |
| Availability | DoS and DDoS attacks | DoS or DDoS attacks are based on network congestion, making the system unusable. The traditional DoS attack is a way that an attacker occupies a large amount of system resources to make the network service abnormal, so as to prevent legitimate users from using the service normally. Similarly, for UAVs, the same idea can be adopted, but this has a greater impact in FANETs, and even leads to the fall and destruction of UAVs, which will threaten ground facilities and the masses. There are three ways to carry out such an attack: flooding, smurfing, and buffer overflow. |
| | Black hole attacks | The malicious node uses the flaws in the routing protocol to pretend to be the optimal path node, so that the malicious node will be preferentially used as a relay node during the route of the data packet is forwarded, which means the routing table is changed. However, after a normal node in the UAV network forwards a data packet to a malicious node, the malicious node does not transmit the data packet forward to the next node in the routing table, but directly discards it, resulting in the transmission of the control command failed. |

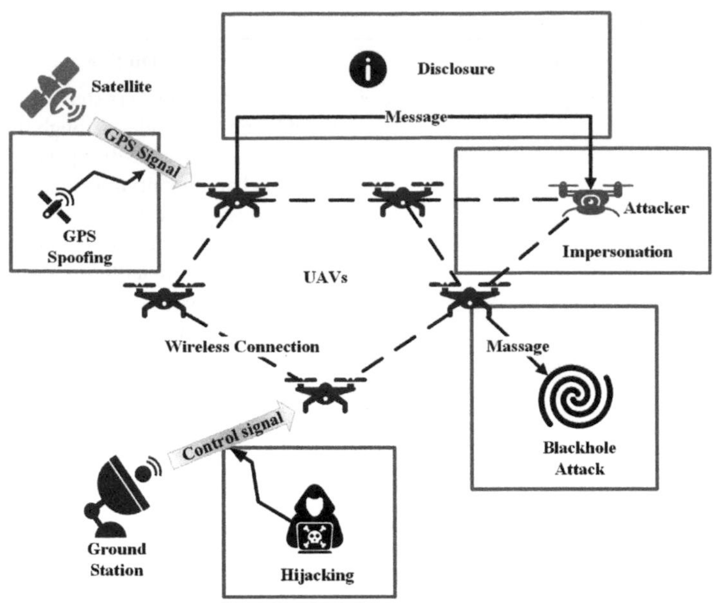


FIGURE 6.2 Types of attacks.

Distributed Denial of Service (DDoS) attacks, and black hole attacks. Furthermore, scenarios of various attacks are depicted in Figure 6.2.

## 6.3 INTRUSION DETECTION SYSTEM

The intrusion detection plays a significant role in maintaining the security of FANETs, which is designed to improve the UAV network security and performance to prevent UAVs from being attacked by attackers inside or outside the network. Figure 6.3 shows the normal unmanned aircraft system safety framework, and it shows clearly that three processes are included in IDS architecture. The first is data collection, generally speaking, from different components, such as sensors, communications links, power supply units, antennas, transceiver units, navigation systems and the UAV control system, IDSs collect monitor signals, command traffic, control instructions, working behavior, energy consumption, and operations of UAV components. In the second step, it simply preprocesses and analyzes the data flow, such as data cleaning, data integration, data transformation, and data reduction. After that, the next process finds abnormal nodes in FANETs based on the processed data and makes a decision on how to deal with these nodes. The third step is a response to the analysis results. Different nodes are treated in different ways: (1) for normal nodes, keep them in a normal state of communication; (2) for abnormal nodes, drive them out of the network and prohibit the communication; (3) for uncertain nodes, leave them in the network for continued observation, but prohibit other nodes communicate with them.

Given that integrity is attacked at high frequency, it is reasonable to take integrity as the basis for categorizing the IDS. Thus, the IDS could be able to sort the following intrusion types: (1) integrity attack; (2) integrity and availability attack; and (3) integrity, availability, and confidentiality attack.

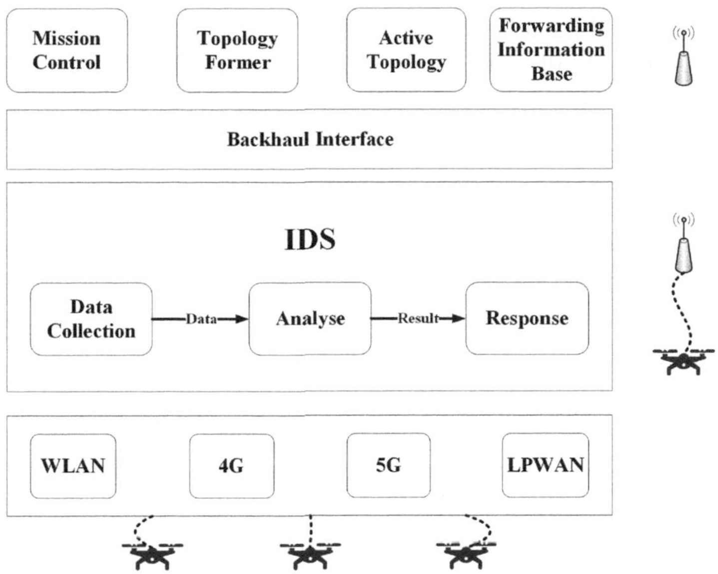


FIGURE 6.3 IDS architecture.

### 6.3.1 IDS Based on Integrity Attacks

Arthur [10] pointed out that the UAV autonomous driving system only had a safety mechanism to avoid collisions with objects on the trajectory, and couldn't deal with malicious hacker attacks and signal spoofing. Attacks are generally launched by aircraft on the ground or nearby flying vehicles. Through these attacks, the attacker can control the drone's autopilot system or modify the drone's flight settings. Therefore, the authors proposed a deep learning-based and adaptive intrusion detection system (DLAIDS) to identify intruders and ensure drones' safe return-to-home. The proposed IDS uses self-taught learning with a multiclass support vector machine (SVM) to ensure the detection accuracy, even applying to unknown attacks. In particular, a self-healing method based on deep Q network is used in the IDS recovery phase, which is a deep reinforcement learning algorithm for dynamic route learning and could promote the safe return of drones. Experimental simulation results showed that the proposed IDS had high accuracy, sensitivity, and specificity in the face of UAV network attacks.

Zhou et al. [11] considered the application of the UAV equipped with an air-to-ground friendly jammer (FJUAV) to improve the physical layer security of a legitimate transmitter-receiver pair if the eavesdropper location was unknown. Authors developed a framework to tackle the problem how to efficiently optimize the jamming power and the UAV location so as to maximize the interference to the eavesdropper while mitigating the jamming impact on the legitimate receiver. They first derived the outage probability of the legitimate receiver and the intercept probability for a given eavesdropper location in terms of the jamming power and the 3D UAV location. Then, inspired by the jamming coverage in Ref. [12], they developed a new area-based security measure which they referred to as the intercept probability security region (IPSR). Based on the derived IPSR and outage probability, they proposed and formulated a non-convex binary integer optimization problem to maximize the IPSR subject to an outage probability constraint. Finally, they developed a computationally efficient iterative algorithm that decoupled the original optimization problem into two-dimensional (2D) horizontal location, jamming power, and UAV altitude subproblems. Simulation results showed that their proposed iterative algorithm achieves an IPSR that was comparable to a high-complexity exhaustive search solution while reducing the execution time by up to 97%.

Miao et al. [13] proposed obstacle recognition and intrusion detection algorithm (ORIDA) for the UAV based on an airborne LiDAR (ALORID). The proposed ALORID algorithm performs data preprocessing and motion detection of intrusions using 2D LiDAR data. In the data preprocessing step, firstly, the coordinate conversion of the LiDAR data is performed according to UAV motion parameters. Then, the LiDAR data graph at the current moment is generated by the image noisy point filtering algorithm. After that, the improved density-based spatial clustering of applications with a noise algorithm is used for image clustering of intrusions to obtain the LiDAR time-domain cumulative graph in a certain detection time. Finally, the motion recognition and location detection of each cluster are completed. The experiment results verified the effectiveness of the proposed algorithm in identifying the moving state of the intrusions.

Muniraj and Farhood [14] proposed a statistical analysis framework (SAF) for identification of cyber-physical attacks on sensors of a small unmanned aerial system (sUAS) to ensure a secure autopilot. A framework is presented wherein techniques from statistical analysis are used in a probabilistic setting to detect sensor attacks. The attack detection approach consists of two types of anomaly detection methods: (1) anomaly detection using attack signatures, which are based on measurements from safe sensors, and (2) anomaly detection based on residuals, which are computed by a residual generator from the outputs of a state estimator. By incorporating knowledge about the physical system and using a probabilistic framework, the proposed approach minimizes the false alarm rates by concentrating on the response of the UAS to attacks but not on the attack mechanism itself.

### 6.3.2 IDS Based on Integrity and Availability Attacks

Sedjelmaci et al. [15] designed a hierarchical detection and response system (HDRS) that ran on drones and ground stations. The hierarchical scheme can detect malicious anomalies that threaten the security of FANETs and respond in time. The scheme mainly targets lethal cyber attacks on UAV networks, such as false information dissemination, GPS spoofing, jamming, and black hole and gray hole attacks. The hierarchical IDS monitors the behavior of the UAV nodes according to the detection rules for intrusion detection, and then uses the SVM learning algorithm on the ground station to verify the detected attack, evaluate the node, and make a final decision. Finally, nodes are divided into four categories: normal, abnormal, suspicious, and malicious. A large number of experimental simulations showed that this scheme performed well in terms of attack detection even though there were many UAVs and attackers because it had a high detection rate, a low false positive rate, and a low communication overhead.

Sedjelmaci et al. [16] proposed a cyber security system (CSS) based on IDS to prevent the most dangerous and fatal attacks that might occur on UAV networks. CSS relies on specific detection policies to detect network attacks against data integrity and network availability in a timely manner. In addition, in order to reduce the false positive and negative rates, CSS includes a threat estimation model based on a Belief approach. In this approach, IDS observes the behavior of each drone, calculates the corresponding belief function, and then assigns a joint threat level to the monitored node. According to the simulation results, CSS had the ability to detect attacks with a high detection rate and a low false positive rate, even though the number of UAVs and attackers was high.

Sedjelmaci et al. [17] proposed a Bayesian game model (BGM), which included two kinds of security agents, namely, intrusion detection system and intrusion ejection system (IES) that ran the monitoring and attacker-ejection process, respectively. Their aim is to secure the UAV-aided network against lethal attackers, in particular, to address the issues that the optimal activation of the monitoring process and the optimal attacker ejection process. A Bayesian game is a game in which information about the characteristics of the other players is incomplete. In other words, in their Bayesian game, the incompleteness of information means that the IDS (or IES) player is unsure of the type of the attacker player, and the attacker does not know whether the neighboring node is an IDS or IES player. By means of Bayesian Nash equilibrium, they proved by simulations that the proposed security game framework incurred a low overhead to detect and deter lethal attacks with high accuracy.

Lauf et al. [18] proposed a decentralized intrusion detection system (DIDS) that used a combination of detection strategies while minimizing resource utilization for use with mobile ad hoc networks (MANET). The anomaly-based intrusion detection is provided by analyzing the context from application-level interactions of nodes, where each interaction corresponds to a specific function or behavior within the operational scenario of the network. A static set of behaviors is determined offline, and these behaviors are tracked dynamically during the operation of the network. There are two stages in the process of detecting. During the first stage of the IDS, the detection strategy employs the analysis of global and local maxima in the probability density functions of the behaviors to isolate deviance at the granularity of a single node. The first stage is used to capture the typical behavior of the network and provides tuning and calibration for the second stage. During the second stage, a cross-correlative component is used to detect multiple threats simultaneously. The approach distributes the IDS among all connected network nodes, allowing each node to identify potential threats individually. Parallelism increases system scalability by removing the burden of analysis from one machine monitoring the entire collective (i.e., centralized) to multiple devices monitoring only their relevant interactions (i.e., decentralized).

### 6.3.3 IDS Based on Integrity, Availability, and Confidentiality Attacks

Sayeed et al. [5] proposed a multi-tier framework (MTF) for safeguarding UAS against malicious attackers and recovering the rogue UAVs. The framework can capitalize and take advantage of 4G, 5G, or any available communication channel and is split into three components: The UAS mission planning, the UAS safety component, and the communication component. The proposed framework enforces a dynamic conceptual grid-based layout over the actual geographical deployment. The dynamically shuffling grid ascertains the security of transmission channels, as every time the grid is shuffled periodically or based on abnormal behavior, the safety paradigm is reinitiated. Public key cryptographic algorithms are deployed for securing communication links and long short-term memory [19]/convolutional neural network [20] are employed for behavior and time-series predictions to detect abnormality in behavioral, statistical, and mobility patterns. Principal component analysis based on multivariate statistical analysis is used for detecting outliers in the aerial network environment. The behavior prediction and outlier detection algorithms significantly improve the overall performance of the network and provide immunity against intruders with reduced false positives, high accuracy, and better detection rate.

Mitchell and Chen [21] proposed a specification-based intrusion detection system (SIDS) to protect FANETs against cyber attacks. In their proposal, they implemented some detection rules with regard to the most dangerous attacks that could occur in FANETs, such as reckless, random, and opportunistic attackers. In this work, the transition of a node from a safe to an unsafe state and vice versa is used to detect the degree of compliance. According to their simulation results, their detection system exhibited a low false negative. However, the false positive rate was quite high when an opportunistic attack unfolds. The main weakness of this system lies in assuming that the monitoring UAV, implementing the IDS agent, is always trusted. Otherwise, such a node could be compromised and provides false detection, i.e., claim that a normal UAV is an attacker. What's more, since only a few behavior indicators are used to frame rules, the system is simple and fails to withstand cyber security attacks on UAVs.

Condomines et al. [22] proposed a hybrid intrusion detection system (HIDS) based on both spectral traffic analysis and a robust controller /observer for anomaly estimation inside UAV networks. The model is a two-step mechanism and it firstly characterizes the traffic by using a statistical signature of the traffic exchanged in the network. Then it selects a precise estimator model to reconstruct the attack traffic. Note that, statistical signatures, based on wavelet analysis, are selected because they offer a wide spectral characterization of the entire traffic process. Each signature provides a unique identification of the current traffic. By looking up this signature in a bank of signatures, it is possible to characterize and make a model of the anomaly in the UAV network. Subsequently, the attack will be analyzed by using a robust control estimation to reconstruct the attack traffic. All types of attacks which do not follow the initial characteristics trigger an alarm and, consequently, the malicious traffic can be analyzed in depth. This approach has the major advantage that it is not associated with a specific type of the attack. Any attacks that do not follow the initial model can be detected, analyzed, and managed.

## 6.4 COMPARISON AND EVALUATION OF IDSs

### 6.4.1 Comparison

In order to figure out the difference of each IDS, the comparison of the time, scenario, technology, and attack is presented in Table 6.3.

From scenarios, we observe that most IDSs are only applicable to specific scenarios. Thus, the shortcoming of the intrusion detection technology is exposed. It is almost impossible to detect all potential attacks by using one IDS. Of course, UAVs are not subject to all kinds of attacks in real life. When choosing an IDS for UAVs, it can be decided according to the scenarios where UAVs are deployed. Compared with the all-around IDSs, an IDS

TABLE 6.3 Comparison of IDSs

| IDS | Time | Scenario | Technology | Attacks |
|---|---|---|---|---|
| DLAID [10] | 2019 | UAVs in autopilot collect data from hostile regions, and safely return home | Self-taught learning Multiclass SVM Deep-Q Network | GPS spoofing Jamming |
| FJUAV [11] | 2018 | UAVs equipped with friendly jammers assist secure communications between legitimate transmitter-receiver pairs to against unknown eavesdroppers | Iterative algorithm | Eavesdropping |
| ORIDA [13] | 2020 | Common scenarios | Light detection and ranging Machine learning Obstacle recognition | Obstacle interference |
| SAF [14] | 2017 | The secure autopilot for the small unmanned aircraft system | Statistical analysis Attack signatures Residuals | Sensor attacks |
| HDRS [15] | 2018 | UAVs are envisioned to carry out explorations in isolated zones to collect and transmit critical information | Rules-based anomaly detection SVM learning algorithm | GPS spoofing Jamming False information dissemination Black hole and gray hole attacks |
| CSS [16] | 2016 | UAVs carry out exploration tasks in remote isolated zones | Specific detection policies Threat estimation | Cyber attacks |
| BGM [17] | 2017 | UAVs provide a safety-oriented vehicular network, in which critical relevant information is exchanged among vehicles and UAVs | Bayesian game Nash equilibrium | False information attacks DoS attacks |
| DIDS [18] | 2010 | Resource-constrained Ad hoc network | Anomaly-based intrusion detection | Jamming Spoofing attacks |
| MTF [5] | 2020 | The framework implements a conceptual grid-based system layout. Every time the grid is shuffled or a node crosses into a different section of the grid, security parameters are re-negotiated. | Cryptographic privacy protection Multivariate statistical analysis Deep learning Predictions | Impersonation attacks Privacy and policy violation UAV capture Spoofing Masquerading Flooding attacks Routing and TCP attacks Jamming Application attacks |
| SIDS [21] | 2012 | IDSs are embedded in UAVs to secure infrastructure (sensors or actuators) | Specification-based intrusion detection | Direct the weapon to resource Energy wasting Activating counter UAV capture Exfiltrate mission data |
| HIDS [22] | 2019 | FANETs | Spectral analysis Robust estimator | Flooding attacks |

designed for a specific scene tends to have a better performance in this scene. In practical applications, it is feasible to combine a variety of IDSs to resist potential attacks.

From the perspective of technology, as time goes by, more and more mature technologies are introduced into IDS, such as deep learning, SVM, and statistical analysis. Therefore, intrusion detection will make progress with the development of other technologies. To be sure, in the future, IDSs will bring in more brand-new technologies to achieve better performance. This also inspires scholars and scientific researchers to keep pace with the times and actively introduce new technologies into IDSs to optimize its efficiency.

Comparing the attacks that IDSs can defend, it is obvious that spoofing attacks, DoS attacks, and jamming are detected at high frequency. This is because these attacks are more common in FANETs. Listing attacks targeted by all IDSs is helpful to select the appropriate IDS in line with attacks that users expect to be defended. In actual use, the probability of occurrence of some attacks is almost zero, so we can ignore these attacks. We should pay more attention to defend against attacks that are very harmful or have a high probability of occurrence.

### 6.4.2 Evaluation

The development of an appropriate intrusion detection system requires a set of characteristics and parameters that should be taken into account. These parameters are based on the challenges in FANETs environment. In Ref. [23], as requirements of IDSs, several parameters were identified. To achieve an IDS that meets required levels of effectiveness (i.e., minimizing detection errors with minimum service interruptions) and efficiency (i.e., reducing computational and communication overhead), they identified the challenges of detection latency, IDS computational cost, implementation overhead, threat and behavior modeling, effective threat assessment, maximum network throughput with minimum cost, effective monitoring and attack response, and lightweight IDS with minimum resource consumption to pave a way to build a cyber-physical hardened IDS.

According to an overview of all IDSs, it shows that all mentioned IDSs implement experiments by using the simulator to evaluate the performance. To perform qualitative comparison, literature in IDSs is summarized in Table 6.4, and the following metrics are selected as indicators: detection rate, false positive rate, false negative rate, accuracy rate, communications overhead, and efficiency:

TABLE 6.4 Summary of Measure Metrics in Proposed IDSs

| IDS | Detection Rate | False Positive Rate | False Negative Rate | Accuracy Rate | Communications Overhead | Efficiency | Others |
|---|---|---|---|---|---|---|---|
| DLAIDS [10] | | | | √ | | | |
| FJUAV [11] | | | | | | | √ |
| ORIDA [13] | √ | | | | | | √ |
| SAF [14] | | | | | | | √ |
| HDRS [15] | √ | √ | | | √ | √ | |
| CSS [16] | √ | √ | √ | √ | | | |
| BGM [17] | | | | √ | √ | | √ |
| DIDS [18] | | | | | | | √ |
| MTF [5] | √ | √ | | √ | | | √ |
| SIDS [21] | √ | √ | √ | | | | |
| HIDS [22] | | | | | | | √ |

- Detection rate: The ratio of correctly identified attackers over the total number of attackers.
- False positive rate: The number of normal UAVs that are incorrectly classified as attackers over the total number of normal UAVs.
- False negative rate: The rate of malicious UAVs that are incorrectly classified as normal over the total number of malicious nodes.
- Accuracy rate: The ratio of correctly sorted nodes over the total number of modes, depending on detection rate, false positive rate, and false negative rate.
- Communications overhead: The amount of bytes generated by the IDS.
- Efficiency: The time required for the IDS to detect attackers, depending on the moment when the attack starts and the detection time of the cyber attack.

It clearly presents that specific values of indicators are not listed in the table, because different IDSs are tested under different simulation environments. Thus, we only focus on whether the IDS considers the above indicators. Generally speaking, the more indicators are valued, the more superior IDSs are. However, some IDSs must be measured by other indicators, and that doesn't mean those IDSs valued by other metrics are ineffective. When evaluating IDS, it is necessary to comprehensively consider the scenario impacted performance.

The inspiration of this section is that when designing IDS, designers must consider the above factors as much as possible, so that IDS can have an outstanding performance to be suitable for more scenarios.

## 6.5 CONCLUSION AND FUTURE

The intrusion detection enables UAVs to distinguish trustworthy nodes and untrustworthy ones in FANETs. It leads to reducing the risk of UAVs being attacked by other malicious UAVs. Due to the importance of security of UAVs and their communications in FANETs, this study has conducted a systematic review of current research that aims at detecting intrusion in FANETs.

In this review, a brief overview of the state-of-the-art UAV-IDS mechanisms has been provided. This review summarized the applications of UAVs and features and corresponding requirements of FANETs. What's more, it listed six common attacks including eavesdropping, identity spoofing, GPS spoofing, jamming attacks, DoS and DDoS attacks, and black hole attacks. By defining a taxonomy based on information security, IDSs were categorized into three types: (1) IDS based on integrity attacks, (2) IDS based on integrity and availability attacks, and (3) IDS based on integrity, availability, and confidentiality attacks. Eleven IDSs were explained in detail in terms of detection mechanism, advantages, disadvantages, etc. Afterwards, a multi-dimensional comparison was presented, and time, scenario, technology, and attacks were all taken into consideration. Lastly, the summary of measure metrics in proposed IDSs made it easy to extract the metrics for selecting and measuring an appropriate IDS.

On the one hand, future works can focus on improving the accuracy, reducing the overhead, and optimizing other indicators. On the other hand, it needs to be designed to detect more kinds of attacks to meet security requirements. With the advent of novel attack methodologies such as APTs (advanced persistent threats), UAVs have become vulnerable to sophisticated cyber attacks that may cause catastrophic damages, thereby necessitating the development of novel IDSs for UAV security [24]. From the technical point, it is necessary to merge more novel technologies in IDS, such as deep learning, neural networks, and other technologies in the field of artificial intelligence. They have the nature of self-learning, so they could identify some unknown attacks and be extended to more scenarios. Meanwhile, due to the limited resources of UAVs, it is necessary to balance the computational overhead and efficiency.

## REFERENCES


1. G. Pajares, "Overview and Current Status of Remote Sensing Applications Based on Unmanned Aerial Vehicles (UAVs)," *Photogrammetric Engineering & Remote Sensing*, vol. 81, no. 4, pp. 281–329, 2015.
2. H. Menouar, I. Guvenc, K. Akkaya, A. S. Uluagac, A. Kadri, A. Tuncer, "UAV Enabled Intelligent Transportation Systems for the Smart City: Applications and Challenges," *IEEE Communications Magazine*, vol. 55, no. 3, pp. 22–28, 2017.
3. E. Biermann, E. Cloete, L. M. Venter, "A Comparison of Intrusion Detection Systems," *Computers & Security*, vol. 20, no. 8, pp. 676–683, 2001.
4. A. Kim, B. Wampler, J. Goppert, et al., "Cyber Attack Vulnerabilities Analysis for Unmanned Aerial Vehicles," in Infotech@Aerospace, Garden Grove, CA, June 2012, pp. 2012–2438.
5. M. A. Sayeed, R. Kumar, V. Sharma, "Safeguarding Unmanned Aerial Systems: An Approach for Identifying Malicious Aerial Nodes," *IET Communications*, vol. 14, no. 17, pp. 3000–3012, 2020.
6. A. Y. Javaid, W. Sun, V. K. Devabhaktuni, et al., "Cyber Security Threat Analysis and Modeling of an Unmanned Aerial Vehicle System," in *2012 IEEE Conference on Technologies for Homeland Security*, Waltham, MA, USA, Nov. 2012, pp. 585–590.
7. W. Wang, S. Yan, H. Li, et al., "Cross-Layer Attack and Defense in Cognitive Radio Networks," in *IEEE Global Telecommunications Conference*, Miami, FL, USA, Dec. 2010, pp. 1–6.
8. J. Alves-Foss, "Multi-Protocol Attacks and the Public Key Infrastructure," in *Proceedings of 21st National Information Systems Security Conference*, 1998.
9. H. König, "Network Attack Detection and Defense," *PIK - Praxis der Informationsverarbeitung und Kommunikation*, vol. 35, no. 1, pp. 1–2, 2012.
10. M. P. Arthur, "Detecting Signal Spoofing and Jamming Attacks in UAV Networks Using a Lightweight IDS," in *2019 International Conference on Computer, Information and Telecommunication Systems*, Beijing, China, Aug. 2019, pp. 1–5.
11. Y. Zhou, et al., "Improving Physical Layer Security via a UAV Friendly Jammer for Unknown Eavesdropper Location," in *IEEE Transactions on Vehicular Technology*, vol. 67, no. 11, pp. 11280–11284, Nov. 2018.
12. J. Vilela, M. Bloch, J. Barros, S. W. McLaughlin, "Wireless Secrecy Regions with Friendly Jamming," *IEEE Transactions on Information Forensics and Security*, vol. 6, no. 2, pp. 256–266, 2011.
13. Y. Miao, Y. Tang, B. A. Alzahrani, et al., "Airborne LiDAR Assisted Obstacle Recognition and Intrusion Detection Towards Unmanned Aerial Vehicle: Architecture, Modeling and Evaluation," *IEEE Transactions on Intelligent Transportation Systems*, vol. 22, no. 7, pp. 4531–4540, 2020.

14. D. Muniraj, M. Farhood, "A Framework for Detection of Sensor Attacks on Small Unmanned Aircraft Systems," in *2017 International Conference on Unmanned Aircraft Systems*, Miami, FL, USA, Jun. 2017, pp. 1189–1198.
15. H. Sedjelmaci, S. M. Senouci, N. Ansari, "A Hierarchical Detection and Response System to Enhance Security Against Lethal Cyber-Attacks in UAV Networks," *IEEE Transactions on Systems, Man, and Cybernetics: Systems*, vol. 48, no. 9, pp. 1594–1606, 2017.
16. H. Sedjelmaci, S. M. Senouci, M. A. Messous, "How to Detect Cyber-Attacks in Unmanned Aerial Vehicles Network," in *GLOBECOM 2016-2016 IEEE Global Communications Conference*, Washington, DC, USA, Dec. 2016, pp. 1–6.
17. H. Sedjelmaci, S. M. Senouci, N. Ansari, "Intrusion Detection and Ejection Framework Against Lethal Attacks in UAV-Aided Networks: A Bayesian Game-Theoretic Methodology," *IEEE Transactions on Intelligent Transportation Systems*, vol. 18, no. 5, pp. 1–11, 2017.
18. A. P. Lauf, R. A. Peters, W. H. Robinson, "A Distributed Intrusion Detection System for Resource-Constrained Devices in Ad-hoc Networks," *Ad Hoc Networks*, vol. 8, no. 3, pp. 253–266, 2010.
19. S. Hochreiter, J. Schmidhuber, "Long Short-Term Memory," *Neural Computation*, vol. 9, no. 8, pp. 1735–1780, 1997.
20. Y. LeCun, L. Bottou, Y. Bengio, P. Haffner, et al., "Gradient-Based Learning Applied to Document Recognition," *Proceedings of the IEEE*, vol. 86, no. 11, pp. 2278–2324, 1998.
21. R. Mitchell, I. R. Chen, "Specification Based Intrusion Detection for Unmanned Aircraft Systems," in *Proceedings of the First ACM MobiHoc Workshop on Airborne Networks and Communications*, pp. 31–36, 2012.
22. J. P. Condomines, R. Zhang, N. Larrieu, "Network Intrusion Detection System for UAV Ad-hoc Communication: From Methodology Design to Real Test Validation," *Ad Hoc Networks*, vol. 90, pp. 101759.1–101759.14, 2019.
23. G. Choudhary, V. Sharma, I. You, et al., "Intrusion Detection Systems for Networked Unmanned Aerial Vehicles: A Survey," in *2018 14th International Wireless Communications & Mobile Computing Conference*, Limassol, Cyprus, June 2018, pp. 560–565.
24. J. Jithish, S. Sankaran, "A Game-Theoretic Approach for Ensuring Trustworthiness in Cyber-Physical Systems with Applications to Multiloop UAV Control," *Transactions on Emerging Telecommunications Technologies*, vol. 35, no. 5, e4042, 2021.

CHAPTER 7

# Critical Analysis of Security and Privacy Challenges for UAV Networks

Hassan Jalil Hadi, Xueru Du, and Yue Cao

*Wuhan University*

CONTENTS



## LIST OF ABBREVIATIONS

| | |
|---|---|
| **ACA** | Arctic control area |
| **ADS-B** | Automatic dependent surveillance broadcast |
| **AES** | Advanced encryption sandard |
| **API** | Application programming interference |
| **BER** | Bit Error Rate |
| **CNN** | Convolutional neural network |
| **DDoS** | Distributed denial of service |
| **DoS** | Denial-of-service |
| **ECC** | Elliptic curve cryptography |
| **GCA** | Group collision avoidance |
| **GCS** | Ground control station |

DOI: 10.1201/b22998-7

| | |
|---|---|
| **IDS** | Intrusion detection system |
| **IoD** | Internet of drones |
| **IoT** | Internet of things |
| **L-PPS** | Lightweight privacy-preserving scheme |
| **MAV-link** | Micro air vehicle link |
| **ML** | Machine learning |
| **MLIDS** | Machine learning intrusion detection system |
| **NDN** | Named data networking |
| **OTP** | One time pad |
| **PLS** | Physical signaling sublayer |
| **SAA** | Sense and avoid |
| **SDN** | Software defined network |
| **SHA-1** | Secure hash algorithm |
| **TCALAS** | Temporal credential-based anonymous lightweight authentication scheme |
| **UAS** | Unmanned aircraft system |
| **UAV** | Unmanned aerial vehicle |
| **VU** | Vulnerabilities |

## 7.1 INTRODUCTION

The use of mobility and locality aware vehicles has increased in nearly every domain. This has led to the significant increase in the number of these vehicles and associated applications. A popular class of such vehicles is Unmanned Aerial Vehicles (UAVs) or drones. These vehicles are capable to fly with or without any pilot. These can be controlled remotely through radio or Wi-Fi. Advance features of these vehicles are capability of carrying heavy payloads, mine's detection, and unethical activities' scanning in specific areas [1]. In recent years, the usage of UAVs by military for crucial operations has intensified. Other applications include rescue missions, search missions, courier services, fire-fighting surveillance, ecological surveys, and damage assessments. Figure 7.1 shows the applications of UAVs.

The applications of UAVs are dramatically increasing as a result of their quick movement, low upkeep cost, and capacity to monitor the real-time environment. However, the majority of such application domains have caused a great threat to human lives because of unidentified weather and other environmental factors. As these vehicles are susceptible to numerous security risks, they might be attacked in different ways. The results of many attacks might be overwhelming. For example, some of these attacks are intended to steal information and affect confidentiality, integrity, and availability [2]. With the help of their communication channels, the UAVs carry a lot of information which must be secured. The information carried by UAVs is of multiple forms like text, images, videos, and audios. Although many encryption algorithms have been proposed to secure sensitive information, the most professional UAVs are not secure [3]. This has raised many security concerns about the communication protocols of UAVs. The main security concern related to

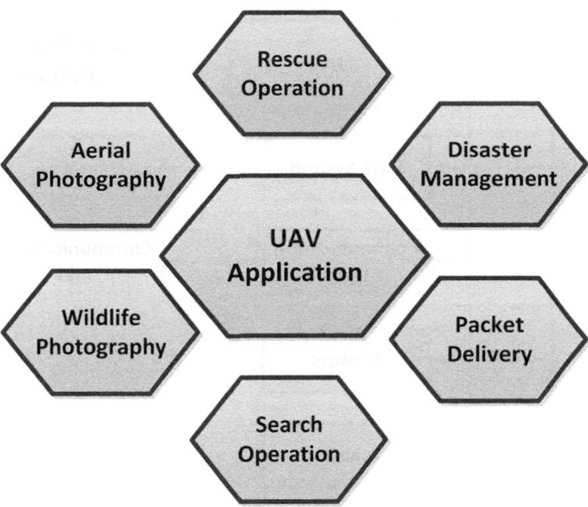


FIGURE 7.1 Applications of UAVs.

these communication protocols is securing information sent using connections like Wi-Fi. Usually, to send data to base stations, UAVs use a wireless link which is an easy target for attackers. So, it is necessary to prevent data interception by intruders. For this purpose, one widely used mechanism is encryption like Advanced Encryption Standard (AES) being used commonly at present. However, its use is not efficient when it comes to real-time applications due to communication overhead, particularly in case of extremely high data transfer rate [3].

Another important security concern for UAVs is the attackers interfering with UAVs. Their intention is to either take control of the aerial vehicle to disable communication between Ground Control Station (GCS) and the aerial vehicle. For this purpose, numerous attacks such as spoofing [4], jamming, and false data injection are being used. In this chapter, we will analyze the security and privacy challenges faced by UAV networks, but it is necessary to know first the UAV system's components and information flow among these components in order to effectively analyze these challenges.

### 7.1.1 UAV System

The base system of a UAV is manufactured for aerial vehicles which have the responsibility to link together all components. It is utilized for communication among these components and to control sensors as well as navigation or communications systems as shown in Figure 7.2. The base system of the UAVs is also used to integrate the optional components like weapon systems, etc. The sensors system in UAVs is made up of sensory equipment having integrated functionalities for pre-processing, for example, sensors having Global Positioning System (GPS), sensors along with cameras, and sensors with radars. Next, avionics systems in these vehicles have the responsibility to execute the control commands issued by controller like spoilers, engine commands, stabilizers, and flaps. The UAVs

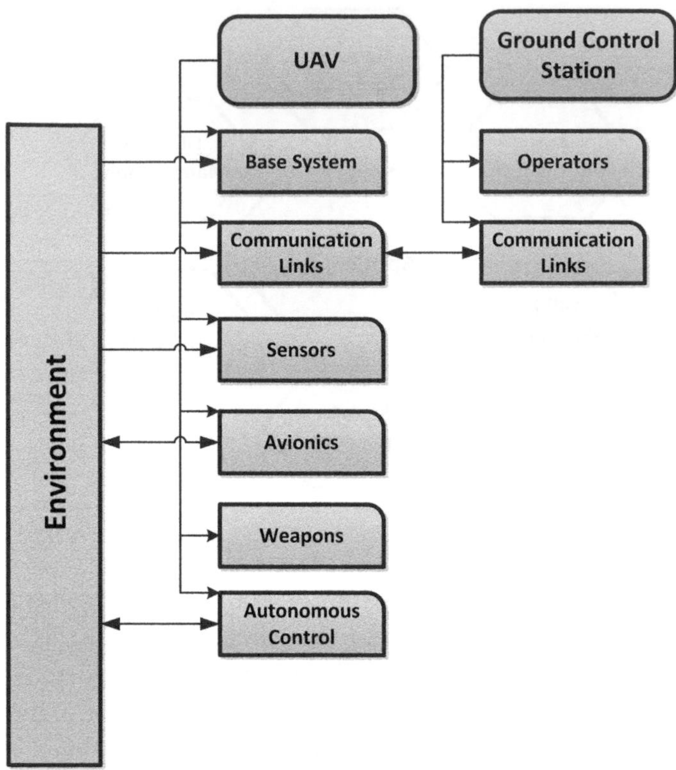


FIGURE 7.2 Components and information flow.

mainly depend on a wireless communication channel being a direct communication line or indirect communication line by using satellites.

For attacking a UAV, an intruder must affect the UAV externally except the intruder is able to access the UAV system physically. Because of UAV communication systems' wireless nature, these systems rely heavily on the external input. This offers numerous input channels to an attacker in order to attack UAV systems. The information in UAVs streams between the UAVs and external environment over different channels. The two-way communication between the GPS and UAV communication system is the highly exposed channel which can be utilized for attacking the UAVs. The other most critical and sensitive component of the UAVs is flow of information from the external environment to UAVs sensors. Such links are highly prone to manipulations. Moreover, reliability of these sensors is not trustworthy. The knowledge of the host about receptiveness of components to commands is a key for controlling UAVs during cyber-attacks.

### 7.1.2 Motivation

In the last few decades, numerous review papers about UAVs' security have been published [5–7]. Many of these reviews are incomprehensive and address just a few security concerns.

Moreover, protocols in these reviews are not properly examined when it comes to security and vulnerabilities. Thus, there is a need for an exhaustive survey that contains all UAVs' vulnerabilities and security threats. This survey investigates the following two aspects of UAVs:

- Security and privacy protocols for UAVs.
- Vulnerabilities present in these protocols.

Initially, drones were mostly used by military in relief operations like rescue operations. Currently, UAVs are also common in civil applications like capturing videos and images of many areas. Because of their massive increase, UAVs have become an easy target for attackers. It is insecure to carry out sensitive operations through UAVs unless there are strong security protocols. These protocols for UAVs should be very strong. These protocols should have the ability to resist cyber-attacks like integrity, confidentiality, denial of services (DoS), etc. Vulnerabilities present in UAVs' security protocols cause many cyber threats. For securing UAVs from these threats, these security protocols must be free of vulnerabilities. In recent years, different UAV security protocols are proposed. Some of those provide a great sense of security. Still, it is important to investigate vulnerabilities in UAV security protocols. The main motivation behind this survey is to explore these vulnerabilities and provide solutions for them.

### 7.1.3 Contributions of This Survey

In the published literature, only a few survey papers have discussed security vulnerabilities present in security protocols of UAVs [7–11]. For example, UAVs' security vulnerabilities have been discussed in Ref. [9]. This paper has discussed vulnerabilities that might be used for attacking UAVs. These include attacks on control and communication systems, jamming and spoofing attacks. However, there is no discussion about how to secure UAVs from these attacks. In Ref. [9], security and privacy issues present in UAVs were discussed by Zhi et al. Only two categories of attacks have been discussed. The first is spoofing attacks and the second is Wi-Fi attacks. Still, this paper is incomprehensive because it lacks in-depth exploration of attacks and vulnerabilities of UAVs. Another study has highlighted UAVs communication protocols and their vulnerabilities [11]. However, the study did not discuss the prevention of attacks. Table 7.1 presents comparison of literature in this chapter and recent literature relevant to security of UAVs. This chapter has covered all crucial aspects like UAVs' security, vulnerabilities, and countermeasures.

Following are the contributions of this survey:

- In this chapter an in-depth review of existing literature related to UAVs' security is provided.
- Vulnerabilities existing in security protocols are explored and analyzed.
- Some solutions possible to eliminate the identified vulnerabilities are also offered in this chapter.

TABLE 7.1 Comparing Chapter with Existing Review Paper

| Security and Privacy Classification in UAVs | Sub-Classification | Paper [16] | Paper [17] | Paper [18] | Paper [11] | Paper [19] | Current Survey |
|---|---|---|---|---|---|---|---|
| Secure Communication in UAVs | Symmetric cryptography protocols for security | | | ✓ | | | ✓ |
| | Asymmetric cryptography protocols for security | | | ✓ | | | ✓ |
| | Lightweight authentication protocols for UAV | ✓ | | | | ✓ | ✓ |
| Intrusion Detection System (IDS) | Machine learning-based IDS | | | | | ✓ | ✓ |
| | Signature-based IDS | | | | | ✓ | ✓ |
| Packet Forwarding | | | | | | | ✓ |
| Spoofing Attacks | GPS spoofing | ✓ | ✓ | ✓ | ✓ | | ✓ |
| | False data injection | | ✓ | ✓ | ✓ | | ✓ |
| Wi-Fi Insecurity | | | ✓ | ✓ | ✓ | | ✓ |
| Jamming Attacks on UAV | | ✓ | ✓ | ✓ | ✓ | | ✓ |
| Attacks on Control Surface | | ✓ | | | | ✓ | ✓ |
| Fuzzing Attacks on UAVs | | | | | | ✓ | ✓ |
| Evil UAVs Detection | | | ✓ | | ✓ | | ✓ |

- It is discussed how numerous vulnerabilities like Wi-Fi insecurity can be used for attacking UAVs.
- Vulnerabilities in routing protocols of UAVs and how they threaten security are explored in the chapter.
- Finally, the chapter provides possible future directions for research in order to increase UAVs' security.

This research paper is arranged in this way: Section 7.2 presents survey of current security protocols designed to increase UAVs' security. In Section 7.3, possible future research directions are proposed. Finally, Section 7.4 concludes the study.

## 7.2 SECURITY AND PRIVACY IN UAVs

No doubt, drones have brought different benefits including commercial and personal benefits. Yet, there are many securities, privacy and safety drawbacks that need to be addressed before fully depending on them [12].

Cyber criminals can use these drones to invade the safety and privacy of individuals as well as general public. Many attributes of drones are used in attacks including unauthorized inspections as well as high-level operations. Drones should not be used for capturing images and for recording the videos of individuals without their consent [13]. The drones should not be allowed in residential and public properties as images taken by drones might be used to carry out illegal activities like scamming. Majority drones, at present, are Wi-Fi enabled for broadcasting the captured videos to smart devices. Many drones have Wi-Fi for remote control using smart devices. Because Wi-Fi connections have weak passwords, attackers easily access Wi-Fi. After that, they can interfere with the communication, and particularly Wi-Fi passwords are not encrypted [14].

Attackers may use unauthorized UAVs for destroying authorized UAVs with the help of physical collisions. As unauthorized and authorized UAVs run into each other frequently, it is important to prevent any collision among them. Different modes are investigating in existing literature for preventing collision among UAVs [15]. The goal was to give a Sense-and-Avoid algorithm for sensing UAV and referring to the hurdles placed by an attacker. Barfield has presented another model for SAA in which he proposed an independent collision prevention system that has the ability to protect UAVs from accidents. There was no failure during recorded during practical trails of tested flights.

In addition to above, many algorithms to avoid collision have been designed. These algorithms have helped in accomplishing some crucial challenges including Individual Collision Avoidance (ICA) as well as Group Collision Avoidance (GCA).

Yang et al. have presented another method in Ref. [16]. This method is based on a 3D path organization for UAV that comprises tracking down a collision-free path. Several methods to avoid collision are also present to get rid of any hurdle facing UAVs. Ueno et al. have given a novel algorithm that is capable of correctly locating the objects in UAVs vicinity. In another research work, Brandt and Colton [17] provided that drones are more appropriate for operating indoors because of their flexible as well as well-controlled operations in confined and small areas. Israelsen et al. [18] presented an algorithm for manually controlling UAVs applying automated Obstacle Collision Avoidance.

Additionally, avoiding UAVs' collisions, this is also crucial to secure the communication among UAVs as well as GCS. To secure this communication, different security protocols were proposed. The nature of UAVs' application decides the use of these security protocols. This will be discussed later. There are three main categories of UAVs security protocols. The first is secure communication. The second is security of the physical layer and the third is intrusion detection system. Figure 7.3 shows the block diagram for these security protocols.

### 7.2.1 Secure Communication in UAVs

UAVs have ability for observing a wider area without taking any aid from network. While in flight, UAVs continuously exchange sensitive information by communicating with GCS. Because of dynamic topology, new challenges are created by the exchange of information. UAVs are often used to transmit data from a node to GCS. Attackers can attack the transmitted data in different ways. In maximum military application, important information is transmitted between authorized users via wireless communication channels. As these communication channels are insecure medium, it is easy to steal information by launching cyber-attacks. These may include confidentiality attacks, integrity attacks, and availability

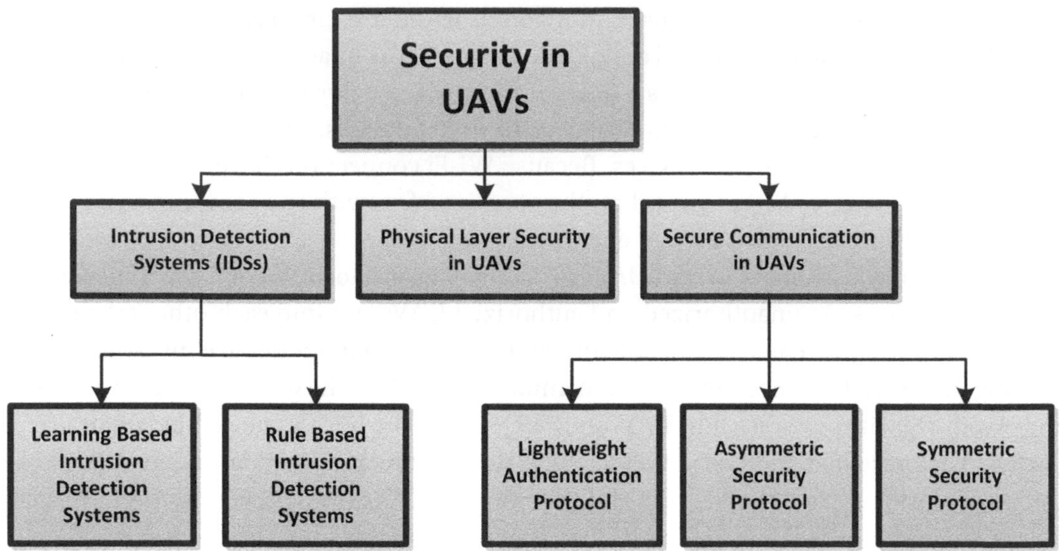


FIGURE 7.3 Security classification in UAVs.

attacks. For protecting the information from attackers, numerous security protocols are being used. These security protocols help in securing the transmission as well as in authentication of users. Examples of these protocols include symmetric protocols for UAV security and asymmetric protocols for security. These two types of protocols are used for securing communication among UAVs and GCS. Just a private key is used for encrypting and decrypting data in symmetric cryptography security protocols. However, asymmetric cryptography protocols use pair keys (one public and one private). Public key is used for data encryption while private key is used for decryption. Symmetric and asymmetric protocols are scribed in the following sections. After that, authentication protocols that are lightweight are discussed. These protocols are used in case of low computational complexity and less memory required.

#### *7.2.1.1 Symmetric Cryptography Protocols*

These days, cryptographic protocols are used for ensuring confidentiality, integrity, and availability. Particularly, symmetric cryptographic protocols are being used for protecting sensitive data like text, audio, video, and images. In these protocols, sender and receiver both share the same key for encryption and decryption. For example, One Time Pad (OTP) is used frequently for securing transmission. For this purpose, OTP needs having the same size as the size of data. For example, when an image containing pixels of N rows and M columns is transmitted, key should be the same as the size of the image. Security MAV-link (Micro Air Vehicle Link) is increased by applying OTP encryption in Ref. [19]. A decryption plus encryption method is used for securing the data transmission. Many commands are used to control UAVs, like start and takeoff commands. All commands are given in bit format which can be shown by 0–1. A long text, secured by using encryption, is created with the combination of these bits. OTP encryption mechanism also has some disadvantages. For example, the size of key and data should be equal. In order to send large data, it is necessary to share key at the receiver side. Thus, key sharing is a problem because it consumes most of the bandwidth.

Furthermore, this key might be used only once. Hence, for every transmission a new key is required [19]. Algorithm in Ref. [19] can be improved by implementing robust transformation mechanisms like discrete wavelet and discrete cosine transforms. These mechanisms initially convert original data into several frequency coefficients that are fully different from real data. Further, this type of transformation is quicker than transformation performed on the original message directly [20]. A chaotic Lorenz system is presented in Ref. [21]. This system can encrypt as well as decrypt actual and transformed messages. The system has long-term unpredictability. Chaotic Lorenz may generate higher randomness with minimum differences in seed values. UAVs collect data from cameras, sensors, and later pass data to the Lorenz encoder. No direct encryption is applied to plain message. Every information is changed into bits and then encrypted. Following encryption, UAVs send encrypted information to receive. The receiver then decrypts information by using reversing the Lorenz system. This encryption algorithm lies in the category of symmetric protocol. Yet, this protocol [21] also has some drawbacks. For example, no data scrambling mechanism is present in this system. Actually, the security of encryption algorithms relies on both diffusion and confusion [22].

#### *7.2.1.2 Asymmetric Cryptography Protocols*

Asymmetric protocols for security fuse two separate keys including one public and one private key. These keys are used for encryption and decryption of data by sender and receiver, respectively. The confidentiality of a public key is not very important since the information encrypted through public key cannot be decrypted using same key. There is always need for private key to decrypt data. The author has proposed a mechanism in Ref. [23], for checking the originality of the data. This protocol verifies whether data is sent by the original ground station or eavesdroppers. This protocol falls in category of asymmetric protocols. These protocols are used for ensuring the integrity of data being transmitted among different devices or sensors. With the help of X.509, the length of generated signatures has become shorter. This made the authentication process faster.

The scheme presented in Ref. [24] verifies signatures once the data is received by receiver. Here, the verification process is executed by UAV (after receiving the data from GCS) to confirm confidentiality before executing the last action. A 164 bits hash is created through SHA-1 Algorithm on sender sides. This hash is encrypted through a public key before it is sent to the other side. Private key is utilized for decrypting the hash on receiving side. Later, hash is calculated by receiver for the original message. Finally, both hashes are compared by the receiver to check the originality of message.

Security of Automatic Dependent Surveillance Broadcast (ADS-B) can be enhanced by asymmetric protocols [23]. The ADS-B, being a protocol for air traffic surveillance, is insecure. It is used for detecting the other surrounding UAVs. ADS-B has some disadvantages like it lacks built-in mechanisms for security, for example, an encryption module to protect from eavesdroppers and data tampering. It is important that ADS-B technology meets security requirements. Wesson et al. [23] have investigated either an asymmetric protocol can increase security in ADS-B or not. For this purpose, they have evaluated the existing security protocols based on their ability for making ADS-B more secure. Then after evaluation, they have contended that Key Elliptic Curve Digital Signature is more practical.

Further, the application of asymmetric cartography through ADS-B is costly as well as time-consuming because here different keys are provided for encrypting and decrypting the actual text. Communication security protocols may also be used for ensuring data integrity being transmitted. Diffie as well as Hellman presented a protocol for key exchange between the sender and receiver [25]. This protocol has been used in past decades. Unfortunately, both parties have no prior information whether keys sent over the communication channel are shared by authorized person or not.

In Ref. [26] authors have proposed a protocol for exchanging public keys. In this protocol, keys are exchanged by sensor nodes and communication between nodes starts after authenticating the surrounding nodes. In this framework, sensor nodes are actual communication parties. Here, one sensor sends a message encrypted with public key and other senor decrypts message using private key and random numbers generated by the sensor itself. Likewise, the other party transmits an encrypted message using public key. The cipher message is then decrypted by the first party through private key as well as random number. If both parties have the same messages, then the sensor is considered an original entity and will be allowed to continue communication.

Valentin-Alexandru et al. developed trust-based security protocol to secure UAVs [27]. In this protocol, UAV determined sensors are assigned some trust values to check accuracy and correctness of data. This methodology has three modules: (1) direct determination of trust value, (2) indirect determination of trust value, and (3) final determination of trust value calculated by UAVs. Each sensor determines its own trust value. UAVs use their own trust value plus indirect values produced by sensors. If an attacker places a sensor, that sensor can also produce trust value. However, UAV will determine the ultimate trust value. For this, trust values saved in the UAV log file will be compared with those received by a UAV. If the result of comparison is a negative trust value, then sensors are not trustworthy and UAV will not take more data from sensors. Thus, an attacker will not have any access to UAV. Contrary to this a positive trust value indicates that sensors are trustworthy and UAV can take further communications with sensors. While zero trust value shows that more information is required to determine either sensors can be trusted or not.

Yoon et al. in [28] have implemented a protocol that can verify whether data received by a UAV is sent from an attacker or original ground station. In this scheme, a UAV sends a randomly encrypted message to GCS. Once data is received by GCS it is decrypted using a private key then again encrypted using a public key. At last, it is sent to UAV where it is compared to values stored in UAV log file. If the verification is successful, the UAV can take off. If verification fails, this tells that UAV attackers are trying to get control over UAV. Hence, connection with that communication channel is terminated. No further communication between sender (GCS) and receiver (UAV) will take place. In this mechanism, only random text encryption and comparison are used for checking if the UAV is hijacked or secure. The scheme provides authentication of data. Yet, in case of larger data transmission for UAV authentication, high bandwidth, processing time, and cost are required. Instead of whole text encryption a small hash can be used to overcome these issues.

In addition to the above, Steinmann et al. have proposed key negotiation framework for providing security and confidentiality to data stored in UAV chip. The major theme in this method is to develop a scheme that changes random keys continuously. For example, One-Time PAD technique can be used to encrypt data because in this technique the size of original message and key is the same. Here, if one key is explored, the whole message can be revealed easily. Therefore, random keys generation can increase the security of original message and keys. In this mechanism the sender sends a message to the receiver along with a hash code after encrypting with a public key. The receiver uses a private key for data decryption, calculates value of hash, and later compares both received and calculated hashes. If both hash codes match, it ensures the integrity of the message. Authentication is implemented through this message. Yet it is not good to keep keys as secret.

#### *7.2.1.3 Lightweight Authentication Protocols for UAV*

Lightweight encryption protocols belong to another class of security protocols used to hide confidential information from intruders. These lightweight protocols can help to reduce the information encoding time. Lightweight protocols do not take much memory which permits UAVs to perform faster. A lightweight protocol is implemented in Ref. [29]. The protocol is more suitable for switching context frequently in a highly multi-task environment. In Ref. [30] Wang et al. developed a blockchain-based solid routing scheme for Unarmed Aerial System (UAS) networking. In this algorithm, lightweight blockchain is used like a bargaining chip for strengthening routing of UAS, using 5G cellular technology. This algorithm is not the same as traditional algorithms for routing. As it can easily identify spiteful UAS, reduce attack intensity, and avoid the attackers. The proposed algorithm is intended to secure UAS and expand the deployment of swarm UAS network over a wider range. Integration of low-cost devices into UAVs can help to secure data with the use of Internet of Things. For minimizing the effects of cyber-attacks, sessions keys known by both participating parties should be used for data encoding.

Conversely, it is difficult to get required abilities for both encoding and session keys with low-cost IoT integration, due to some performance limitations. Demeri et al. [31] have implemented an integrated secure data transfer mechanism by using low-cost aerial policy that integrates numerous cryptographic accelerators. The components are combined using moldable Application Programming Interference (API) through a hardware and software design yielded costless drones. With recent advancements in wireless communication systems as well as electronic devices, great relaxation is being offered to public by UAVs.

Moreover, UAV security is getting higher attention because of emerging security issues, financial and strategic information, and significance associated with UAV applications. For providing privacy and security to communicating parties, lightweight protocol for authentication is proposed in Ref. [32]. This protocol offers safe communication between the ground station and UAVs. This framework has also used packet capture (PCAP) for ensuring secure communication between both parties. The prime idea behind PCAP is a UAV as well as a ground station using seed values from chaotic maps which can shuffle actual message randomly based on chaotic sequence [32]. Yet, UAVs are mistrusted when it comes to dabble and device capturing attacks due to the advancement of remote-control environments and the presence of lower resources. This enhances the risk of data stealing. Haque and Chowdhury [33] have concentrated just on secure transmission of data sent by UAV to GCS. The concept of lightweight along with data security is discussed and a novel framework was developed for achieving the desired task. To make the system lightweight, selective encryption is performed. In this algorithm, other than cryptographic method watermarking is also used for increasing data confidentiality and integrity. The reason behind using specific encryption is to offer stabilization among UAVs within limited resources. However, selective encryption has advantage, particularly in case of real-time application where quick processing is necessary. To point out authentication problems and data insecurity, a two-step lightweight correlative authentication protocol is implemented in Ref. [34]. In this method, a Software Defined Network (SDN) is assisted with the UAV network that is installed in necessary spying areas. Additionally, evidence for security of the protocol is also provided to highlight security features present in it. For UAVs, Smart Internet of Drones assisted framework is proposed in Ref. [35]. This framework collects all needed information independently. To reduce computational cost, Lightweight Privacy-Preserving Scheme (L-PPS) is provided in Ref. [35]. This scheme provides robustness among IoT devices along with a proper authentication time. Moreover, because of limited resources as well as substantial risk around UAV, attackers can perform numerous attacks including confidentiality, wireless, and man in middle attacks. To protect UAVs from these attacks, an authentication is necessary to be implemented prior to UAVs start communication. It is also urgent that implemented authentication mechanism guarantees that original drone is prime concern of UAVs network security.

As standardized authentication mechanism having a username and a password, the secret key cannot be very secure. The RSA certification needs long-lasting keys that cannot fulfill lightweight demand in UAVs infrastructure. A lightweight recognition mode for authentication based on Elliptic Curve Cryptography (ECC) is developed in Ref. [36]. It has three phases: (1) Initiation of ECC certification, (2) identity authentication, and (3) verification of key compatibility. The authors have mentioned that phases one and two are completely matched with the two-way authentication, while the third phase validates the consistency of verification keys. Contrary to traditional modes of authentication in a UAV network, the method proposed in Ref. [36] depends upon short keys as well as reduced computing workload. Barka et al. [37] proposed lightweight communication technique for aerial NDN. This technique can assist NDN security, also has 80% prediction accuracy. This technique has reduced end-to-end pause. With the help of this technique the pause is less than a second in worst condition. Average energy use is also reduced by this technique. A new authentication mechanism is proposed in Ref. [38]. As UAVs have small batteries and consist of limited memory, the lightweight security protocols suit them well. Srinivas et al. [38] presented a Temporal Credential-Based Anonymous Lightweight Authentication Scheme (TCALAS) for IoD networks. However, the scheme proposed in Ref. [38] can work in one condition only where there is a single flying zone that cannot be expanded. Further, the exploration provided in Ref. [6] indicates that the scheme proposed by Srinivas et al.

does not prove to be efficient for availability and traceability attacks. In Ref. [6] an upgraded version of the TCALAS scheme, uTCALAS, is provided by Ali et al. This method provides security against different attacks including availability and traceability attacks. It increases extendibility and can work in numerous areas where more different flying zones are available. Also, they have successfully achieved fast computation for authentication taking 2.29 ms for accomplishing the whole authentication process.

### 7.2.2 Security of UAV Physical Layer

One frequently adopted performance measure in security of physical layer is rate of secrecy at which data is transmitted securely [39]. Conventional encryption protocols are vulnerable due to key distribution as well as higher processing time. Examination of physical layer properties of communication channels can assist secure transmission. To get maximum secrecy rate for transmitted data, physical layer security is used frequently. It is crucial for all communication devices and security controls mounted in a UAV. Unlike traditional approaches of cryptographic security, PLS takes benefits from the attributes of cellular channels, for example, interference, fading, and nose for boosting signals reception at receiver end and reducing the quality of signals for eavesdroppers [40]. PSL can be ensured with the incorporation of cryptographic protocols. There are numerous cryptographic protocols presented in literature. These protocols provide significant security level, but unfortunately there is no protocol that can provide ideal security. Hence, PSL is getting more attention. To maximize secrecy rate and increase security of UAV communication, many works have been proposed on PSL [41–43]. In recent years, static relay-based systems for communication have gained importance for improving existing PSL algorithms. A novel framework based on relying on technique called UAV empowered mobile relay is considered the valuable technology. The authors have developed an upgraded version of PSL algorithm based on UAV enabled mobile relying [44]. For improving communication system security, buffer-aided mobile-based relay is implemented that permits quick and independent data arrival.

### 7.2.3 Intrusion Detection System

#### *7.2.3.1 Intrusion Detection Using Machine Learning*

Digital machines can perform multiple tasks depending on user instructions. For automation of tasks, different mechanisms and techniques are used. For example, Machine Learning (ML), Neural Networks, and deep learning are used commonly. There are two phases in ML algorithms including training and testing. In training, model is trained with the help of data to predict future events depending on training. Testing phase is used to measure the accuracy of the model with the help of different strategies. These ML techniques can also be applied to UAVs to detect the intrusion. When UAVs are trained, it can recognize the intrusion patterns. A novel algorithm was presented in Ref. [45]. This algorithm is based on deep reinforcement as well as weighted least square [46]. It is used to calculate jamming signal's using Convolutional Neural Network (CNN) [47]. Bit Error Rate (BER) is used to select a relay power element. Once UAV receives values of weights and that from ground station it compares the values of parameters for relay selection. If it is greater than zero, UAV uses reinforcement and randomly selected values to send message.

#### *7.2.3.2 Signature-Based Intrusion Detection*

For making devices intelligent, instructions must be given to these devices. There is a need to define rules in rule-based tasks. Devices send messages to base station and act depending upon these rules. When it comes to UAVs, different rules are stored in the chips for and levels for rule acceptance are created. A novel IDS is proposed in Ref. [48]. This is based on rules of behavior for minimizing false negative predictions. Seven attacks related to integrity, confidentiality, and availability were discussed in this detection method. All these types of attacks are considered, and after their detection, UAVs take defense steps to protect themselves from these attacks. Thus, IDSs remove the impact of attacks to prevent system from hazards.

## 7.3 CHALLENGES AND FUTURE DIRECTIONS FOR RESEARCH

Due to advancement in UAV technology, their use and application has increased exponentially. There are numerous research areas that lack proper investigation and solutions, so there is a need to tackle these areas in efficient way. This can help to generate more secure and trustworthy UAVs in future. The innovative solutions should be restricted with specific needs and limitations, for example, unwavering quality, intricacy, etc. This section provides some probable future solutions for security of UAVs.

### 7.3.1 Connectivity Insecurity

There are two frequently used methods for connecting UAVs to GCS. First is connecting through Wi-Fi and second is through radio signals. No doubt wired connections are more secure as compared to wireless, but these cannot be used due to range restrictions. Thus, wireless connections are mostly used. Security of these wireless connections can be increased through encryption [49] and watermarking methods [50]. The novel solution can be proposed by combining the watermarking and cryptography in single protocol as watermarked text can retain itself. Thus, it can be assumed that if password is hacked, the cracked password might be used by attackers before applying reverse watermarking.

### 7.3.2 Authentication Drawback in UAVs

Authentication is most important in UAV networks as most of the time eavesdroppers use false information for hijacking UAVs. Spoofing is mostly used for distracting UAVs. Also, authentication is crucial for identifying the right signals. Researchers are trying to tackle this problem. No doubt there are numerous authentication protocols, but still there is a necessity to improve these protocols. A brute force attack can predict single random numbers. Thus, instead of using these numbers, a chaotic map can be used with a key to generate multiple random numbers [51–54]. However, low-dimensional chaotic map is not good. A high-dimensional map can solve this problem [55].

### 7.3.3 Intrusion Detection Systems (IDS) in UAVs

Numerous ML algorithms are present for intrusion detection. These algorithms predict the coming events with the help of already learned data. For increasing the accuracy of these, algorithms' features must be relevant. For example, prediction results cannot be correct if the features in the dataset are irrelevant. For improving ML-based intrusion detection techniques there is a need to include deep learning. This will help to make prediction more accurately. More number of features for intrusion detection can increase model accuracy.

## 7.4 CONCLUSION

In this chapter, issues in UAVs security and privacy protocols were discussed. Existing protocols for UAV security were reviewed and vulnerabilities in these protocols were identified as well. It concludes that all protocols need improvement for securing UAVs. Moreover, vulnerability should be addressed for securing the coming generation of UAVs. Next, the future direction for research is summarized. Strong UAV security is a prime concern, and researchers should focus on the suggested research direction. In this way, the security of UAVs can be improved.

## REFERENCES


1. E. G. AbdAllah, H. S. Hassanein, and M. Zulkernine, "A survey of security attacks in information-centric networking," *IEEE Communications Surveys & Tutorials*, vol. 17, no. 3, pp. 1441–1454, 2015.
2. H. Zhu, M. L. Cummings, M. Elfar, Z. Wang, and M. Pajic, "Operator strategy model development in UAV hacking detection," *IEEE Transactions on Human-Machine Systems*, vol. 49, no. 6, pp. 540–549, 2019.
3. J. Li, S. Kamin, G. Zheng, F. Neubrech, S. Zhang, and N. Liu, "Addressable metasurfaces for dynamic holography and optical information encryption," *Science Advances*, vol. 4, no. 6, p. eaar6768, 2018.
4. M. P. Arthur, "Detecting signal spoofing and jamming attacks in UAV networks using a lightweight IDS," in *2019 International Conference on Computer, Information and Telecommunication Systems (CITS)*, Beijing, China, 2019, pp. 1–5: IEEE.
5. H. Shakhatreh, et al., "Unmanned aerial vehicles (UAVs): A survey on civil applications and key research challenges," *IEEE Access*, vol. 7, pp. 48572–48631, 2019.
6. A. Sharma, et al., "Communication and networking technologies for UAVs: A survey," *Journal of Network and Computer Applications*, vol. 168, p. 102739, 2020.
7. J. M. Hamamreh, H. M. Furqan, and H. Arslan, "Classifications and applications of physical layer security techniques for confidentiality: A comprehensive survey," *IEEE Communications Surveys & Tutorials*, vol. 21, no. 2, pp. 1773–1828, 2018.
8. A. Abbaspour, K. K. Yen, S. Noei, and A. Sargolzaei, "Detection of fault data injection attack on UAV using adaptive neural network," *Procedia Computer Science*, vol. 95, pp. 193–200, 2016.
9. Y. Zhi, Z. Fu, X. Sun, and J. Yu, "Security and privacy issues of UAV: A survey," *Mobile Networks and Applications*, vol. 25, no. 1, pp. 95–101, 2020.
10. C. L. Krishna and R. R. Murphy, "A review on cybersecurity vulnerabilities for unmanned aerial vehicles," in *2017 IEEE International Symposium on Safety, Security and Rescue Robotics (SSRR)*, 2017, pp. 194–199: IEEE.
11. N. A. Khan, S. N. Brohi, and N. Jhanjhi, "UAV's applications, architecture, security issues and attack scenarios: A survey," in *Intelligent Computing and Innovation on Data Science*: Springer, Singapore, 2020, pp. 753–760.
12. M. A. Kafi, Y. Challal, D. Djenouri, M. Doudou, A. Bouabdallah, and N. Badache, "A study of wireless sensor networks for urban traffic monitoring: Applications and architectures," *Procedia Computer Science*, vol. 19, pp. 617–626, 2013.
13. P. A. Peterson, "Cryptkeeper: Improving security with encrypted RAM," in *2010 IEEE International Conference on Technologies for Homeland Security (HST)*, Waltham, MA, United States, 2010, pp. 120–126: IEEE.

14. A. Jones and G. L. Kovacich, *Global Information Warfare: The New Digital Battlefield*. CRC Press, Boca Raton, FL, 2015.
15. A. Zeitlin, A. Lacher, J. Kuchar, and A. Drumm, "Collision avoidance for unmanned aircraft: Proving the safety case," MITRE CORP MCLEAN VA 2006.
16. L. Yang, J. Qi, J. Xiao, and X. Yong, "A literature review of UAV 3D path planning," in *Proceeding of the 11th World Congress on Intelligent Control and Automation*, Shenyang, 2014, pp. 2376–2381: IEEE.
17. A. M. Brandt and M. B. Colton, "Haptic collision avoidance for a remotely operated quadrotor UAV in indoor environments," in *2010 IEEE International Conference on Systems, Man and Cybernetics*, Istanbul, Turkey, 2010, pp. 2724–2731: IEEE.
18. J. Israelsen, M. Beall, D. Bareiss, D. Stuart, E. Keeney, and J. van den Berg, "Automatic collision avoidance for manually tele-operated unmanned aerial vehicles," in *2014 IEEE International Conference on Robotics and Automation (ICRA)*, Hong Kong, China, 2014, pp. 6638–6643: IEEE.
19. S. Atoev, O.-J. Kwon, C.-Y. Kim, S.-H. Lee, Y.-R. Choi, and K.-R. Kwon, "The secure UAV communication link based on OTP encryption technique," in *2019 Eleventh International Conference on Ubiquitous and Future Networks (ICUFN)*, Zagreb, Croatia, 2019, pp. 1–3: IEEE.
20. Y. Ma, X. Wu, G. Yu, Y. Xu, and Y. Wang, "Pedestrian detection and tracking from low-resolution unmanned aerial vehicle thermal imagery," *Sensors*, vol. 16, no. 4, p. 446, 2016.
21. V. V. Kirichenko, "Information security of communication channel with UAV," *Electronics and Control Systems*, no. 3, pp. 23–27, 2015.
22. C. E. Shannon, "Communication theory of secrecy systems," *The Bell System Technical Journal*, vol. 28, no. 4, pp. 656–715, 1949.
23. K. D. Wesson, T. E. Humphreys, and B. L. Evans, "Can cryptography secure next generation air traffic surveillance?" *IEEE Security and Privacy Magazine*, 2014.
24. W.-J. Pan, Z.-L. Feng, and Y. Wang, "ADS-B data authentication based on ECC and X. 509 certificate," *Journal of Electronic Science and Technology*, vol. 10, no. 1, pp. 51–55, 2012.
25. E. Bresson, O. Chevassut, and D. Pointcheval, "Provably secure authenticated group Diffie-Hellman key exchange," *ACM Transactions on Information and System Security (TISSEC)*, vol. 10, no. 3, pp. 10-es, 2007.
26. O. K. Sahingoz, "Multi-level dynamic key management for scalable wireless sensor networks with UAV," in *Ubiquitous Information Technologies and Applications*: Springer, Dordrecht, 2013, pp. 11–19.
27. V. Valentin-Alexandru, B. Ion, and P. Victor-Valeriu, "Energy efficient trust-based security mechanism for wireless sensors and unmanned aerial vehicles," in *2019 11th International Conference on Electronics, Computers and Artificial Intelligence (ECAI)*, Pitesti, Romania, 2019, pp. 1–6: IEEE.
28. K. Yoon, D. Park, Y. Yim, K. Kim, S. K. Yang, and M. Robinson, "Security authentication system using encrypted channel on UAV network," in *2017 First IEEE International Conference on Robotic Computing (IRC)*, Taichung, Taiwan, 2017, pp. 393–398: IEEE.
29. K. Driscoll, "Lightweight crypto for lightweight unmanned arial systems," in *2018 Integrated Communications, Navigation, Surveillance Conference (ICNS)*, Herndon, VA, USA, 2018, pp. 1–15: IEEE.
30. J. Wang, Y. Liu, S. Niu, and H. Song, "Lightweight blockchain assisted secure routing of swarm UAS networking," *Computer Communications*, vol. 165, pp. 131–140, 2021.
31. A. Demeri, W. Diehl, and A. Salman, "Saddle: Secure aerial data delivery with lightweight encryption," in *Science and Information Conference*, London, United Kingdom, 2020, pp. 204–223: Springer.
32. C. Pu and Y. Li, "Lightweight authentication protocol for unmanned aerial vehicles using physical unclonable function and chaotic system," in *2020 IEEE International Symposium on Local and Metropolitan Area Networks (LANMAN)*, Orlando, FL, USA, 2020, pp. 1–6: IEEE.

33. M. S. Haque and M. U. Chowdhury, "A new cyber security framework towards secure data communication for unmanned aerial vehicle (UAV)," in *International Conference on Security and Privacy in Communication Systems*, Canada, 2017, pp. 113–122: Springer.
34. T. Alladi, V. Chamola, and N. Kumar, "PARTH: A two-stage lightweight mutual authentication protocol for UAV surveillance networks," *Computer Communications*, vol. 160, pp. 81–90, 2020.
35. B. D. Deebak and F. Al-Turjman, "A smart lightweight privacy preservation scheme for IoT-based UAV communication systems," *Computer Communications*, vol. 162, pp. 102–117, 2020.
36. L. Teng, et al., "Lightweight security authentication mechanism towards UAV networks," in *2019 International Conference on Networking and Network Applications (NaNA)*, Daegu, Korea (South), 2019, pp. 379–384: IEEE.
37. E. Barka, C. A. Kerrache, R. Hussain, N. Lagraa, A. Lakas, and S. H. Bouk, "A trusted lightweight communication strategy for flying named data networking," *Sensors*, vol. 18, no. 8, p. 2683, 2018.
38. J. Srinivas, A. K. Das, N. Kumar, and J. Rodrigues, "TCALAS: Temporal credential-based anonymous lightweight authentication scheme for Internet of drones environment," *IEEE Transactions on Vehicular Technology*, vol. 68, no. 7, pp. 6903–6916, 2019.
39. Q. Li, Y. Yang, W.-K. Ma, M. Lin, J. Ge, and J. Lin, "Robust cooperative beamforming and artificial noise design for physical-layer secrecy in AF multi-antenna multi-relay networks," *IEEE Transactions on Signal Processing*, vol. 63, no. 1, pp. 206–220, 2014.
40. N. Yang, L. Wang, G. Geraci, M. Elkashlan, J. Yuan, and M. Di Renzo, "Safeguarding 5G wireless communication networks using physical layer security," *IEEE Communications Magazine*, vol. 53, no. 4, pp. 20–27, 2015.
41. Y. Zeng, R. Zhang, and T. J. Lim, "Throughput maximization for UAV-enabled mobile relaying systems," *IEEE Transactions on Communications*, vol. 64, no. 12, pp. 4983–4996, 2016.
42. D. H. Choi, S. H. Kim, and D. K. Sung, "Energy-efficient maneuvering and communication of a single UAV-based relay," *IEEE Transactions on Aerospace and Electronic Systems*, vol. 50, no. 3, pp. 2320–2327, 2014.
43. K. Li, R. C. Voicu, S. S. Kanhere, W. Ni, and E. Tovar, "Energy efficient legitimate wireless surveillance of UAV communications," *IEEE Transactions on Vehicular Technology*, vol. 68, no. 3, pp. 2283–2293, 2019.
44. Q. Wang, Z. Chen, W. Mei, and J. Fang, "Improving physical layer security using UAV-enabled mobile relaying," *IEEE Wireless Communications Letters*, vol. 6, no. 3, pp. 310–313, 2017.
45. X. Lu, L. Xiao, C. Dai, and H. Dai, "UAV-aided cellular communications with deep reinforcement learning against jamming," *IEEE Wireless Communications*, vol. 27, no. 4, pp. 48–53, 2020.
46. W.-S. Ra, I.-H. Whang, and J. B. Park, "Robust weighted least squares range estimator for UAV applications," in *2008 SICE Annual Conference*, Chofu, Japan, 2008, pp. 251–255: IEEE.
47. M. B. Bejiga, A. Zeggada, A. Nouffidj, and F. Melgani, "A convolutional neural network approach for assisting avalanche search and rescue operations with UAV imagery," *Remote Sensing*, vol. 9, no. 2, p. 100, 2017.
48. H. Sedjelmaci, S. M. Senouci, and N. Ansari, "Intrusion detection and ejection framework against lethal attacks in UAV-aided networks: A Bayesian game-theoretic methodology," *IEEE Transactions on Intelligent Transportation Systems*, vol. 18, no. 5, pp. 1143–1153, 2016.
49. V. Kriz and P. Gabrlik, "UranusLink-Communication protocol for UAV with small overhead and encryption ability," *IFAC-PapersOnLine*, vol. 48, no. 4, pp. 474–479, 2015.
50. M. P. Marcinak and B. G. Mobasseri, "Digital video watermarking for metadata embedding in UAV video," in *MILCOM 2005-2005 IEEE Military Communications Conference*, Atlantic City, NJ, USA, 2005, pp. 1637 Vol. 3–5: IEEE.
51. Z. Cheng, Y. X. Tang, and Y. L. Liu, "3-D path planning for UAV based on chaos particle swarm optimization," *Applied Mechanics and Materials*, vol. 232, pp. 625–630, 2012.

52. M. Rosalie, G. Danoy, S. Chaumette, and P. Bouvry, "Chaos-enhanced mobility models for multilevel swarms of UAVs," *Swarm and Evolutionary Computation*, vol. 41, pp. 36–48, 2018.
53. M. Rosalie, G. Danoy, S. Chaumette, and P. Bouvry, "From random process to chaotic behavior in swarms of UAVs," in *Proceedings of the 6th ACM Symposium on Development and Analysis of Intelligent Vehicular Networks and Applications*, Malte, 2016, pp. 9–15.
54. J. Sun, et al., "A data authentication scheme for UAV ad hoc network communication," *The Journal of Supercomputing*, vol. 76, no. 6, pp. 4041–4056, 2020.
55. P. S. Gohari, H. Mohammadi, and S. Taghvaei, "Using chaotic maps for 3D boundary surveillance by quadrotor robot," *Applied Soft Computing*, vol. 76, pp. 68–77, 2019.

CHAPTER 8

# Security in Connected Drones Environment

Saim Shahid

*Nottingham Trent University*

Shiv Prakash

*University of Allahabad*

Mukesh Prasad

*University of Technology Sydney*

## CONTENTS



## LIST OF ABBREVIATIONS

**ADS-B** Automatic dependent surveillance broadcast
**AP** Access point
**ARP** Address resolution protocol
**BVLOS** Beyond visual line of sight
**DHCP** Dynamic host configuration protocol
**DNS** Domain name system

DOI: 10.1201/b22998-8

| | |
|---|---|
| **ESP8266** | Espressif systems wifi chip |
| **GCS** | Ground control station |
| **GPIO** | General-purpose input output |
| **GPS** | Global positioning system |
| **HTTP** | Hypertext transfer protocol |
| **MAC** | Medium access control |
| **Rx** | Receiver pin |
| **SSID** | Service set identifier |
| **TCP/IP** | Transmission control protocol/internet protocol |
| **Tx** | Transmitter pin |
| **UART** | Universal asynchronous receiver-transmitter |
| **UAVs** | Unmanned aerial vehicles |
| **VLOS** | Visual line of sight |

## 8.1 INTRODUCTION

As the advancement in technology progresses, the ever-increasing need for robotic vehicles arises. Hundreds of types of robotic vehicles are developed by humans for their specific purposes. These vehicles effectively decrease expensive human labor. Many robots are also built to work in extreme difficult situations and environments where humans cannot. The rovers are built to explore other planets in the solar system, e.g., Mars is the prime example of robotic vehicles made by humans. These are designed to work in the atmospheres of the planets in which it is impossible for humans to breathe or survive [1].

Unmanned Aerial Vehicles (UAVs) have already proved their significance in this modern world either by delivering groceries to a person's home or by helping military in defense purposes, etc. The drones need to be efficient in design especially in terms of communication security because of their great number of applications nowadays [2,3]. Incompetent security protocols in drones will end up in data breaches and other dangerous consequences which can put human lives in danger [4,5]. Machines, e.g., drones, are prone to hacking, and less robust programming techniques make them more vulnerable [6].

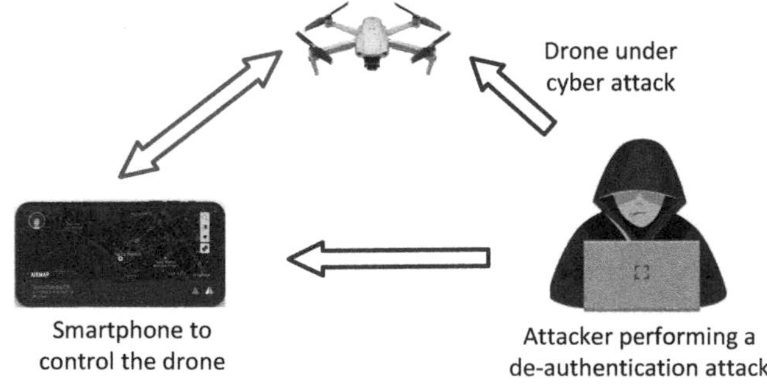


FIGURE 8.1 Drone de-authentication attack example.

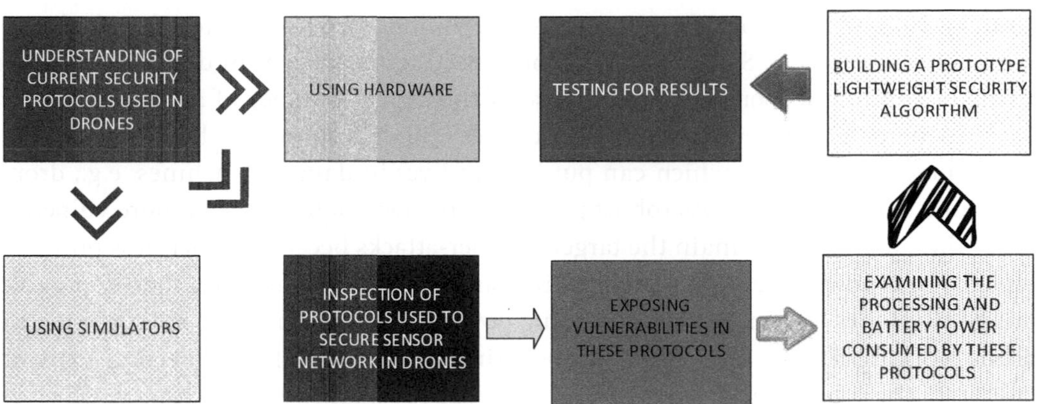


FIGURE 8.2 Flowchart showing different phases to be considered.

Hacking and compromising of drones nowadays is more common than ever before (see Figure 8.1). Drones are used in every field of life, and if somebody hacks or gain controls of a drone illegally, it may put many people's lives in danger [7]. The purpose of this research is to expose those vulnerabilities and try to eliminate them by using both software simulations and hardware. To learn more about drones, their extensive uses, vast communication types and protocols they use, and the vulnerabilities which come with them are the main influencers of doing this task. The need to design an effective lightweight security algorithm which will use less processing power and hence resources is of utmost importance when it comes to drones because of their size and power limitations [8]. So, the idea behind this research is to make a basic version of a security protocol to show how security works in drones.

The aim of this chapter is to learn and implement communication protocols used in drones. Current drones in the industry either use insufficient security protocols or do not use security at all. This leads to high-level risks. So, this chapter is focused on exploiting those risks and then providing suggestions on further improvements (see Figure 8.2). Even a small drone has very high-speed rotors which if comes in contact with a person's body, will result in extreme damage. That is why a robust communication among UAV, smartphone if used, and transmitter is tremendously important.

## 8.2 BACKGROUND STUDIES ON DRONE SECURITY

This document is composed of wireless communication security analysis of drones. There are a wide variety of drones used nowadays for civilian, commercial, government, and military purposes [8]. These drones are not just limited to flying machines, but anything unmanned, for example, an unmanned submarine, is also considered as a drone. The UAVs have already proved their significance in this modern world either by delivering groceries to a person's home or by helping military in defense purposes. The data in

January 2021 showed that there were over 1.7 million drones registered with the government of the United States [9]. The drones need to be efficient in design especially in terms of communication security because of their great number of applications nowadays. Incompetent security protocols in drones will end up in data breaches and other dangerous consequences which can put human lives in danger. Machines, e.g., drones, are prone to hacking, and less robust programming techniques make it more vulnerable. From the start, drones remain the target of cyber-attacks because of their less processing capacity and power issues. The trade off in using less robust, unsecure, shorter programming techniques is reduction of processing power consumption which will then increase drone working/flying time. But in the meantime, the drone will then become extremely vulnerable to cyber-attacks. Hence, there is a need for a lightweight technique which will focus on high-level security so as to reduce the excessive processing consumption to save power for a compact design.

### 8.2.1 General Drone Structure

A drone has three structural components. The first is flight controller which relates to all the processes happening in the drone itself. The second is ground control station (GCS) which supervises the drone functionality. The third component is the wireless link which is established between GCS and the drone [8].

### 8.2.2 Communication Types

The drones can communicate via two types of methods. The first system is based on using radio channels for acquiring and transmitting data from/to the drone which is called Visual Line of Sight (VLOS) transmission. In VLOS connections, the drone does not travel very far from the GCS and remains in eye-sight distance. The second method focuses on using satellites for Beyond Visual Line of Sight (BVLOS) transmissions. In this type of set up, drone may travel to very far remote places and go beyond eye-sight range. So, the GCS then sends the commands to the satellite which then relays the GCS commands to the drone [1]. Drones also utilize satellites to know, transmit, and/or broadcast their location, e.g., in a call to base scenario. Drones can communicate with the ground stations as well as they can connect to other drones depending upon their type and functionality. Drones can also be connected to a mobile network, but this makes the drone network extremely vulnerable. Drones mainly work on two frequencies which are 2.4 and 5 GHz. There are some pros and cons of using 5 GHz over 2.4 GHz which are described in Table 8.1.

### 8.2.3 UAV Infection Methods

It is a piece of code specifically designed for a computer program to alter the working of an overall system [9]. It resembles a kind of illegitimate access to a machine without taking the consent from the administrator. This piece of code uses different strategies to hide itself [8]. It includes fake message transmission utilizing identical protocols as the original message to mislead the drone or ground station or Automatic Dependent Surveillance Broadcast (ADS-B) [10]. ADS-B is a system used for real-time locating the position of air traffic to avoid any incidents [11].

TABLE 8.1 Technical Comparison of 2.4 and 5 GHz Service

| Variables | 2.4 GHz | 5 GHz |
|---|---|---|
| Bandwidth | Short as compared to 5 GHz | Greater |
| Channel noise | Prone to noise | Reduced noise |
| Effective area | Greater | Short as compared to 2.4 GHz |
| Price budget | Moderate | Big |
| Interference | Elevated | Scarce |

#### *8.2.3.1 Three-Way Handshake*

Machines use TCP/IP protocols to send data among each other. Before starting the transmission, client sends a request to the server and the server acknowledges the request and notifies the client back. Then client broadcasts that it has received the request from the server. This is called three-way handshake, and hence, data transmission begins. Hackers abuse this strategy and flood the server with requests by changing the IP of their machines repeatedly. This will make the server busy by occupying all the memory into half-open connections [6].

#### *8.2.3.2 Traffic Analysis*

This process relates to intercepting and analyzing traffic from different devices which evades privacy of users. There are software packages like Wireshark which captures traffic from the data link, and if that traffic is not or weakly encrypted, then all the private data of users will go into wrong hands [12]. Man in the Middle attack also belongs to traffic analysis category and is used for eavesdropping or modification of data.

#### *8.2.3.3 Replay*

In this type of attack, the attacker sniffs the network and when a client sends its credentials to a server for establishing a connection, the attacker gains control of that information. The attacker sends that forged information again to the server in some time future, hence making the server think that the attacker is the original sender. So, the server starts communicating the private data of the client with the attacker. In this way, the hacker can send previous sensor values of a machine in future which results in a very devastating effect on the machine, e.g., "Stuxnet worm" is the name of a replay attack used on the uranium enrichment equipment in Iran [13].

#### *8.2.3.4 De-Authentication*

This is a wireless disassociation attack. The wireless access point disappears from the list of available networks and the host tries to look for that network repeatedly. The connection establishes and de-establishes frequently. This attack uses management frames to make the host vulnerable. Management frames are used to help find an access point, connect to an access point, etc. These frames are not encrypted when sent by the device so there is no protection of the management frame's data, and there is no authentication where this data is coming from. So, the attacker forges this management frame and sends a de-authentication frame to the affected device recurrently [14].

#### *8.2.3.5 Buffer Overflow*

Buffer is a sector in a machine which stores data. Every buffer has a certain limit of storing data. Buffer overflows occur when an attacker tries the use more memory than allocated by that buffer for the programs, hence overwriting the system's data. In this way, the attacker modifies the memory of the computer which it should not. The attacker adds malicious codes in buffer memory which points to an area of the machine which attacker wants to gain access. Thus, resulting in taking control of the machine programs illegitimately [15].

#### *8.2.3.6 ARP Cache Poison*

Domain Name System (DNS) is a system which we use to connect domain names, e.g., www.google.com to an IP address which might be changing from day to day, and it helps us in easily writing human readable alphabets and getting a specific website. DNS Cache poisoning is a flaw in DNS where an attacker can inject a malicious IP address into a name server. Similarly, Address Resolution Protocol (ARP) allows the data to travel from one machine to other by converting the IP addresses of the machines to their MAC addresses. When a device on the network searches for a MAC address by its IP address, the attacker misleads the device that this MAC address belongs to the attacker's device. It is a kind of man in the middle attack. The attacker also does this phenomenon with the access point. Hence all data flows from the attacker's device which makes users on a network extremely vulnerable [16].

#### *8.2.3.7 GPS Spoofing*

In this kind of cyber-attack, the attacker jams the legitimate signal of Global Positioning System (GPS) and broadcast a forged signal which provides false location co-ordinates to the machine, e.g., a UAV. This attack severely damages the integrity of the computer and results in the crash of UAVs [17].

## 8.3 PROTOTYPE DESIGN

### 8.3.1 ESP8266

This device is very versatile in its uses and definitions. It is a networking device. It can connect to any network or can create its own network. It comes in many packages. The one which is used here is ESP01S and has 8 pins. Input power is provided on pin 8. Pin 4 is the enable pin which is also known as CH_PD pin. This pin also needs to be high to turn on the device otherwise if only pin 8 is high, then the device will have power but will not turn on. Pin 1 is for connecting ground. Pins 2 and 7 are Transmitter (Tx) and Receiver (Rx) pins, respectively. Pin 6 is for reset which is active low. The board is powered up by no more than 3.3V. All General-Purpose Input Output (GPIO) and Tx, Rx pins work at 3.3V. The minimum working voltages to power up a ESP8266 board are 2.5 V. The remaining 3 and 5 numbered pins in ESP01S board are for GPIO 2 and GPIO 0, respectively. This module has an onboard LED which can be controlled via programming the microcontroller. ESP01S works in two modes. The first is programming mode and the second is normal UART mode in which the microcontroller executes the uploaded program.

### 8.3.2 Arduino

It is a microcontroller based on ATMEGA328 chip. It is multipurpose having all types of GPIO pins. It has a total of 14 digital pins from pin number 0 to 13. It also has 6 analogue to digital converter pins. It has 16 MHz on board crystal oscillator. It works on 5 V with a voltage regulator on board. A separate DC power jack is provided for powering it up using 12V power supply. This microcontroller supports I2C communication as well as Serial Communication. Two LEDs are connected to Tx and Rx pins to show the data communication. One LED is connected to pin 13 of the board to be controlled via programming. It supports 8-bit processing in its microcontroller. It can connect to other microcontrollers, e.g., ESP8266 using multiple baud rates.

### 8.3.3 USB to ESP01S Programmer

The used device is a USB to TTL converter is specifically designed for ESP01S models. It has female headers on board, so ESP01S male headers can directly connect to it without any wiring procedures. It has a sliding switch on board. When the switch is moved left to the programming side, the ESP01S becomes ready for uploading program from Arduino software. After uploading the program, the switch can be moved left to UART to run the program. In UART mode, the ESP01S can also communicate via UART AT commands with the Arduino UNO software. This programmer uses CH340 chip for USB to TTL conversion purposes.

### 8.3.4 UAV Station

Figure 8.3 shows an UAV station device workflow. At the start, if the UAV is not connected to a network, it will try to connect to an available trusted wireless network. After that, the station UAV will search for the key from the data arrived from AP. The key will be in one of the very first packets of data received from AP. As a sample, the key has been limited to 0 to 100 using C language programming. This key can go up to any number, e.g., 400, 800, 2000, etc., and this will also improve the security because higher number of bits used means more security. The sample algorithm used in station UAV for dynamic key generation for each session is

$$\text{encrypted data} = \frac{\text{key} + \text{data}}{\text{random number}} \tag{8.1}$$

After conversion of data, it is ready to be sent over the network. This dynamic key will change every time when the device resets its connection. There is an onboard LED used for showing the connection status. If it is blinking, it means that the device is not connected. If it is high constantly, it means that the device is connected to a network and data communication is taking place.

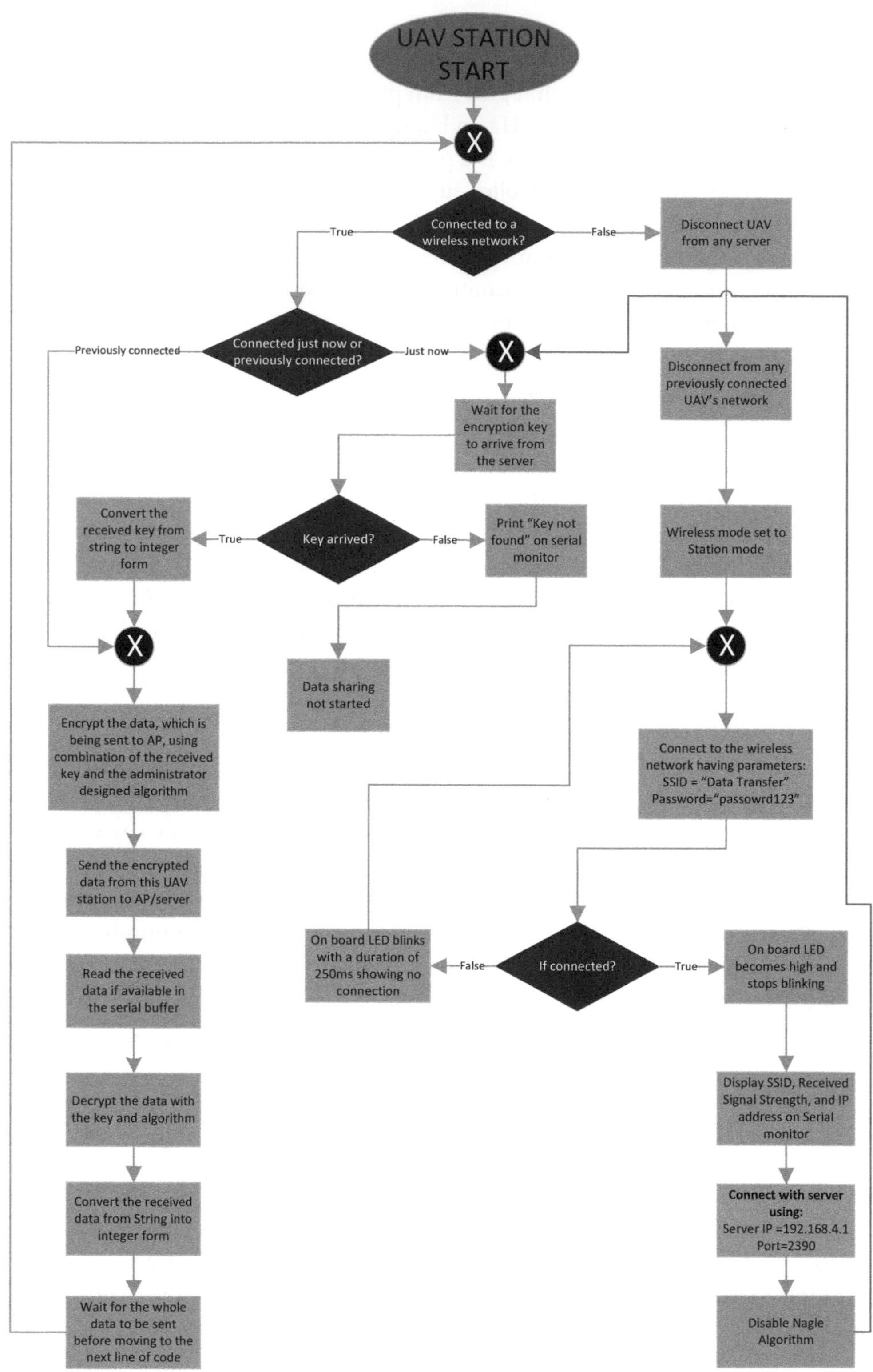


FIGURE 8.3 UAV station workflow part 1.

### 8.3.5 Access Point/ Server

Figure 8.4 workflow diagram shows the functions performed in the AP communication device. On bootup, it will reset all the parameters and set up a server. A network will also be set up so all the UAV stations can find this AP. This Access Point device can be a ground station, or it can be any UAV itself. If this AP is a ground station, then every UAV Station connected to it will be securely communicating the data with this ground station using high security protocols. Furthermore, if this AP is used as a UAV, then by keeping it near the other UAV station drones, it can be used as a server for processing

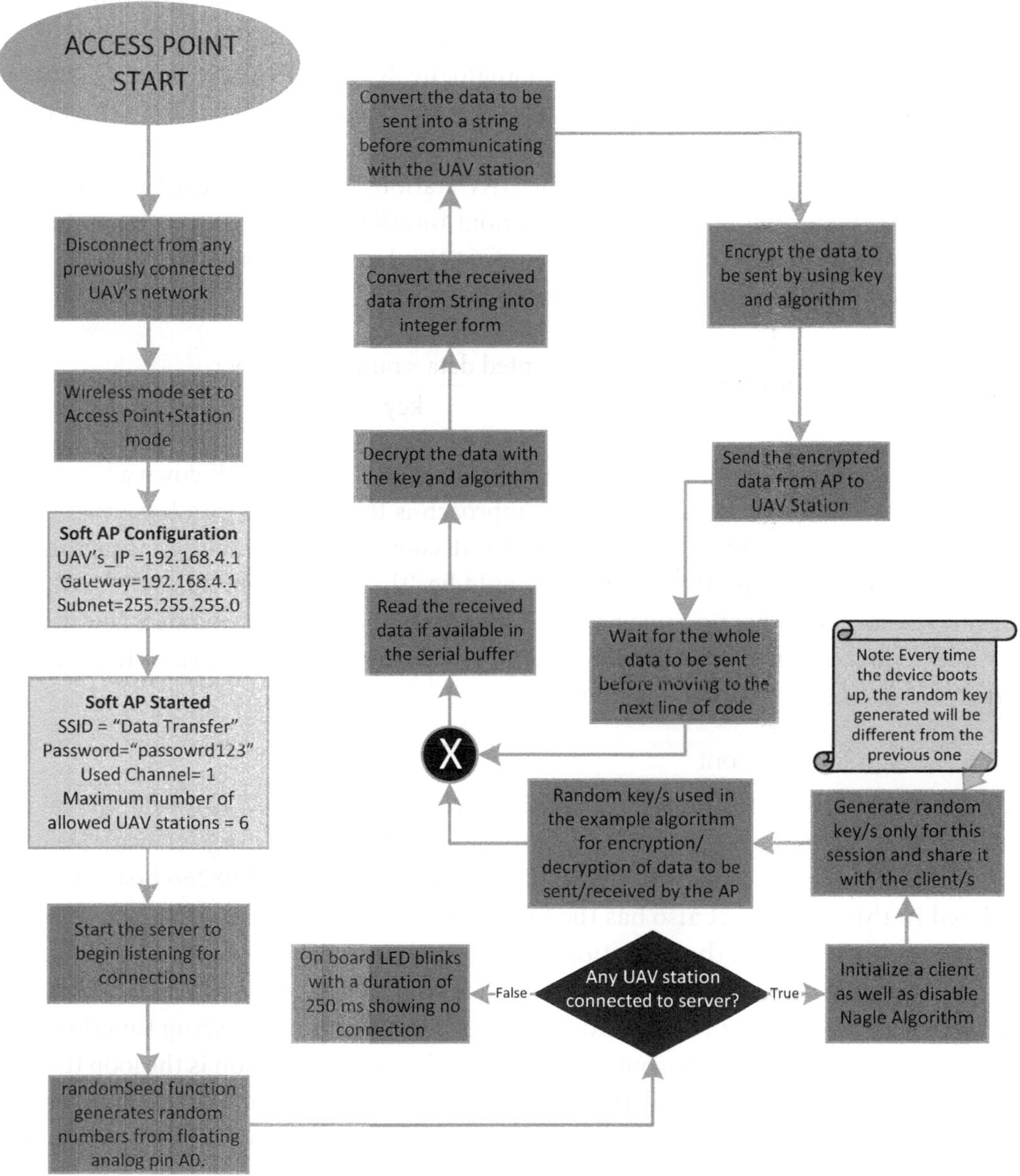


FIGURE 8.4 Access Point design flow.

UAV station's data thus reducing power consumption of UAV stations. In this way, station UAV will just be communicating with an AP drone in very close proximity and not to the far-off ground station, thus increasing data and power consumption efficiency. Security issues arise when the data transmission distance increases. In AP as a UAV method, communication distance is relatively less and thus only the AP and not UAV stations can be programmed to connect with the far-off ground stations by lengthy security protocols, saving all the power and processing in UAV stations. In the C programming language, a key is generated from the random seed function which is then communicated with the UAV stations so they can encrypt their data effectively. This random key is generated from the floating values of the analogue pin available on ESP8266 board. This is done so that the generated random numbers could not be guessed by any human because when an analogue key is at floating input, it generates very different numbers and then those numbers are further processed by the random seed function to provide added security. A buffer zone is placed in the coding after communicating the random key with the UAV stations so to make sure that the key can reach the UAV stations effectively and without interference from other data. The same previous algorithm is used in the AP to encrypt the data as UAV stations. The algorithm used to decrypt data in the AP is expressed in Eq. (8.2).

$$\text{decrypted data} = \frac{\text{encrypted data} \times \text{random number}}{\text{key}} \tag{8.2}$$

For fast communication, Nagle algorithm is disabled which will break down a large packet into small packets. The advantage of this approach is that the data can leave the AP as it arrives in the serial buffer, without waiting. The disadvantage of doing this is that the channel might not be used as effectively as it should be. The key generated in AP is only valid for one session. After booting up or any reconnection, the AP will generate a new dynamic key and send it over to other UAV devices for new dynamic key encryption. This changing key increases the security very effectively.

### 8.3.6 Design Environment

Arduino Genuino is a platform for designing electronics. It is used as the design environment in this chapter. It is an open-source software. It has a wide range of available libraries, and it supports wide variety C language microcontrollers. ESP8266 board library is utilized in this research. It also has the function of serial monitor in which the administrator can directly check the execution of the coded program. Serial commands can also be sent from the serial monitor to any devices in the vicinity. This software is composed of two basic functions, i.e., setup and loop. The first function is the setup function which just runs once when the microcontroller boots up. The other function is the loop function. This function runs continuously if the microcontroller is connected to a power supply. It provides live information about the compilation and uploading of the coded program in the software window.

## 8.4 EVALUATION OF DRONE SECURITY

### 8.4.1 Hardware Wiring and Connections

Two ESP01S devices are used to show the encrypted communication among the UAVs. In the start, Arduino UNO was used to program and execute both ESP01S devices. It was easily connected to Arduino software without any complex procedures. The circuit which is designed for programming ESP01S device is different from that used for the execution of the code. So, it became the biggest hurdle in using the ESP01S with Arduino UNO because the circuit needs to be updated every time after coding and, due to that, it became prone to errors. For programming purposes, Transmitter pin (Tx) of Arduino UNO should be connected to Tx of ESP01S as well as Receiver (Rx) of Arduino UNO should be connected to Rx of ESP01S. GPIO 0 of ESP01S must be connected to ground for the chip to be in programming.

In the code execution mode, the Tx of ESP01S should be connected to Rx of Arduino UNO and Rx of ESP01S should be connected to Tx of Arduino UNO. GPIO 0 and GPIO 2 are not connected to anything in this mode.

The ESP01S microcontroller chip is very delicate and interchanging wires again and again increases the probability of random errors. It is strictly limited to voltage input of 3.3V on all the pins including Tx and Rx pins. Arduino UNO, being working on 5 V, poses a danger to the Serial communication pins of ESP01S if not handled correctly. That is why the hardware was later moved to USB ESP-01 adapter. It has one slide switch which when turned to programming mode enables ESP01S to be programmed without interchanging any wires. On the other side of the sliding switch, there is UART switch. When ESP-01 adapter is set to this side of the switch, it starts working in code execution mode. This chip is voltage limited to 3.3 V so there is no need to change anything.

### 8.4.2 UAV Station Results on Serial Monitor

When a UAV station boots up, it waits to be connected. This waiting time can be seen, inside the UAV station's screenshot, in the form of dots within the first line of serial monitor of Arduino. When the UAV station gets connected to the AP, it notifies on the serial monitor that it is connected and shows the parameters of connection, e.g., the SSID of AP to which it is connected to, Received Signal Strength, Server IP address to which it is connected to and the IP address it received from the DHCP pool of the server. Then it shows the generated random key on the serial monitor which in this case is 54. It is visible in the UAV station's output screenshot that for a short period of time after getting the key, it does not receive any data from the AP and that is because of the inclusion of buffer zone in the code. In the buffer zone, the AP does not send any data so that the key might reach safely the UAV station which is why a server timeout is shown in the UAV station's output. After the buffer zone ends, encrypted communication starts. For example, the sent data which is marked as 1 in the black colored box in Figure 8.5 inside the UAV station's output has the value of 12039 on the UAV station's side but on the AP's side, the encrypted version of the value can be derived from Equation 8.1 which is following with random number taken as 99 and key size 54.

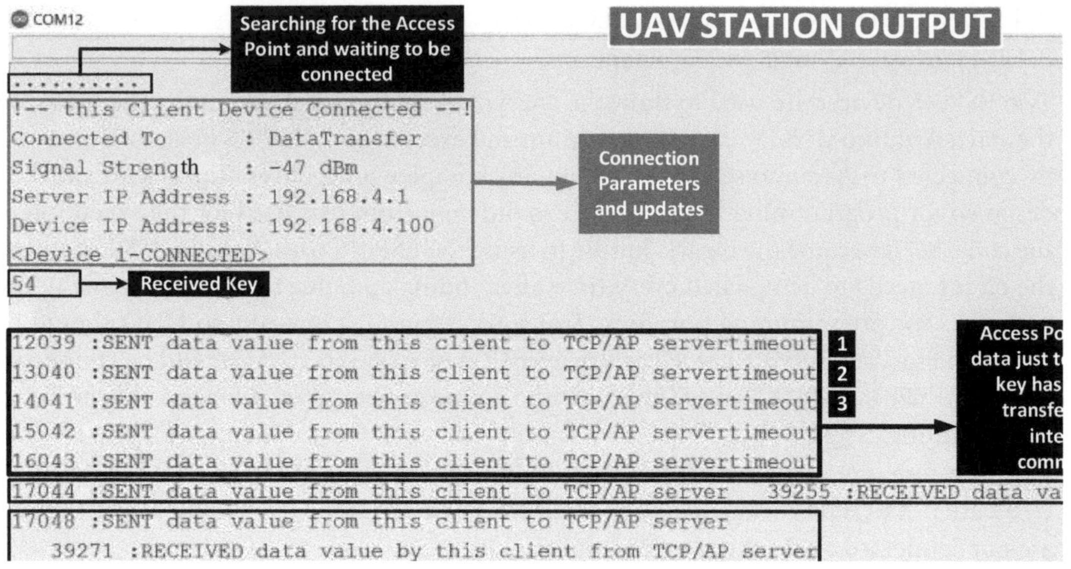


FIGURE 8.5 Unmanned Aerial Vehicle Station's Output in Arduino's serial monitor.

$$\text{encrypted data} = \frac{\text{key} + \text{data}}{99} = \frac{54 + 12{,}039}{99} = 6566 \tag{8.3}$$

For data marked as number 2 in black colored box, the encrypted output will be calculated as follows:

$$\frac{54 + 13{,}040}{99} = 7112 \tag{8.4}$$

For data marked as number 3 in black colored box, the encrypted output will be calculated as follows:

$$\frac{54 + 14{,}041}{99} = 7658 \tag{8.5}$$

### 8.4.3 UAV AP/Server Results on Serial Monitor

In the first line of AP's serial monitor's output (see Figure 8.6), the device identifies itself as an Access Point. In the second line, it displays its SSID as "DataTransfer" and notifies that this AP has been started and listening for connections. In the third line, it apprises its internal IP and gateway. Just before the beginning of the data transmission, the AP starts its server and prints on the serial monitor that the server has been started. The initial data received, shown in the first black box, informs about the number of packets received which in this case is 12 due to disabling of the Nagle algorithm. When Nagle algorithm is disabled, the packets will be communicated as soon as they become available without any delay, and because of this, the packet size decreases. After that, the serial monitor shows the IP and

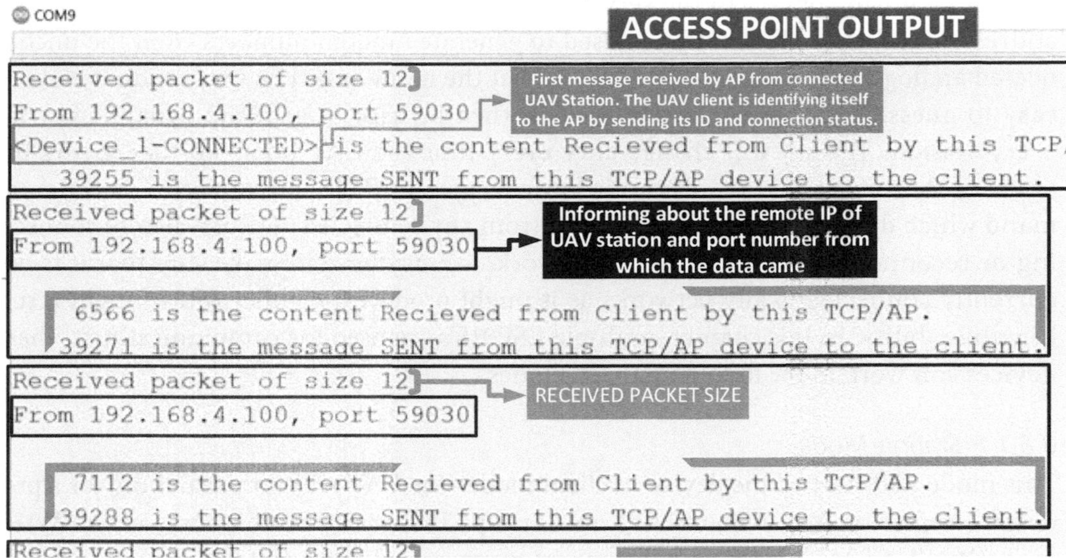


FIGURE 8.6 Output of the Access Point in Arduino's Serial Monitor.

port from which it has received the packet. Then the content of the first message arrives in the AP which in this case is the identification of the device connected. The connected UAV station device is sending out its identity to the AP. In this case the device is identifying itself as "Device_1". After this identification, the normal data transfer begins. It can be seen from Equations 8.3 to 8.5 that the received encrypted data from UAV station/client is the same as calculated which shows that the encryption is successful. The key which is used to encrypt the data will change for every reconnection; hence, it is called dynamic key encryption.

## 8.5 SOFTWARE-BASED TESTING

### 8.5.1 Access Point

"ESP8266WiFi.h" is the library used extensively in programming the hardware. It is the main library employed by ESP8266 devices and includes all the sub-functions which are used to control the Wireless microcontroller chip to its full extent. This library holds "IPAddress" command which converts a string into an Internet Protocol (IP) address, e.g., drone's access point IP, gateway IP, and subnet IP address. Every Transmission Control Protocol (TCP) uses a port number to recognize the destination application, e.g., WhatsApp application in smart phones use TCP 443 port to communicate the data among devices. The sender device will attach the port number with the payload so the receiver device will read the port number with the payload and provide the data to the application which is associated with this port number. A random port number of 2390 is picked because this UAV is not designed to connect to internet. "WiFiServer" command is used to initialize the server with the port number. It's like giving name to a server which will be used further in the coding. After initialization, the server will start listening for connections on the defined port number. "WiFiClient" command initializes a client so it can connect to the server with the aforementioned port and a specific IP address. “randomSeed” function is used to generate random numbers from the unconnected analog pin 0 of esp8266 module so that the number will be very random and not easy to guess. These random numbers will then be used to generate dynamic key for every session. The key will change after every time the UAV boots up thus giving the name of the encryption, i.e., dynamic key encryption. “**WiFi**.disconnect()” is the command which disconnects the drone device from any connected network. Before connecting or reconnecting the UAV to any network, we may have to make sure that it is not currently connected to any networks as it might produce a conflict which might result in vulnerability. In this chapter, multiple ESP01S’s are used for communications. These devices can work in the following three modes.

#### *8.5.1.1 Station Mode*

This mode will start up the device in client mode. The UAV will start searching for a predefined server or Access Point (AP) on start-up. The syntax for this function is “**WiFi**.mode(WIFI_STA)”. The Client will connect to server and gets an IP address for data link. In this mode, this station UAV is not connected to internet at all but to its server/AP. If the AP is connected to the internet, then all the data of this UAV station device will go through the connected AP to the internet if desired [10].

#### *8.5.1.2 Access Point Mode*

On bootup, AP will start looking for connections from client/station devices. When a client sends a request of connection, it will provide an IP address from the Dynamic Host Configuration Protocol (DHCP) pool. The syntax for this command is “**WiFi**.mode(WIFI_AP)”. In this mode, this AP device is not connected to the internet; hence, only communication among Stations and APs is taking place. For this, it is called soft Access Point (soft AP). From this method, multiple station-based UAVs can talk to each other through AP, i.e., another UAV or ground station. Neither stations nor APs can launch HTTP requests to internet in these mentioned methods [10].

#### *8.5.1.3 Station and Access Point Simultaneously*

This procedure can start the device in both station and AP modes together. The AP can be connected to further stations which might not be connected to internet. It will also be connected to a router or another AP as station. After getting the data from its stations, it will provide the data to its own AP as station itself. The syntax used for this command is “**WiFi**.mode(WIFI_STA_AP)”. In this method, if the AP, to which this device is behaving as Station, is connected to internet, then this device can launch HTTP requests to internet and get/send data from/to internet. It can be used when an AP has to get data from other connected UAVs as well as to publish or compare data from the internet [10].

The line “**WiFi**.softAPConfig(Access Point IP address, AP gateway, subnet of AP)” will provide Internet Protocol address, gateway, and subnet to the Access Point or Server for configuration. Subnet tells about the number of networking devices which can connect to this AP. Gateway is the path through which a device can connect to internet if asked. It is mostly the first IP address from the IP pool of DHCP. For example, if subnet provided to an AP UAV is 255.255.255.0 which is equal to /24, then 254 devices can connect to this UAV because first position, i.e., 0 and last position i.e., 255 is system reserved. It means that if IP list from a DHCP pool starts at 192.168.10.1, then the last available IP for any device will be 192.168.10.255. Normally the AP gateway is the first available IP address from the DHCP pool which in this case is 192.168.10.1 [11].

The line of code "**WiFi**.softAP (const char* Service Set Identifier (SSID), const char* passphrase, int channel, int SSID hidden?, int maximum number of connections?)" provides device setup parameters to the wireless shield. In the first part, the command will set the SSID of the wireless shield of UAV or ground station. All the devices which might want to connect with this UAV will search it by its SSID. In the second part, the command will set the password of the wireless shield for a secure network. All the clients or stations connected will have this password preloaded into them by the network administrator. The third part of the command will specify about the channel to be used by the UAV's network as there are multiple channels present in a frequency band. This is done to avoid any possible interference with any other channel. The data is also prone to theft if using a compromised channel. The fourth part of the command will specify that either the SSID of UAV should be visible by any client/station in the vicinity or not. If this choice is set to 0, then every station in the vicinity can search the SSID of this AP, but if this setting is set to 1, then the AP's SSID will be hidden, and then no station will be able to find this SSID automatically. So, to connect to this hidden AP, every station which wants to connect to it must have the AP's password as well as the SSID stored in it. The last part of the command shows the total number of allowed connections with this UAV's AP. In the command "TCP_SERVER.begin()", TCP_SERVER is the name given to the server and begin is used so the server may begin listening for connections from stations. The statement "TCP_SERVER.hasClient()" will check that if a client is connected to TCP_SERVER or not. This statement will return true if a client is connected to this specified server and will return false if no client is connected to the server.

The command "TCP_SERVER.available()" is used to check the data already received from a station or client device and present in the serial port. The buffer of serial port can retain 64 bytes of data in it. This command will return the quantity of bytes which is ready to be read from the serial buffer [18].

"millis()" is used to count time since the ESP microcontroller has been started. Its value goes back to 0 after 50 days of being continuously turned on [19]. The command "random (int x)" will generate random numbers from 0 to integer x. These numbers are used to make sure that the generated dynamic keys are fully random and not guessable. This command "TCP_Client.readStringUntil('\r')" will read the data sent by the clients. In this case, the client's name is TCP_Client. "readStringUntil('\r')" will search for the character mentioned in the brackets and terminates the string when it reads that character. This command will look through the data string directly from the serial buffer [20].

"sizeof(String Message)" is used to find the size of message received. In this case, the message was composed of 12 bytes. The command "TCP_Client.remoteIP()" returns the IP address of the client which is connected to this server/AP [21] while "TCP_Client. remotePort()" returns the port number which is received along with the data to direct the incoming packet towards accurate destination [22]. The line of code "TCP_Client. flush()" will wait for the data to be completely sent from the serial buffer to the client or any other device before moving to the next line of code [23]. The command "TCP_Client. connected()" will check that if the client is connected or not. It will return true if the client is connected and false if it is not connected [24]. "TCP_Client.println(char x [])" command will send the data to the client which, here, is named as TCP_Client. The data will be sent in the form of a character array.

"dtostrf(long/double a, char b, char c, char d)" is the command to make a string from double or long values. In this case, values of "a" will be converted to a character or string and stored into "d" for transmission. "b" tells about the width of the data which is going to be converted from "a" to "d". "b" is the minimum width of the output string, and it includes the possible signs of positive, negative, or decimal points. For example, a value of "-1.11" has the output minimum width of 5, so in this case b=5. "c" tells us about the precision of the converted numbered string. It tells about how many figures after decimal point to keep. In the example of "-1.11", if we want to keep both figures after decimal point into output string, then "c" will be equal to 2 [25].

The command "setNoDelay(1)" deactivates Nagle Algorithm. This algorithm waits for a long enough packet to be built by merging all the little packets and then sending all of them together hence reducing traffic and increasing channel efficiency [26]. The disadvantage of using Nagle approach is potential delays in the data transmission because it takes inputs for a long period of time and bursts arrays of data at once. So, in a Real Time System, this delay can cause errors that's why "setNoDelay" function is set to true [27].

### 8.5.2 Client/Station

The command "toInt()" will convert the string which is coming from the server/AP into an integer. The output data type after conversion is long. For conversion to start, the string which is at input must be a valid integer meaning it must have just numerical numbers in it and not alphabetical words or special characters otherwise the function will return 0 [28]. The function "**WiFi**.status()" will return the link status of wireless network. In total there are 8 different return types if using this function. If "WL_CONNECTED" is returned, it means that the UAV is connected to a network. If "WL_DISCONNECTED" is returned, it denotes that the UAV is decoupled from a network. When no network is available in the vicinity at all, then "WL_NO_SSID_AVAIL" is returned [29].

The command "stop()" is used together with the client name, e.g., "client.stop". This command disengages the client from the server [30]. The command "**WiFi**.begin("Network name", "Network password")" will provide the UAV with a SSID and password to connect to. When this command would run, if the network mentioned in the function is available, the UAV will connect itself to it [31]. The function "TCP_Client.connect(TCP_Server IP address, Port)" is used to connect a client to a specific IP address and port of the server/AP. This command is also used for DNS lookup. For DNS lookup, URL is used instead of IP address. If the client successfully connects with the desired IP address and port, the function will return true, otherwise false [32]. The command "**WiFi**.RSSI()" will ask the UAV about the received signal power of the connected wireless network. The UAV will measure this value at receiver antenna and return a value of type "long" having the unit of dBm [33].

## 8.6 RESULT ANALYSIS

### 8.6.1 Connection Time of AP/Server and Transmission Rate

Figure 8.7 shows the time in seconds which server required to connect to a UAV node. This data is composed of 15 independent readings. It shows how much time the server took to establish a connection. According to this data, most of the connections were established at 7th second with no connections established on or before 3rd second.

Figure 8.8 shows the transmission rate of the AP with the UAV station. This data is based on five simulations. An average of 149 packets communicated among AP and UAV station per second with highest being at 158 and lowest at 130 per second. A packet is composed of 12 bytes.

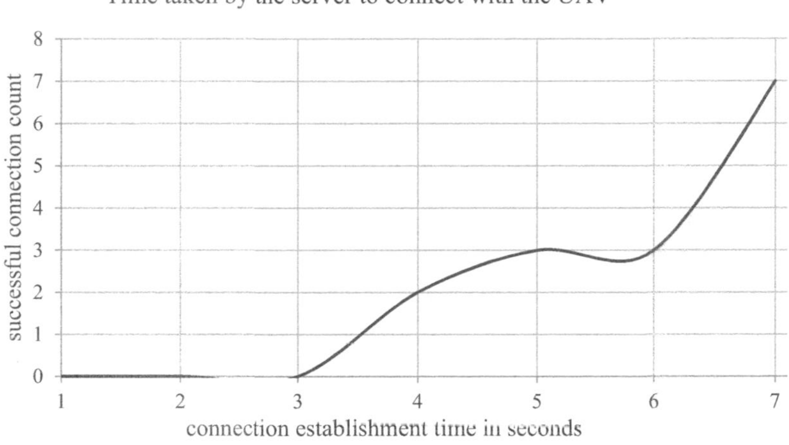


FIGURE 8.7 Connection establishment time.

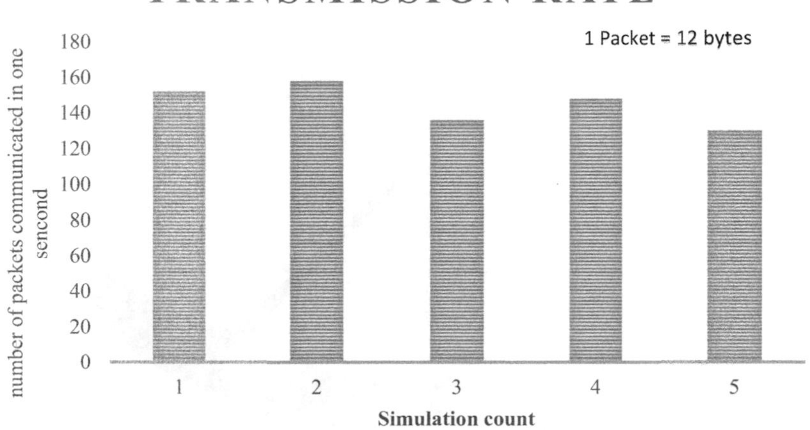


FIGURE 8.8 Transmission speed per second.

### 8.6.2 Data Visibility

The pie charts in Figures 8.9 and 8.10 show the data which was visible before and after the encryption. In unencrypted network, algorithms which are stored inside the server and UAV station were not visible because they were not communicated at all. But if an attacker knows about the input and output data then he/she can guess the algorithm. UAV station, when connected to server, sends its identification to the server. Data being routed to any device depends on this special identification. If this identification is compromised, then all the data being sent gets compromised. Transmitted and received data are also visible when sent through non-encrypted channel. Device credentials that either a device is started as

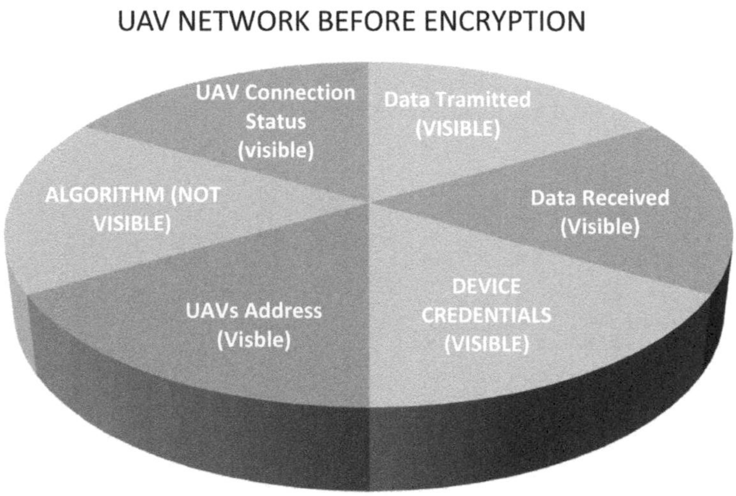

UAV NETWORK BEFORE ENCRYPTION

UAV Connection Status (visible)
Data Tramitted (VISIBLE)
Data Received (Visible)
DEVICE CREDENTIALS (VISIBLE)
UAVs Address (Visble)
ALGORITHM (NOT VISIBLE)

FIGURE 8.9 Unencrypted channel.

UAV NETWORK AFTER ENCRYPTION

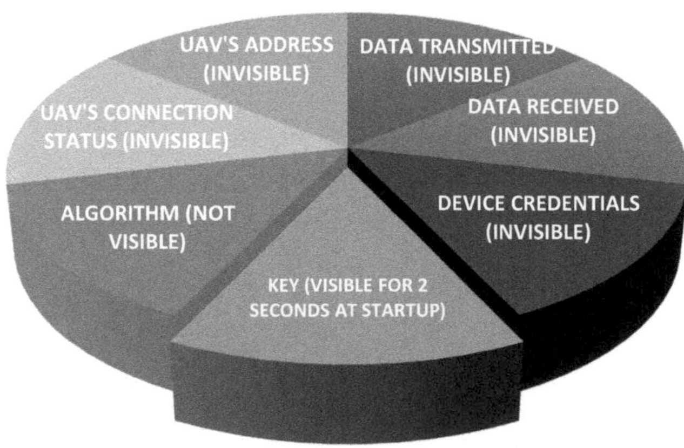


FIGURE 8.10 Protected data after encryption.

AP or UAV station are also visible. The IP addresses of connected devices are also in the open when using a non-encrypted channel.

When the channel is encrypted, everything becomes invisible. Only the random key generated will be visible for a microsecond when communication starts. The exposure of this random key for a very short interval of time is not a very major concern because the system is also secured with a username and password and hence only known devices can connect to the network. Figure 8.10 shows all the data which became invisible after the encryption procedure.

### 8.6.3 Security Analysis

The bar graph in Figure 8.11 shows the security analysis of the accomplishments in this chapter. It compares the amount of security provided by the used method in this section with other techniques. In the first one, there is no security at all because the channel is open and unencrypted. The second one has the security of username and password only and all the data transfer is unencrypted which is dangerous. The fourth one is based on a research journal in which different dynamic keys were used for each packet [34]. In most recent techniques, Elliptic Curve Cryptography (ECC) is being utilized which makes the random key very unpredictable by using comparatively a smaller number of bits [35].

## 8.7 CONCLUSION AND FUTURE WORKS

This research is aimed at securing the unprotected, unencrypted channels used by the UAVs, and it turned out to be a full success. A dynamic key approach is developed in this chapter. This dynamic key expires after each session and renews before the starting of a new session. The UAV station and AP are also protected by a username and password. Twelve bytes of data in every packet are secured and sent to the UAVs in the vicinity. All the device parameters, updates, and address are secured using this approach. Device identifications are also encrypted so to deter any attacker within the network. The connection establishment times of the UAV stations with the server are also kept in consideration during this job so as not to make any unwanted delays caused by the encryption/decryption processes.

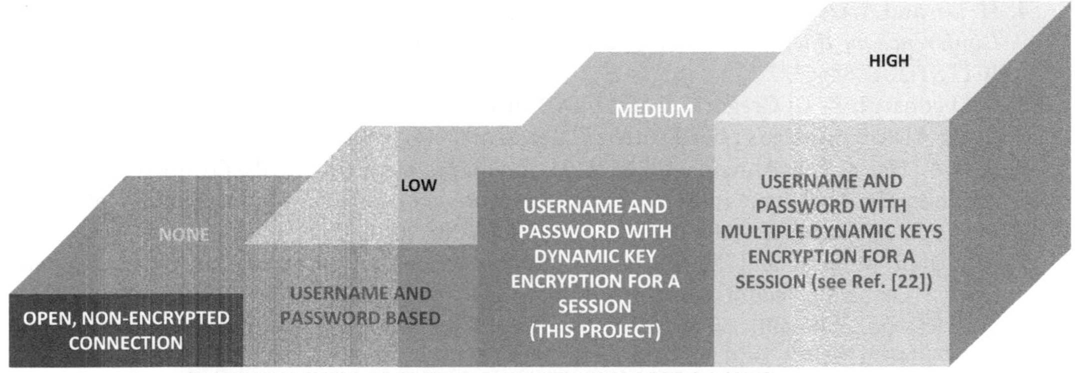


FIGURE 8.11 Security analysis.

### 8.7.1 Strengths

- Provides an acceptable level of security in a network.
- Lightweight security protocol.
- Saves processing and battery power.
- Used TCP/IP protocols to communicate among the UAV networks.
- Data transfer rate is also not affected by the encryption and decryption procedures.
- Sensitive data successfully became hidden by the suggested and implemented encryption/decryption technique in this chapter.
- The encryption/ decryption procedures do not affect the connection establishment times of the UAV stations with the server.

### 8.7.2 Limitations of the Study

Due to limited time availability, only a sample of data is encrypted to show that how the proposed idea will fully work if implemented in a required timescale. Due to the limitation of battery and processing power, encryption must be composed of a smaller number of bits which makes the network less secure if the session becomes too long. In future works, every packet of data can be assigned with its own dynamic key rather than using a single key for a whole session. Ultimately, this work depicted how unsecured UAV networks can be guarded by using a lightweight dynamic key, username, and password protection approach.

## REFERENCES

1. G. K., Dey, R. Hossen, M. S. Noor, and K. T. Ahmmed, "Distance controlled rescue and security mobile robot," in *IEEE*, Dhaka, Bangladesh, 2013.
2. R. K. Barnhart, D. M. Marshall, and E. Shappee, *Introduction to Unmanned Aircraft Systems*, CRC Press, Boca Raton, FL, 2021.
3. A. P. Namanya, et al., "The world of malware: An overview," in *2018 IEEE 6th International Conference on Future Internet of Things and Cloud (FiCloud)*, 2018. DOI: 10.1109/FiCloud.2018.00067.
4. H. Loi and A. Olmsted, "Low-cost detection of backdoor malware," in *2017 12th International Conference for Internet Technology and Secured Transactions (ICITST)*, 2017. DOI: 10.23919/ICITST.2017.8356377.
5. M. Leonardi, L. Di Gregorio, and D. Di Fausto, "Air traffic security: Aircraft classification using ADS-B message's phase-pattern," *Aerospace*, 4(4) pp. 3, 2017.
6. S. H. C. Haris, R. B. Ahmad, and M. A. H. A. Ghani, "Detecting TCP SYN flood attack based on anomaly detection," in *2010 Second International Conference on Network Applications, Protocols and Services*, 2010. DOI: 10.1109/NETAPPS.2010.50.
7. M. Riahi Manesh and N. Kaabouch, "Cyber-attacks on unmanned aerial system networks: Detection, countermeasure, and future research directions," *Computer Security*, vol. 85, pp. 386–401, 2019. Available: https://www.sciencedirect.com/science/article/pii/S0167404819300963. DOI: 10.1016/j.cose.2019.05.003.
8. J. P. Yaacoub, H. Noura, O. Salman, and A. Chehab, "Security analysis of drones systems: Attacks, limitations, and recommendations," Elsevier, p. 39, 2020.

9. M. Satell, "Ultimate list of drone stats for 2021," 2021. [Online]. Available: https://www.phillybyair.com/blog/drone-stats/. [Accessed 11 05 2021].
10. randomnerdtutorials, "ESP32 useful Wi-Fi library functions (Arduino IDE)," [Online]. Available: https://randomnerdtutorials.com/esp32-useful-wi-fi-functions-arduino/. [Accessed 28 08 2021].
11. I. Grokhotkov, "Soft access point class," 2017. [Online]. Available: https://arduino-esp8266.readthedocs.io/en/latest/esp8266wifi/soft-access-point-class.html. [Accessed 28 08 2021].
12. A. Dabir and A. Matrawy, "Bottleneck analysis of traffic monitoring using wireshark," in *2007 Innovations in Information Technologies (IIT)*, 2007. DOI: 10.1109/IIT.2007.4430446.
13. M. Hosseinzadeh, B. Sinopoli, and E. Garone, "Feasibility and detection of replay attack in networked constrained cyber-physical systems," in *2019 57th Annual Allerton Conference on Communication, Control, and Computing (Allerton)*, 2019. DOI: 10.1109/ALLERTON.2019.8919762.
14. T. Rakhra, et al., "De authentication attack: A review," in *2020 IEEE International Symposium on Sustainable Energy, Signal Processing and Cyber Security (iSSSC)*, 2020. DOI: 10.1109/iSSSC50941.2020.9358889.
15. D. Fu and F. Shi, "Buffer overflow exploit and defensive techniques," in *2012 Fourth International Conference on Multimedia Information Networking and Security*, 2012. DOI: 10.1109/MINES.2012.81.
16. J. S. Meghana, T. Subashri, and K. R. Vimal, "A survey on ARP cache poisoning and techniques for detection and mitigation," in *2017 Fourth International Conference on Signal Processing, Communication and Networking (ICSCN)*, 2017. DOI: 10.1109/ICSCN.2017.8085417.
17. I. G. Ferrão, et al., "GPS spoofing: Detecting GPS fraud in unmanned aerial vehicles," in *2020 Latin American Robotics Symposium (LARS), 2020 Brazilian Symposium on Robotics (SBR) and 2020 Workshop on Robotics in Education (WRE)*, 2020. DOI: 10.1109/LARS/SBR/WRE51543.2020.9307036.
18. arduino, "available()," 2021. [Online]. Available: https://www.arduino.cc/en/Serial.Available. [Accessed 28 08 2021].
19. Arduino, "millis()," 2021. [Online]. Available: https://www.arduino.cc/reference/en/language/functions/time/millis/. [Accessed 29 08 2021].
20. arduino.cc, "Serial.readStringUntil()," 2021. [Online]. Available: https://www.arduino.cc/reference/en/language/functions/communication/serial/readstringuntil/. [Accessed 29 08 2021].
21. arduino.cc, "ClientRemoteIP," 2021. [Online]. Available: https://www.arduino.cc/en/Reference/ClientRemoteIP. [Accessed 29 08 2021].
22. arduino.cc, "RemotePort," 2021. [Online]. Available: https://www.arduino.cc/en/Reference/ClientRemotePort. [Accessed 29 08 2021].
23. arduino.cc, "Serial.flush()," 2021. [Online]. Available: https://www.arduino.cc/reference/en/language/functions/communication/serial/flush/. [Accessed 29 08 2021].
24. arduino.cc, "connected()," 2021. [Online]. Available: https://www.arduino.cc/en/Reference/ClientConnected. [Accessed 29 08 2021].
25. nongnu.org, "avr-libc," 2016. [Online]. Available: http://www.nongnu.org/avr-libc/user-manual/group__avr__stdlib.html#ga060c998e77fb5fc0d3168b3ce8771d42. [Accessed 29 08 2021].
26. wikipedia, "Nagle's algorithm," 2021. [Online]. Available: https://en.wikipedia.org/wiki/Nagle%27s_algorithm. [Accessed 29 08 2021].
27. I. Grokhotkov, "Client class," 2017. [Online]. Available: https://arduino-esp8266.readthedocs.io/en/latest/esp8266wifi/client-class.html. [Accessed 29 08 2021].
28. arduino.cc, "toInt()," 2021. [Online]. Available: https://www.arduino.cc/reference/en/language/variables/data-types/string/functions/toint/. [Accessed 29 08 2021].
29. arduino.cc, "WiFi.status()," 2021. [Online]. Available: https://www.arduino.cc/en/Reference/WiFiStatus. [Accessed 29 08 2021].

30. arduino.cc, "stop()," 2021. [Online]. Available: https://www.arduino.cc/en/Reference/ClientStop. [Accessed 29 08 2021].
31. arduino.cc, "WiFi.begin()," 2021. [Online]. Available: https://www.arduino.cc/en/Reference/WiFiBegin. [Accessed 29 08 2021].
32. arduino.cc, "connect()," 2021. [Online]. Available: https://www.arduino.cc/en/Reference/ClientConnect. [Accessed 29 08 2021].
33. arduino.cc, "WiFi.RSSI()," 2021. [Online]. Available: https://www.arduino.cc/en/Reference/WiFiRSSI. [Accessed 29 08 2021].
34. H. H. Ngo, X. Wu, P. D. Le, C. Wilson, and B. Srinivasan, "Dynamic key cryptography and applications," *International Journal of Network Security*, vol. 10, p. 15, 2010.
35. L. Teng, M. Jianfeng, F. Pengbin, M. Yue, M. Xindi, Z. Jiawei, C. Gao, and L. Di, "Lightweight security authentication mechanism towards UAV networks," in *2019 International Conference on Networking and Network Applications (NaNA)*, 2019.

CHAPTER 9

# Routing Protocols in Vehicular and Aerial Delay Tolerant Networks

Endong Liu

*University College London*

Yue Cao

*Wuhan University*

## CONTENTS



## LIST OF ABBREVIATIONS

**AFW** Autonomous flight wireless nodes
**AODV** Ad-hoc on demand vistance vector routing
**AOMDV** Ad-hoc on-demand multi-path distance vector
**CEL** Current energy level
**CGR** Contact graph routing
**CGR-ETO** Contact graph routing-earliest transmission opportunity
**CIA** Confidentiality, integrity, and availability

DOI: 10.1201/b22998-9

| | |
|---|---|
| **CNN** | Current neighbours node |
| **CPUP** | Contact plan update protocol |
| **CRP** | Contact review procedure |
| **DGR** | Delegation geographic routing |
| **DP** | Delivery predictability |
| **DSR** | Dynamic source routing |
| **DTN** | Delay tolerant network |
| **E2E** | End-to-end |
| **EAER** | Energy-aware epidemic routing |
| **ECGR** | Enhanced contact graph routing |
| **EMP-DSR** | Enhanced multi-path dynamic source routing |
| **FCFS** | First come first served |
| **FIFO** | First in first out |
| **GPS** | Global location system |
| **IBC** | Identity-based cryptography |
| **LMP-DSR** | Load-balanced multi-path dynamic source routing |
| **MoVe** | Motion vector |
| **MR** | Mobile router |
| **NDN** | Never die network |
| **OSPF** | Open shortest path first |
| **PER** | Predict and relay |
| **PROPHET** | Probabilistic routing protocol using history of encounters and transitivity |
| **QoS** | Quality of service |
| **RD** | Route discovery |
| **REER** | Route error |
| **RIP** | Routing information protocol |
| **RM** | Route maintenance |
| **RREP** | Route reply |
| **RREQ** | Route request |
| **RSU** | Road side unit |
| **RTT** | Round-trip time |
| **SCF** | Store-carry-forward |
| **SV** | Summary vector |
| **TTI** | Time to intercept |
| **UID** | Unique identification |
| **V2I** | Vehicle-to-infrastructure |
| **V2V** | Vehicle-to-vehicle |
| **V2X** | Vehicle-to-everything |
| **WFPP** | Weighted flight path planning |

## 9.1 INTRODUCTION

Traditional networks are mainly divided into two classes: wired networks and wireless networks. Wired networks connect the computer network using coaxial cables, twisted pairs, and fibre, while wireless networks (e.g., cellular networks and Wi-Fi) need to use air as the transmission medium, and electromagnetic waves and infrared rays are used as carriers to transmit data; this kind of networking method is convenient and flexible. With the development of the Internet, nowadays the resources of network contain fantastic multimedia content and people prefer to access them at anytime and anywhere.

However, both wired and wireless networks rely on infrastructure to connect with each other, and their architecture or the topology can be regarded as an undirected graph whose nodes are routers and edges are continuous bidirectional End-to-End (E2E) paths. Because of the high requirement of quality of service (QoS), conversations on the Internet need short round-trip time (RTT) and low error rates. Although the success of this structure has been verified by users all over the world, it is still easy to find its shortcomings. Its advantages indicate that this model only meets ideal states but cannot satisfy extreme cases that have frequent disconnections, long propagation delay, or low transmission reliability, and networks consist of these situations that should be tolerated called *Delay Tolerant Network (DTN).*[1]

DTN architectures are only suitable for intermittent connectivity, where conventional Internet routing protocols (e.g., RIP and OSPF) would fail. Internet protocols need to check whether an established E2E path has been created and calculate the best path from sources to destinations using routing tables, then start to transmit packages to the best next hop. While in DTNs, the connectivity between pairwise nodes exists only when these two nodes come into the transmission ranges of each other. The contact types are opportunistic and routing protocols cannot know when the next conversation happens exactly, so nodes in DTNs need to store bundles (like packages in traditional routing protocols) for a long period of time until they meet another node and then begin to transmit bundles again. This kind of mechanism is called *Store-Carry-Forward (SCF).* SCF emphasizes the significance of persistent storage and mobility of nodes when developing protocols in DTNs, rather than memory storage and stable structures in existing Internet routing protocols.

This chapter will focus on the improvements and applications in the Vehicular and Aerial Delay Tolerant Networks. The rest of this chapter is organized as follows. Section 9.2 introduces some basic routing protocols in DTNs that are divided into two strategies: Single-Copy based and Multi-Copy based. Improvements in several aspects in vehicular and aerial DTNs will be presented in Section 9.3 and Section 9.4 separately. Section 9.5 is mainly about the applications of vehicular and aerial DTNs nowadays and in the future while the application fields include some applications in Smart City. Moreover, some works of literature that combine other technologies in computing science, like the social network, machine learning, security, and privacy, will be discussed in Section 9.6. Finally, this chapter is concluded in Section 9.7.

## 9.2 BASIC ROUTING PROTOCOLS IN DELAY TOLERANT NETWORKS

While many routing protocols have been developed for Vehicular and Aerial DTNs in recent years, most of them and their applications are derived from some basic routing protocols. Basic routing protocols usually have extreme performance in the ideal cases, while advanced ones prefer to make a trade-off or do some improvements in some areas for their specific applications. Therefore, before introducing the advanced routing protocols, basic routing protocols need to be illustrated. These basic routing protocols can be classified into two clusters based on the number of message copies in Sections 2.1 and 2.2.

### 9.2.1 Single-Copy-based Routing Protocols

Single-Copy-based routing protocols[2] represent that only a single copy of each message exists in the network at any time. The main feature of single-copy algorithms is the lower number of transmission and lower contention for shared resources but cannot increase the delivery ratio in the system. Common routing protocols consist of Direct Delivery Routing Protocol and First Contact Routing Protocol.

#### *9.2.1.1 Direct Delivery*

Direct Delivery Routing Protocol is the simplest strategy using the Storage-Carry-Forward mechanism in DTNs. In this algorithm, each node in the network carries the self-created message and moves continuously until it encounters the destination node. The entire communication process never uses other nodes to help transmission. The advantage of this method is the lowest overhead ratio. However, the disadvantages are also obvious since its efficiency is the lowest. As an extreme of DTN routing protocol, it is usually compared with other protocols as a benchmark.

#### *9.2.1.2 First Contact*

First Contact Routing Protocol defers from Direct Delivery Routing Protocol when one node meets another node. Whenever two nodes come into the transmission range of each other, one node will build a conversation with its neighbour and start to forward a new copy of every message it carries. If the transmission is finished successfully, the node will drop old messages it obtains. First Contact can be regarded as a simple improved version of Direct Delivery due to the increasing delivery probability, but it has deleting operations during the whole transmission so it may cause more resources to be wasted.

### 9.2.2 Multi-Copy-based Routing Protocols

Multi-copy-based routing protocols[3] indicate multiple copies of a message may exist concurrently in the network. These algorithms are flooding-based schemes and usually have lower delivery delay and higher robustness compared with single-copy-based algorithms. However, problems are also caused by flooding messages in the network. Supposed that each node tries to give copies of each message in its buffer memory to any nodes it meets, there will be an exponential number of messages in the whole system and the value of overhead will be increased into an unimaginable number. Because of limited buffer memory and bandwidth, it is hard or impossible to deploy in the real world and some algorithms are only based on infinite buffer size or high transmission speed. Here, the most common routing algorithm will be shown first, while two other basic algorithms using historically and geographically topological information are introduced separately.

#### *9.2.2.1 Epidemic*

Direct Delivery routing protocol is an extreme routing protocol, that is, messages are never copied, and messages are delivered only when they reach the destination node. Epidemic[4] routing protocol is the other extreme. It uses the flooding mechanism to pass messages to neighbouring nodes whenever it has a chance. As its name implies, its behaviour is like the "contact-infection" of an epidemic virus.

Detailed implementation of Epidemic routing protocol is based on the *Unique Identification (UID)* and a bit vector called *Summary Vector (SV)*. Every created message will have a *UID* and this *UID* will be associated with all its copies which means all replicated messages have the same *UID*. This *UID* will never change until the message has been sent to the destination node or it needs to be dropped because the Time to Live expires or the node does not have enough buffer to store the message. All the message *UID* in memory form a hash table and the *SV* is the summary of the list of messages in one node.

As illustrated in Figure 9.1, when two nodes build connections, they will exchange their *SVs* and compare the received *SV* with its own *SV*. After comparison, they will request, send, or receive all unknown messages to the paired node. Since all the messages in the buffer are ordered on a First-Come-First-Served (FCFS) strategy, messages will be sent one by one until the transmission is finished. If the conversation keeps for a quite long time and both nodes have enough buffer size, these two nodes will have the same hash table for lists of messages.

Because there always exists one message copy on the shortest path, it is not difficult to find that Epidemic algorithm has the highest delivery rate if the buffer of each node is large enough. However, it also has a terrible drawback that is it needs huge consumption

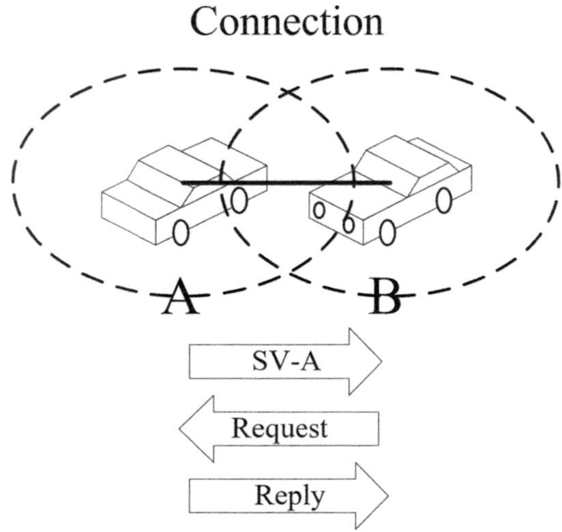


FIGURE 9.1 *Summary Vector (SV)* exchange in epidemic.

of network resources, so epidemic routing protocol is an idealized method, and it is also usually compared with other protocols as a benchmark.

#### 9.2.2.2 PROPHET

Probabilistic Routing Protocol using History of Encounters and Transitivity (PROPHET)[5] is designed for choosing the most suitable relay nodes to meet the goal above. It uses historically topological information of previous encounters between nodes to dynamically update a key probabilistic metric called *Delivery Predictability (DP)*.

The process of calculation and update of *DP* has been divided into three parts. The first one is the encounter part where *DP* will be updated whenever two nodes meet:

$$P_{(A,B)} = P_{(A,B)\text{old}} + \left(1 - P_{(A,B)\text{old}}\right) \times P_{\text{init}} \tag{9.1}$$

where $P_{(A,B)}$ is *DP* at every node *A* for each known destination *B*. $P_{(A,B)\text{old}}$ is the previous *DP* and $P_{\text{init}}$ is an initialization constant that takes a value between [0,1]. From Equation 9.1, more frequent encounters between nodes mean to have higher *DP*.

However, if two nodes have not met for a while, *DP* should also be updated, and this is the second part about aging:

$$P_{(A,B)} = P_{(A,B)\text{old}} \times \gamma^k \tag{9.2}$$

where $\gamma \in (0,1)$ is an aging constant while $k$ is the number of time units that have elapsed since the last time that *DP* was updated. The conclusion derived from Equation 9.2 is that *DP* would decrease by the time passed without communication between two nodes.

The final part is the significance of considering transitivity relationships between nodes:

$$P_{(A,C)} = P_{(A,C)\text{old}} + \left(1 - P_{(A,C)\text{old}}\right) \times P_{(A,B)} \times P_{(B,C)} \times \beta \tag{9.3}$$

where $\beta$ is scaling constant and its range is from 0 to 1. Therefore, Equation 9.3 shows if node *A* often encounters node *B* and node *B* often encounters node *C*, then *DP* for connection of node *A* and node *C* should also be higher due to the transitive property.

The process of message transmission in PROPHET routing protocol is close to which in Epidemic routing protocol except for the order of messages in memory. PROPHET routing protocol uses Priority Queue and prefers to transmit messages with higher *DPs* and drop messages with lower ones. For nodes that have the same *DP*, the order still follows the FCFS rule. This strategy will guarantee the delivery probability suppose that there is not enough bandwidth or buffer size during one conversation.

#### 9.2.2.3 MoVe

The Motion Vector (MoVe)[6] predicts the *closest distance* to the location of the destination node using the information of relative velocities of one *Mobile Router (MR)* and its neighbouring nodes. Because all the information can be achieved from Global Location System (GPS), MoVe is regarded as the geographically topological routing protocol.

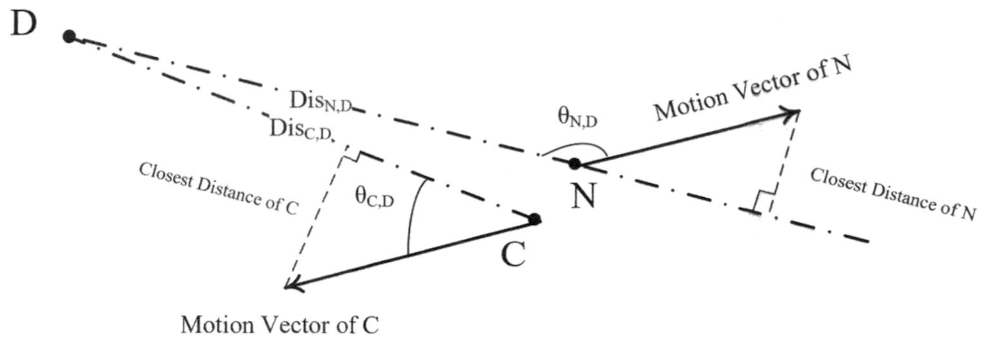

FIGURE 9.2 MoVe.

Additionally, MoVe not only defines the closest distance but also gives three moving states of each node. As shown in Figure 9.2, given the current node $C$, the neighbouring node $N$, and the destination node $D$:

The angle between $C$ and $D$ can be computed by

$$\theta_{C,D} = \arccos \frac{\overrightarrow{CD} \cdot \overrightarrow{MV_C}}{\left|\overrightarrow{CD}\right|\left|\overrightarrow{MV_C}\right|} \tag{9.4}$$

According to this angle, if $\theta > \frac{\pi}{2}$, the moving state is *Away*. If $\theta = \frac{\pi}{2}$, the moving state is *Still*. If $\theta < \frac{\pi}{2}$, the moving state is *Towards*. Moreover, the predicted *closest distance* is $d_C = \sin\theta_{C,D} \cdot \left|\overrightarrow{CD}\right|$. When the moving state and the closest distance are given, the rules of forwarding decisions (current node to the neighbouring node) can be summarized in Table 9.1.

However, MoVe routing protocol does not consider the state of the destination node, since if the destination node moves towards the relay node, this relay node is considered to have a higher delivery probability. In addition, forwarding messages to the relay node using MoVe is always the local optimum but may not be global optimal solution. Therefore, there are still various improvements to the MoVe routing protocol.

## 9.3 IMPROVEMENTS IN VEHICULAR DTNs

### 9.3.1 Flooding Restriction

Although Epidemic routing protocol has a high delivery probability because of its flooding scheme, the resource problem is caused by numerous messages in the network. This problem may greatly degrade its performance, especially in overhead and latency. Some routing protocols have been proposed to complete the flooding restriction.

#### *9.3.1.1 Spray-and-Wait*

Spray-and-Wait[7] routing protocol is one efficient method to solve the resource. It has two versions: one is the normal mode while the other one is the binary mode. Each

TABLE 9.1 Rules of Forwarding Decisions in MoVe

| Current State | Neighbouring State | Forwarding Decision |
|---|---|---|
| *Away* | *Away* | if $\lvert\overline{CD}\rvert > \lvert\overline{ND}\rvert$ Yes |
| *Away* | *Still* | Yes |
| *Away* | *Towards* | Yes |
| *Still* | *Away* | No |
| *Still* | *Still* | No |
| *Still* | *Towards* | Yes |
| *Towards* | *Away* | No |
| *Towards* | *Still* | No |
| *Towards* | *Towards* | if $d_C > d_N$ Yes |

mode has two phases: Spray Phases and Wait Phases, and the primary difference is in the first phase.

The main idea in the spray phase is to generate $L$ copies for each created message in its source node and then spray these copies into the whole network. The basic idea of normal spray mode sends only one of $L(L>1)$ message copies. When node $N_i$ meets node $N_j$, they will exchange their *SVs* to their differences. Then, $N_i$ will send one copy of each missing message to $N_j$. After this connection is done, take the delivered message $Msg_i$ as an example, there are $L=L-1$ copies of $Msg_i$ in the node $N_i$ and there is only $L=1$ copy of $Msg_i$ in the node $N_j$. Another advanced mechanism in the spray phase is the binary mode. Similarly, suppose that node $N_i$ has $L(L>1)$ message copies of $Msg_i$ in the beginning and it meets node $N_j$. After successful transmission, there are $L=L/2$ copies of message $Msg_i$ in $N_i$ and $L=L/2$ copies in $N_j$. The simple example of binary spray mode is displayed in Figure 9.3 if $L=4$ copies in the beginning.

When there is only $L=1$ copy of $Msg_i$ in one node, this node comes into the Wait Phase. In this phase, nodes will apply Direct Delivery routing protocols which means they will forward the message only to the target destination.

Spray-and-Wait combines the high delivery probability of Epidemic, lower average delivery delay, and fewer transmissions per delivered message, which is close to the optimal mechanism. Additionally, it has good scalability when cooperated with other DTN routing protocols.

#### *9.3.1.2 Spray-and-Focus*

Spray-and-Wait already has good performance in scenarios with high mobility, but it still has a serious problem. Supposed that nodes have low mobility or run in a sparse network in the first phase, the diffusion area may not be large enough to have good performance. An enhanced algorithm named Spray-and-Focus[8] is designed for this problem. The solution is quite simple. It turns the Wait Phrase which applies Direct Delivery protocol to Focus Phase.

The Focus Phase applies another Single-Copy-based routing protocol. Different from Direct Delivery and First Contact, it uses a utility-based function to set and calculate *last encounter timers* in Spray-and-Focus. Furthermore, this utility-based function can use any suitable parameters that appear in transmissions. Hence, the rule of forwarding decision can be generally concluded:

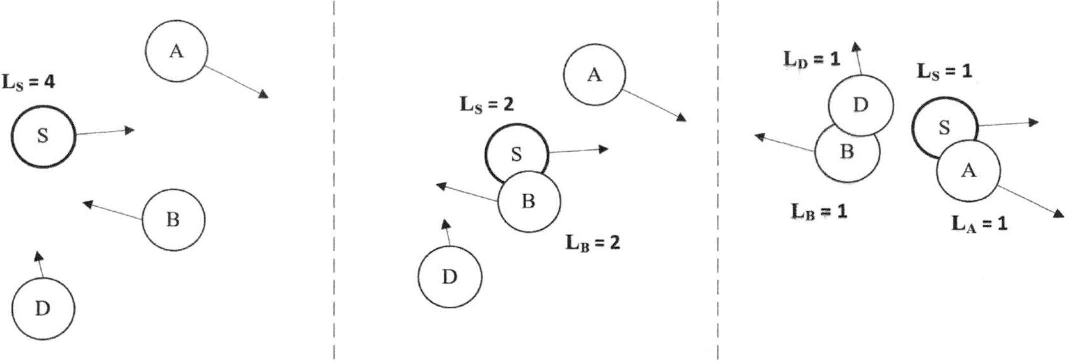


FIGURE 9.3 Binary mode in spray phase.

$$C \rightarrow N \Leftrightarrow U_N(D) > U_C(D) + U_{th} \tag{9.5}$$

where $C$ is the current node, $N$ is the neighbouring node, and $D$ is the destination node, and $\rightarrow$ is the forwarding action, $U$ is the utility-based function, and $U_{th}$ (utility threshold) is a parameter in this algorithm. The change of Focus Phase in Spray-and-Focus will deliver messages much faster despite the low mobility and diffusivity and lead to the reduction of the delivery latency.

### 9.3.2 Queue Management

Queue management defines the order of messages in the queue to deliver if the buffer is limited or the connection time is not enough. It is common in routing protocols of DTNs, and there are several kinds of schemes that have been developed. For example, Epidemic routing protocol uses First-In-First-Out (FIFO), which is usually the default queue management policy, and PROPHET applies Priority Queue based on its *Delivery Predictability*. Some of these policies are destination independent which means only parameters that are not related to the information of the destination node are used while some are destination-dependent. Here, one more complex queue management policy will be introduced.

#### 9.3.2.1 MaxProp

MaxProp[9] discovers the optimal delivery paths by considering the DTN as a weighted directed graph. It defines its own metric called *Estimating Delivery Likelihood* which computes the total cost of the shortest path to the destination using a variation of Dijkstra's algorithm. This information is stored in a routing table in each node to select the next hop like in normal routing protocols on the Internet.

Moreover, MaxProp also uses another information which is *hop count*. MaxProp separates two methods using one parameter named *threshold*. However, the constant *threshold* is not suitable for any situation, so a conditional expression is needed to define the dynamic *threshold*:

$$\text{threshold} = \begin{cases} x, & x \in (-\infty, b/2) \\ \min(x,(b-x)), & x \in [b/2, b] \\ 0, & x \in (b, +\infty) \end{cases} \tag{9.6}$$

where $x$ is the average number of bytes that transferred per transfer opportunity while $b$ is the byte size of the buffer in one node. Combined with two parameters, the queue management of the MaxProp is illustrated in Figure 9.4.

Since MaxProp uses Dijkstra to compute the shortest path to the destination node, it can be considered as the routing protocol uses destination-dependent queue management. Moreover, its queue management is also destination independent because it accumulated the number of hops that messages have mentioned. To conclude, the queue management policy in MaxProp is a combination of destination-dependent and destination independent.

### 9.3.3 Social Network

In recent years, social-based methods tried to transfer DTN into the social network. They preferred to analyse the social behaviours of DTN nodes to make better forwarding decisions in routing protocols. To improve the performance of these routing algorithms, they usually used some negative features, like node selfishness, or contemplate the influences of positive characters including community, centrality, and similarity.

#### *9.3.3.1 SimBet*

SimBet[10] turns the routing problem in DTNs into an information flow problem in the social network. It evaluates three parameters: *Betweenness Centrality*, *Social Similarity*, and *Tie Strength* to analyse the interaction of nodes in the social network.

In graph theory, nodes with higher *Betweenness Centrality* have more connections with other nodes in the graph. In the social network, nodes with higher *Betweenness Centrality* can build more relationships with others. Moreover, in DTNs, the nodes with higher *Betweenness Centrality* are regarded as better relays when making forwarding decisions. In SimBet, its utility calculation of *Betweenness Centrality* of the current node C that compared with the neighbouring node $N$ is shown below:

$$BetUtil_C = \frac{Bet_C}{Bet_C + Bet_N} \tag{9.7}$$

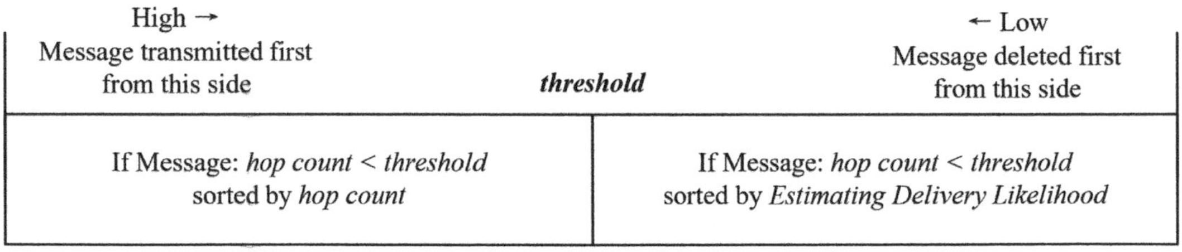

| High → Message transmitted first from this side | *threshold* | ← Low Message deleted first from this side |
|---|---|---|
| If Message: *hop count* < *threshold* sorted by *hop count* | | If Message: *hop count* < *threshold* sorted by *Estimating Delivery Likelihood* |

FIGURE 9.4 Queue management in MaxProp.

*Social Similarity* is another metric of the separation degree. When two nodes have more common neighbouring nodes, *Social Similarity* between them will be higher. If *Social Similarity* is higher, in DTNs, messages will be more difficult to transfer outside their social network. In SimBet, its utility calculation of *Social Similarity* of the current node *C* to the destination node *D* that compared with the neighbouring node *N* is shown below:

$$SimUtil_C(D) = \frac{Sim_C(D)}{Sim_C(D) + Sim_N(D)} \tag{9.8}$$

Due to the variation of topology in DTNs, *Betweenness Centrality* and *Social Similarity* are not enough to describe the complex relationships between nodes. *Tie Strength* is defined as the capability of the interaction behaviour that can activate the information flow. If *Tie Strength* is higher, it will be more likely to activate the information flow. In a previous study, the calculation of *Tie Strength* can be computed based on Frequency,[11] Intimacy/Closeness,[12] and Recency.[13] In SimBet, its utility calculation of *Tie Strength* of the current node *C* to the destination node *D* that compared with the neighbouring node *N* is shown below:

$$TSUtil_C(D) = \frac{TieStrength_C(D)}{TieStrength_C(D) + TieStrength_N(D)} \tag{9.9}$$

Combined with three metrics (set $U = \{\ BetUtil, SimUtil, TSUtil\ \}$), the final utility function of SimBet is as followed.

$$SimBetTSUtil_C(D) = \sum u_C(D), \text{ where } u \in U \tag{9.10}$$

SimBet will use this utility function to make the forwarding decision with a new parameter *Replication*, represented as *R*. Each message has its own *R*, and the number of copies of each message transferred from *C* to *N* is equal to the value of *R*. The computational process of *R* is in Equation (9.11).

$$R_C = \left\lfloor R_{C_{\text{pre}}} \times \frac{SimBetTSUtil_C(D)}{SimBetTSUtil_C(D) + SimBetTSUtil_N(D)} \right\rfloor \tag{9.11}$$

When $R_C < 1$, the current node *C* will not deliver messages to the neighbouring node *N*. When $R_C = 1$, the node *C* will apply Single-Copy mechanism. In addition, when $R_C > 1$, Multi-Copy mechanism will be used. However, SimBet will lead to some other problems: Messages in one community are difficult to transmit into the other communities because of the limited local knowledge, especially since these communities have few communications with each other.

### 9.3.4 Advanced Historical Knowledge Based

In Section 9.2.1, one basic routing protocol based on historical knowledge in DTNs called PROPHET has been introduced. After its publishment, many researchers have done some improvements to enhance the performance of PROPHET: ADPBSW[14] tries to combine the PROPHET with the Binary Spray-and-Wait, AdvPROPHET[15] solves the jitter problem in naïve version in DTNs, PROPHET+ [16] not only includes the *Delivery Predictability* but also provides some other variables to select (e.g., buffer, energy...) which is aimed to construct an adaptive routing protocol, and PROPHET-CLN[17] supplements previous work by considering the congestion level of the relay node. Moreover, some researchers transfer the historical knowledge in DTNs into a mathematical problem.

#### *9.3.4.1 PER*

Predict and Relay (PER)[18] finds that nodes in DTNs, especially in Vehicular DTNs, prefer to connect with some groups of nodes that are frequently visited. These groups are called *Landmarks* in PER. Furthermore, researchers also observe another phenomenon that the node mobility behaviours are semi-deterministic. For instance, students always move around between Classroom, Dormitory, Lab, and Gym. According to two discoveries, PER designs a time-homogeneous semi-Markov model, named TH-SMP Model, to illustrate node mobility between *Landmarks* which is based on the assumption that future states depend on the current state. Therefore, node mobility can be predictable based on reasoning and computing.

Based on this semi-Markov model, PER still requires two parameters to calculate the probability metrics. The first parameter is $P^C$, which represents the transition probability matrix of the embedded Markov chain for the current node $C$. The second parameter is the sojourn time probability distribution matrix $S_{ij}^C(k)$, which means the probability of the current node $N$ who will move from *Landmark* $i$ to *Landmark* $j$ before or at time $k$.

When making forwarding decisions, the current node will check all the connecting neighbouring nodes and the current node itself. After calculating their probability metrics, the current node will select the node with maximum delivery probability. If the selected node is the current node itself, the forwarding behaviour will be failed. If not, the current node will forward the target message to the selected node and then drop the original ones since PER is a Single-Copy-based routing protocol.

## 9.4 IMPROVEMENTS IN AERIAL DTNS

### 9.4.1 Connection Planning

In the early development of DTNs, most routing protocols are designed for the satellite network in deep space. The common routing protocols on the ground usually have continuous and stable connections between nodes. Different from these traditional routing protocols used for the Internet, there may not be an End-to-End path between any two satellites at any time since satellites are dynamically scheduled. Due to this feature, the satellite network can be regarded as a typical DTN.

However, the satellite network is not exactly the DTN in a broad sense. Routing protocols introduced above are usually based on the factor that the movement of nodes in the network is random or semi-deterministic, so the connection selecting is zero knowledge or partial knowledge based (e.g., historical knowledge or geographical knowledge). In the satellite network, all links are pre-planned, so that each satellite can achieve the previous, current, and future information of the network topology, and these link connections are periodically and sequentially bullied between satellites. Therefore, a specific routing protocol and its enhanced versions are developed to meet the requirements in this situation.

#### *9.4.1.1 CGR*

Contact Graph Routing (CGR)[19] is a dynamic routing protocol based on one concept named contact graph. The contact graph is a directed acyclic graph where its root is the contact between the current node and itself while other vertices are contacts between the current node and other nodes in the network.

This contact graph is generated by the *contact plan*. This plan includes the contact information and the range information, which constructs the planned changes in the network topology. CGR uses this plan to update the possible connection opportunities, for example, insert the contact chances in the future or remove the expired contacts.

After the foundation of the contact graph, nodes in the graph will use the Contact Review Procedure (CRP) algorithm to compute the router and select the optimal neighbouring nodes to finish the forward behaviour. The CRP algorithm is one kind of heuristic algorithm which applies a recursion mechanism. For the destination node *D*, CRP will check its closest neighbouring node, and then find the route to the original node *M* that creates the targeted message step by step. For the intermediate transmission node *S*, the potential situations have been concluded into three types: *S* is the local node (*D* is the neighbouring node of the local node), *S* is in the list of Excluded Nodes (node in this list will not participate in the routing calculation process), and all other situations. For each type, the final forwarding decision is also influenced by the starting time of the message, the stopping time of the message, the expiring time of the message, the real-time transmission rate, the residual capacity of the contact, and so on. The details of CRP are illustrated in the original paper.[19]

#### *9.4.1.2 CGR-ETO*

Although CGR performs well in the satellite network and has high reliance due to the accurate information of the network, it still has some problems. The naïve CGR does not consider the queueing delay for the transmission in the buffer. To solve this problem, CGR-ETO[20] is proposed. CGR-ETO replaces the expired time of the message with a new parameter called Earliest Transmission Opportunity (ETO).

ETO is the earliest time for messages to do the transmission behaviours. Its value is associated with the priority of the message which means messages with different priorities will have different ETOs. The initial value of ETO is set to the starting time of the message. When the message is in the transmission queue, CGR-ETO will calculate the estimated transmission time based on the residual capacity and the transmission rate. This estimated transmission time will be regarded as the queueing delay that will be added into the ETO or the current time (the later one will be chosen). Then, all messages that have lower priority will update their ETOs to get more accurate values.

As summarized in the paper,[21] CGR-ETO is considered as a short-term evolution that adds the *Contact Plan Update Protocol (CPUP)*. Another short-term improvement is Enhanced CGR (ECGR)[22] which applies Dijkstra's algorithm for the path selection. This change solves some security problems in the naïve CGR to prevent the route loop and possible route oscillation. Due to this achievement, ECGR has become the core part of the latest CGR version. Moreover, researchers in this paper[21] also put forward two long-term evolutions which are "Path Encoding CGR Extension" (attaches encoded paths into messages) and "Opportunistic CGR Extension" (forwards messages in a probabilistic way), which has a certain leading role in the development of CGR.

### 9.4.2 Passive Routing

To solve different kinds of problems that exist in DTNs, it is necessary to design efficient and stable routing protocols for different realistic conditions and application scenarios. According to the characteristics of the formation method, these routing protocols can be divided into two categories: Active Routing and Passive Routing (also called On-Demand Routing).

Active routing protocols maintain consistent and latest routing information for all nodes in the network. Each node maintains a routing table to store its routing information, and any changes in the configuration of any node will be noticed through the whole network to ensure the reliance of the routing table.

In passive routing protocols, the router will be built only when one message needs to be transferred from the source node to the required destination node. Although they still need the routing table, this routing table will not be consistent for each node in the network, which only contains the information on the transmission path. Therefore, the cost or overhead is much less than active routing protocols, which is the reason that they are commonly used in DTNs. The key processes of passive routing protocols are Route Discovery (RD) and Route Maintenance (RM), and the differences between these passive routing protocols are the methods of RD and RM.

#### *9.4.2.1 DSR*

Dynamic Source Routing (DSR)[23] uses *Route Request (RREQ)* message to request each neighbouring node to check the connected one is the destination node or the optimal relay. Each node that receives *RREQ* will send back *Route Reply (RREP)* message to respond to the previous requests. Due to this design, *RREQ* needs to cover all useful information of nodes that already existed in the current path. To conclude, the RD process is that *RREQ* builds up the path through the network and stores the needed information while *RREP* routed back to each node in the path using the recursion method.

The RM process is based on *Route Error (RERR)* message. To satisfy different data transmission requirements, the maintenance process can be roughly divided into two strategies. The one is Point-to-Point. When one node changes or dies, this node or its neighbouring nodes will send *RERR* to the surrounding nodes and the source nodes. Every node that receives *RERR* will delete the information of the corresponding node from its buffer, then restart the RD process. The other method is End-to-End, which is usually applied when the source node is far away from the destination node. In this situation, the link cannot determine the number of problematic nodes and their location information, so the RD process will be restarted directly from the source node.

Because of the features of passive routing protocols, DSR can tolerate rapid changes in network topology and have less overhead in RM compared with active routing protocols. Additionally, DSR can also support asymmetric communication which means it can forward messages in the unidirectional link only from the source to the destination. However, it still has some disadvantages. With the development of the size of the network, the required information in the buffer will become larger which will lead to more occupation in the cache of each node and increase the cost of RD. Moreover, all control messages are delivered using a flooding mechanism which may cause message collision or information loss, and even cause competition between different routing paths.

#### *9.4.2.2 AODV*

Ad-hoc On Demand Distance Vector Routing (AODV)[24] is like DSR when in RD process. It also uses *RREQ* and *RREP* to find an optimal path to the destination node. Each node records the location of the previous hop and the next hop; it cannot know the entire topology of the network. However, the method of storing node information is different. DSR uses route cache to maintain multiple entries for each destination while AODV uses a traditional routing table to store one current optimal routing path to one corresponding destination. Another different point is that AODV relies on the routing table to send *RREP* back to the source since AODV does not have the source routing.

In the RM process, the sequence number is used for each destination which is aimed to know the freshness of the routing information and avoid possible routing loops. When the network topology changes, one special *RREP* will be sent back to the corresponding source nodes. These changes happen when the destination or intermediate node moves, one next hop becomes unreachable, or other link failures. Compared with DSR, *RERR* in AODV is used to send to the previous hop to delete the useless information till *RERR* is transferred to the source node, therefore, AODV applies the End-to-End method that the RD process will restart directly from the source node.

Because of the features of the Distance-Vector routing protocol, the implementation of AODV is simpler than DSR, and the cost is lower when the network is stable. More comparisons between DSR and AODV are discussed in this paper.[25] However, due to the characteristics of the routing table, AODV only supports one routing path to one destination, and the destination node only can reply to one request. Additionally, although naïve DSR can support multiple paths with its routing buffer, the performance is not ideal. Since multiple paths can support many functions including redundancy mechanism, load balancing, and link aggregation, some research (e.g., EMP-DSR,[26] LMP-DSR,[27] and AOMDV[28]) tried to solve this problem to improve the quality of service.

### 9.4.3 Energy Awareness

Epidemic routing protocol has been mentioned in Section 9.2.2 as a normal benchmark which has high delivery probability when the resources are unlimited. Unfortunately, energy consumption is a significant influencing factor in DTNs when routing protocols are applied in the real world, especially for the Aerial DTNs. Due to the flooding mechanism in Epidemic, the forwarding behaviours will cause huge energy waste and finally lead to the temporary death of nodes in the network. For Unmanned Aerial Vehicle (UAV), its death means not only the loss of communication between other relay nodes or infrastructures but also a possibility of an out-of-control crash. Hence, a reasonable energy limitation policy is required in the development of DTNs.

#### *9.4.3.1 EAER*

Energy-Aware Epidemic Routing (EAER)[29] is the enhanced version of Epidemic routing protocol in energy awareness. In fact, its improvement is not done in one step. The researchers got inspiration from another routing protocol which is *n*-Epidemic.[30] *n*-Epidemic controls the forwarding behaviours that the current node *C* forwards only when there are at least *n* neighbouring nodes in its transmission range. This main idea is illustrated in Figure 9.5 when the threshold is set to $n = 3$.

Although *n*-Epidemic already has a more efficient energy policy compared with the original Epidemic, it covers the possible energy consumption only in the transmission stage. EAER takes some more stages into consideration. In the common transmission stage, EAER includes not only the size of the messages but also the distance between paired nodes. In the function stage, the energy consumption is associated with operations in each node. Furthermore, in the receiving stage, the size of the messages should also be considered. Finally, EAER proposes the *Current Energy Level (CEL)* that combines these parameters.

Additionally, EAER also controls the value of *n* with the node density which is presented by the *Current Neighbours Node (CNN)*. Therefore, EAER proposes an energy-aware heuristic function $f_H(CEL, CNN)$ to determine the value of *n*, which means *n* is dynamically managed and adaptive to the network environment. The main idea of this heuristic

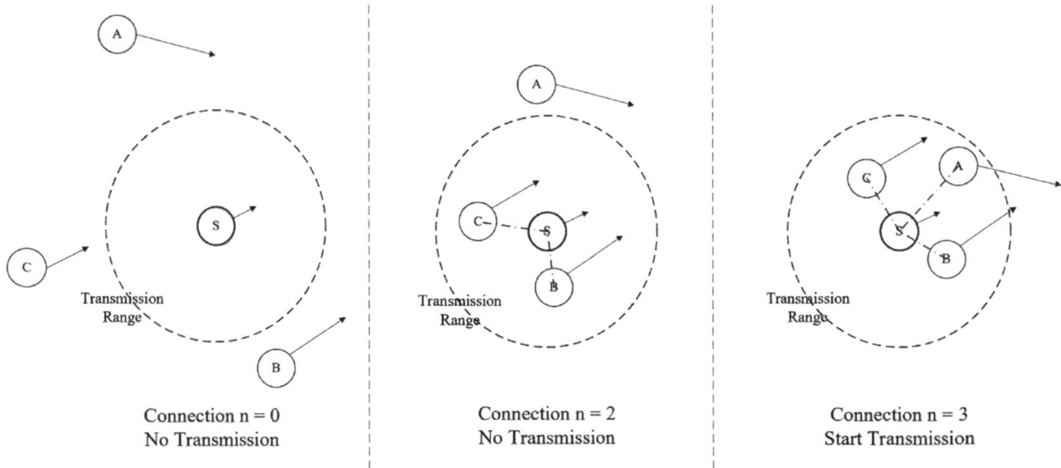


FIGURE 9.5 The *n*-Epidemic routing protocol.

function is to reduce the number of involved nodes during the transmission of messages, which leads to the reduction of energy consumption and less deterioration in delivery probability.

### 9.4.4 Advanced Geographical Knowledge Based

In the routing protocols based on historical knowledge, it is necessary to obtain the up-to-date network topological information, and then use the information to predict the possibility that nodes can encounter in the future. However, the information is not easy to access when in DTNs and might be not timely. Therefore, research on geographical routing protocols is prosperous recently due to their reliability when facing the topological variation. Especially in Vehicular and Aerial DTNs, the motivation of nodes is random or semi-deterministic and the frequency of topology changes is high, so knowing the geographically topological information will perform better[31,32] compared with simple historical routing protocols.

During recent studies, researchers[33] define three classes of geographical routing protocols in DTNs based on the information of the destination node: Destination Unawareness Class, Destination Awareness Class, and Hybrid Class. They also find three challenges that need to be solved: Reliability for Message Relaying, Locating Mobile Destination, and Difficulty for Handling the Local Maximum Problem. Here, one example is given to explain the basis of classification, and one possible solution to the last challenge is put forward for discussions.

#### *9.4.4.1 AeroRP*

AeroRP[34] is a typical example in the Destination Awareness Class. Its main idea is like MoVe but includes the transmission range $R$ and the relative velocity to the destination node. Additionally, AeroRP can be extended to adopt a three-dimensional coordinate environment. Its forwarding decision is made by the metric called *Time to Intercept (TTI)*, and its illustration is in Figure 9.6.

Because of the awareness of the destination node, the imaginary vector between the current node $N_a$ and the destination node $N_d$ can be represented by $\overrightarrow{IV_{a,d}}$. Furthermore, the movement vector of $N_a$ can also be accessed easily and represented by $\overrightarrow{MV}_a$. Hence, the space angle between these two vectors can be calculated:

$$\theta_{a,d} = \arccos \frac{\overrightarrow{MV}_a \cdot \overrightarrow{IV_{a,d}}}{\left|\overrightarrow{MV}_a\right| \left|\overrightarrow{IV_{a,d}}\right|} \tag{9.12}$$

Then, the speed of $N_a$ relative to $N_d$ can be accessed:

$$v_{a,d} = v_a \times \cos\theta_{a,d} \tag{9.13}$$

Based on the distance $Dis_{a,d}$ and relative speed $v_{a,d}$ between $N_a$ and $N_d$, the *TTI* will be computed:

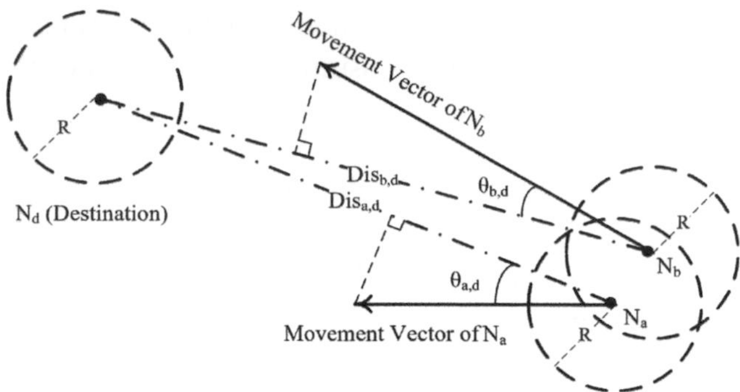


FIGURE 9.6 Moving scenario in AeroRP.

$$TTI = \begin{cases} 0 & \text{when } Dis_{a,d} \leq R \\ \dfrac{Dis_{a,d} - R}{v_{a,d}} & \text{when } Dis_{a,d} > R \text{ and } \theta_{a,d} < \dfrac{\pi}{2} \\ \text{Infinite} & \text{Otherwise} \end{cases} \tag{9.14}$$

AeroRP also defines three modes when the current node contains the lowest *TTI*. The first mode is *Ferry* which stores all messages indefinitely until a node with a lower *TTI* is encountered. The second mode is *Buffer* which queues all messages in a finite buffer with their own expired time until a node with a lower *TTI* is encountered. The last one is *Drop* which will drop all messages with a higher *TTI*. Therefore, when the node with lower *TTI* is found, the forwarding behaviour is done during their connections.

However, AeroRP will not have ideal routing performance if the Aerial Network is sparse. The main reason is its forwarding decision that only transmits messages to the node only with lower *TTI*. When the space angle $\theta_{a,d} \geq \frac{\pi}{2}$, and the *TTI* will be infinite, so that, all message carriers that are moving away from the destination node will not participate in the forwarding process, which will greatly reduce the performance of AeroRP.

#### 9.4.4.2 DGR

Delegation Geographic Routing (DGR)[35] tries to improve the performance of AeroRP and solve its Local Maximum Problem. DGR uses Delegation Forwarding[36] mechanism which prefers to compare the *TTIs* of previously encountered nodes stored in the buffer, instead of the current node itself, with the *TTI* of the presently encountered node. This Delegation Forwarding mechanism overcomes the decision problem that part of encountered nodes might be moving away from the destination node.

The Local Maximum Problem usually happens in sparse DTNs. Because of few connection opportunities in this environment, the request and reply of information in nodes will not be frequent to update. Hence, the outdated information will cause inaccuracy in making forwarding decisions. DGR solves the Local Maximum Problem using a heuristic inequality when forwarding behaviour can be:

$$T_M^{ela} + T_M^{TTI} > T_M^{ini} \tag{9.15}$$

where $T_M^{ela}$ is the elapsed time after the moment that the message $M$ was generated, while $T_M^{TTI}$ is the current $TTI$ of the message $M$, and $T_M^{ini}$ is the initial lifetime of the message $M$.

This inequality applies when message carriers are moving away from the destination node. When the situation meets this inequality, these non-participating nodes will have other chances to forward their messages to their neighbouring nodes. Therefore, each message will be more possible to access the destination node in its lifetime, so that the delivery probability will be increased.

## 9.5 APPLICATIONS

The most significant and close-to-lives application of DTNs is Smart City.[37] Smart City is an advanced form of urban informatization based on a new generation of information technologies and innovative concepts. This helps to alleviate city problems, accelerate the process of urbanization, enhance dynamic management, and improve the quality of life of residents. With these requirements, there are expanding demands on urban infrastructure of all kinds, including transport. Based on these interests, more attention has been paid to the Vehicle-to-Vehicle (V2V) communication in DTNs because of its wide usage in transport applications. V2V communication can enable neighbouring cars to exchange basic safety information such as position, speed, and direction with each other, thus greatly reducing the occurrence of car crashes and alleviating traffic congestion. For electronic vehicles, their information can be used for charging management and charging recommendations. For drones or UAVs, these can be regarded as relays in DTNs to monitor the environment, support disaster recovery, or provide new services to users, especially when the telecommunication infrastructure is in shortage.[38] Furthermore, DTNs support the network in Smart City which is not only limited to the communication between vehicles or drones, but also will be extended to the network between vehicles, infrastructures like Road Side Units (RSUs), and pedestrians to form Vehicle-to-Infrastructure (V2I) and Vehicle-to-Everything (V2X) in the future process of building Smart City.

### 9.5.1 Electric Vehicle Charging in Smart City

Due to the environmentally friendly attribute and the policy support, the increasing number of electric vehicles in the urban city becomes common. However, the slow development of battery capacity leads to not only limitation of the application range but also the decreasing of the experience of users. They may need to find the charging station during their long trip, and even need to wait for a long service time if the found charging station does not have an empty space.

This survey[39] introduces three classes of methods for electric vehicle charging recommendation in the communication aspect. The first category is the Centralized Charging System which uses the cellular network or the Internet. The second one is the Distributed Charging System which focuses on the V2V and V2I networks. The third one is the Hybrid Charging System that combines the features of Centralized and Distributed systems. This survey also mentions the possible urban data for the charging recommendation, so that providers can change the supply and demand relationship to minimize traffic congestion and maximize their interest. It also points out that one strategy with the minimum waiting time can be regarded as the best choice for resource scheduling and user experience.

According to the survey, researchers try to apply the DTN routing protocols for electric vehicle charging management in Smart City.[40] They set the goal for the shortest expected waiting time for charging, including the time from requesting to arriving at the charging station and time consumption in the charging station. Compared with the traditional applications using the cellular network, routing protocols in DTNs for communication have advantages in flexibility and cost-efficiency.

### 9.5.2 Autonomous Logistics Network in Smart City

Another application of the drones in future Smart City is the autonomous logistics network. Many retail companies try to use drones to deliver their goods for higher efficiency, less human resource expenditure, and more interest. Drones in the autonomous logistics network should have their primary function of loading and unloading and be extended to have the capacity to scan and collect data, then carry and transfer messages. The newly designed routing protocol for the requirements should make some improvements to the traditional routing protocols in DTNs.

Hence, researchers[41] propose a heuristic method called Weighted Flight Path Planning (WFPP). Because the working processes require that drones need to deliver goods before their deadline, and back to the stations of the logistics company, and the whole power consumption cannot exceed their battery capacity or can find the charging station before running out of energy, there are many scenes and conditions should be considered.

The weight of each drone is based on the priority of the current carried good, the remaining delivery time of all goods in the storage, the time to live of each message in the buffer, the distance between the destination, and the energy level of the drone. Combined with these influencing factors, WFPP can achieve the minimize the message delivery delay and maximize the message delivery probability, with the success of parcel delivery before the deadline and enough energy left when drones come back to the stations.

### 9.5.3 Disaster Recovery in Smart City

If a large scale of disaster happens, the city will be isolated from other areas due to the disconnection of the network and the transport. Both connections based on the infrastructure cannot be rebuilt in a short time, therefore, the first-line intelligence related to the disaster recovery will be difficult to transmit to the central agency for urgent and accurate decisions and instructions.

Because of the *Store-Carry-Forward* feature of DTNs, connections between nodes do not rely on the existing infrastructure, so routing protocols in DTNs are suitable for post-disaster situations. Some papers[42,43] developed a new network called Never Die Network (NDN) which can be treated as a variation of the DTN. Researchers proposed a data triage method based on the information from the investigation of the previous disasters to improve the efficiency of the routing protocols in NDN. This method divides the common disaster into several periods, and in each period actions with different priorities will be taken for different conditions.

Furthermore, these studies designed the Autonomous Flight Wireless Nodes (AFWs). AFWs not only connect to the Internet and the cellular network with their IP addresses but also support communication in DTNs. During disasters, AFWs will seek any possible nodes that can be communicated and collect the information for future use. Each AFW will try to exchange information with other AFWs, and when they come to the information centre, they will upload the data. Additionally, AFWs will remember the closest charging station using GPS and land to the station for recharging when their energy level is low.

## 9.6 FURTHER DISCUSSIONS

Vehicular and Aerial DTNs have a huge potential in future development because of the tolerant feature. In this section, some further discussions attract the attention of the researchers, and these can be regarded as the new directions of DTNs:

*Social Factors:* This survey has mentioned the social network in Section 3 which includes the *Betweenness Centrality*, *Social Similarity*, and *Tie Strength*.[10] There are some other social factors that can be considered, and node selfishness is one good entry point for research. Routing protocols in DTN that use opportunistic forwarding are based on the forwarding behaviour is selfless, but most nodes in the real world are selfish because they usually are not willing to share the information or to store the information from others. Hence, researchers[44] use a two-dimensional continuous time Markov chain to evaluate the relationship between the node selfishness and the delivery performance. They find that when nodes in DTNs have strong selfishness, the delivery delay will be higher, but the cost when delivering will be lower. Moreover, this survey[45] analyses and concludes the positive and negative social factors which will be helpful to understand this direction in general.

*Quantitative Evaluation:* As explained in the paper,[46] there are two reasons that a quantitative evaluation of the routing protocols in DTNs. The first one is that most routing protocols in DTNs may only be suitable for their specific area that are difficult to prove that they can be extended to apply in other application scenarios. The second reason is caused by the expensive cost of deployments in the real world so researchers prefer to use simulators (e.g., OPNET, QualNet, NS2/NS3, and ONE[47]) to test the performance of routing protocols. Although these simulators can be easier to build the experiment environments and compare the difference between several routing protocols, the simulation result will be somewhat different from the actual operation result due to the lackey of the necessary realism. Therefore, a quantitative evaluation is necessary after proposing a new routing protocol in DTNs. Finally, studying and comparing different quantitative evaluation methods can also be a challenging research direction.

*Artificial Intelligence:* Artificial Intelligence is the study of making computers to simulate the human thinking process and behaviours which make computers able to achieve higher-level applications. In recent years, it has accessed rapid development and has been widely used in many disciplines, so combining artificial intelligence with routing protocols in DTNs is an ideal focus. Here is an example[48] which uses Decision Tree Classification

Algorithm to choose one node as the most suitable relay. The main process is data collection, training the classifier with the data set, and applying the knowledge of the classifier in routing selection. The main contribution of this technology cooperation is the decreasing of network overhead without degrading the delivery probability.

*Security & Privacy:* When technology mentions information exchange and storage, the security and privacy problems should be carefully considered. Based on the CIA (Confidentiality, Integrity, and Availability) Triad in information security, different technologies can be used for these three factors. Since the traditional cryptographic techniques are based on continuous network communication, methods designed for Confidentiality in DTNs should be changed. In this paper,[49] the first anonymous solution for DTNs is proposed. This solution includes the DTN security architecture and the DTN anonymity architecture based on the Identity-Based Cryptography (IBC). Another research team[50] studies the principle of avoiding proximity malware and applies the main character to Integrity and Availability in DTNs. They try to detect two unique challenges which are "insufficient evidence versus evidence collection risk" and "filtering false evidence sequentially and distributedly" so that they can solve the spoofing attack and DDoS.

## 9.7 CONCLUSION

Different from the traditional network that is based on the well-built communication infrastructure, DTNs are designed for the poor environment, so routing protocols in DTNs are trying to solve the long delivery latency and frequent network partitions. To popularize relevant knowledge and applications, and to point out possible future research directions, this chapter lists and simply explains the related routing protocols in vehicular and aerial DTNs and their possible application in future Smart City. Moreover, this chapter hopes to help new researchers who join this field quickly know and understand the relevant routing protocols and find the research directions they are interested in as soon as possible.

## REFERENCES


1. K. Fall. "A delay-tolerant network architecture for challenged internets," in *Proceedings of the 2003 Conference on Applications, Technologies, Architectures, and Protocols for Computer Communications*, Karlsruhe, Germany, 2003, pp. 27–34.
2. T. Spyropoulos, K. Psounis and C. S. Raghavendra, "Single-copy routing in intermittently connected mobile networks," in *2004 First Annual IEEE Communications Society Conference on Sensor and Ad Hoc Communications and Networks, 2004. IEEE SECON 2004*, Santa Clara, California, USA, 2004, pp. 235–244.
3. T. Spyropoulos, K. Psounis and C. S. Raghavendra, "Efficient routing in intermittently connected mobile networks: The multiple-copy case," *IEEE/ACM Transactions on Networking*, vol. 16, no. 1, pp. 77–90, 2008.
4. A. Vahdat, and D. Becker. "Epidemic routing for partially connected ad hoc networks," 2000.
5. A. Lindgren, A. Doria and O. Schelén, "Probabilistic routing in intermittently connected networks," *ACM SIGMOBILE Mobile Computing and Communications Review*, vol. 7, no. 3, pp. 19–20, 2003.
6. J. LeBrun, C.-N. Chuah, D. Ghosal and M. Zhang, "Knowledge-based opportunistic forwarding in vehicular wireless ad hoc networks," in *2005 IEEE 61st Vehicular Technology Conference*, Stockholm, Sweden, 2005, pp. 2289–2293, vol. 4.

7. T. Spyropoulos, K. Psounis and C. S. Raghavendra, "Spray and wait: An efficient routing scheme for intermittently connected mobile networks," in *Proceedings of the 2005 ACM SIGCOMM Workshop on Delay-Tolerant Networking*, Philadelphia, Pennsylvania, USA, 2005, pp. 252–259.
8. T. Spyropoulos, K. Psounis and C. S. Raghavendra, "Spray and focus: Efficient mobility-assisted routing for heterogeneous and correlated mobility," in *Fifth Annual IEEE International Conference on Pervasive Computing and Communications Workshops (PerComW'07)*, White Plains, NY, USA, 2007, pp. 79–85.
9. J. Burgess, B. Gallagher, D. Jensen and B. N. Levine, "MaxProp: Routing for vehicle-based disruption-tolerant networks," in *Proceedings IEEE INFOCOM 2006. 25TH IEEE International Conference on Computer Communications*, Barcelona, Spain, 2006, pp. 1–11.
10. E. M. Daly and M. Haahr, "Social network analysis for information flow in disconnected delay-tolerant MANETs," *IEEE Transactions on Mobile Computing*, vol. 8, no. 5, pp. 606–621, 2009.
11. H. Dubois-Ferriere, M. Grossglauser and M. Vetterli. "Age matters: Efficient route discovery in mobile ad hoc networks using encounter ages," in *Proceedings of the 4th ACM International Symposium on Mobile Ad Hoc Networking & Computing*, Annapolis, Maryland, USA, 2003, pp. 257–266.
12. M. Everett and S.P. Borgatti, "Ego network betweenness," *Social networks*, vol. 27, no. 3, pp. 31–38, 2005.
13. A. Khelil, P. J. Marron and K. Rothermel, "Contact-based mobility metrics for delay-tolerant ad hoc networking," in *13th IEEE International Symposium on Modeling, Analysis, and Simulation of Computer and Telecommunication Systems*, Atlanta, GA, USA, 2005, pp. 435–444.
14. J. Xue, X. Fan, Y. Cao, J. Fang and J. Li, "Spray and wait routing based on average delivery probability in delay tolerant network," in *2009 International Conference on Networks Security, Wireless Communications and Trusted Computing*, Wuhan, China, 2009, pp. 500–502.
15. J. Xue, J. Li, Y. Cao and J. Fang, "Advanced PROPHET routing in delay tolerant network," in *2009 International Conference on Communication Software and Networks*, Chengdu, China, 2009, pp. 411–413.
16. T. Huang, C. Lee and L. Chen, "PRoPHET+: An adaptive PRoPHET-based routing protocol for opportunistic network," in *2010 24th IEEE International Conference on Advanced Information Networking and Applications*, Perth, WA, Australia, 2010, pp. 112–119.
17. G. Wang, J. Tao, H. Zhang and D. Pan, "A improved prophet routing based on congestion level of nodes in DTN," in *2017 IEEE 2nd Advanced Information Technology, Electronic and Automation Control Conference (IAEAC)*, Chongqing, China, 2017, pp. 1666–1669.
18. Q. Yuan, I. Cardei and J. Wu, "An efficient prediction-based routing in disruption-tolerant networks," *IEEE Transactions on Parallel and Distributed Systems*, vol. 23, no. 1, pp. 19–31, 2012.
19. S. C. Burleigh, "Contact graph routing". No. NPO-45488. 2011.
20. N. Bezirgiannidis, F. Tsapeli, S. Diamantopoulos and V. Tsaoussidis, "Towards flexibility and accuracy in space DTN communications," in *Proceedings of the 8th ACM MobiCom Workshop on Challenged Networks*, pp. 43–48, 2013.
21. G. Araniti, et al., "Contact graph routing in DTN space networks: Overview, enhancements and performance," *IEEE Communications Magazine*, vol. 53, no. 3, pp. 38–46, 2015.
22. J. Segui, E. Jennings and S. Burleigh, "Enhancing contact graph routing for delay tolerant space networking," in *2011 IEEE Global Telecommunications Conference - GLOBECOM 2011*, Houston, TX, USA, 2011.
23. D. B. Johnson, "The dynamic source routing protocol for mobile ad hoc networks." in *IETF Internet Draft*, http://www.ietf.org/internetdrafts/draft-ietf-manet-dsr04.txt, 2000.
24. C. E. Perkins and E. M. Royer, "Ad-hoc on-demand distance vector routing," in *Proceedings WMCSA'99. Second IEEE Workshop on Mobile Computing Systems and Applications*, New Orleans, LA, USA, 1999, pp. 90–100.

25. C. E. Perkins, E. M. Royer, S. R. Das and M. K. Marina, "Performance comparison of two on-demand routing protocols for ad hoc networks," *IEEE Personal Communications*, vol. 8, no. 1, pp. 16–28, 2001.
26. E. K. Asl, M. Damanafshan, M. Abbaspour, M. Noorhosseini and K. Shekoufandeh, "EMP-DSR: An enhanced multi-path dynamic source routing algorithm for MANETs based on ant colony optimization," in *2009 Third Asia International Conference on Modelling & Simulation*, Bundang, Indonesia, 2009, pp. 692–697.
27. L. K. Malviya and D. Tiwari, "LMP-DSR: Load balanced multi-path dynamic source routing protocol for mobile Ad-Hoc network," in *2013 Fourth International Conference on Computing, Communications and Networking Technologies (ICCCNT)*, Tiruchengode, India, 2013, pp. 1–5.
28. M. K. Marin, and S. R. Das, "On-demand multipath distance vector routing in ad hoc networks," in *Proceedings Ninth International Conference on Network Protocols. ICNP 2001*, Riverside, CA, USA, 2001, pp. 14–23.
29. F. De Rango, S. Amelio and P. Fazio, "Enhancements of epidemic routing in delay tolerant networks from an energy perspective," in *2013 9th International Wireless Communications and Mobile Computing Conference (IWCMC)*, Sardinia, Italy, 2013, pp. 731–735.
30. X. Lu and P. Hui, "An energy-efficient n-epidemic routing protocol for delay tolerant networks," in *2010 IEEE Fifth International Conference on Networking, Architecture, and Storage*, Macau, China, 2010, pp. 341–347.
31. V. N. G. J. Soares, J. J. P. C. Rodrigues, J. A. Dias and J. N. Isento, "Performance analysis of routing protocols for vehicular delay-tolerant networks," in *SoftCOM 2012, 20th International Conference on Software, Telecommunications and Computer Networks*, Split, Croatia, 2012, pp. 1–5.
32. V. N. G. J. Soares, J. J. P. C. Rodrigues and F. Farahmand, "Performance assessment of a geographic routing protocol for vehicular delay-tolerant networks," in *2012 IEEE Wireless Communications and Networking Conference (WCNC)*, Paris, France, 2012, pp. 2526–2531.
33. T. Wang, Y. Cao, Y. Zhou and P. Li. "A survey on geographic routing protocols in delay/disruption tolerant networks," *International Journal of Distributed Sensor Networks*, vol. 12, no. 2, p. 3174670, 2016.
34. K. Peters, A. Jabbar, E. K. Çetinkaya and J. P. G. Sterbenz, "A geographical routing protocol for highly-dynamic aeronautical networks," in *2011 IEEE Wireless Communications and Networking Conference*, 2011, pp. 492–497.
35. Y. Cao, Z. Sun, N. Wang, H. Cruickshank and N. Ahmad, "A reliable and efficient geographic routing scheme for delay/disruption tolerant networks," *IEEE Wireless Communications Letters*, vol. 2, no. 6, pp. 603–606, 2013.
36. V. Erramilli, M. Crovella, A. Chaintreau and C. Diot. "Delegation forwarding," in *Proceedings of the 9th ACM International Symposium on Mobile Ad Hoc Networking and Computing*, Hong Kong, Hong Kong, China, pp. 251–260, 2008.
37. A. Zanella, N. Bui, A. Castellani, L. Vangelista and M. Zorzi, "Internet of things for smart cities," *IEEE Internet of Things Journal*, vol. 1, no. 1, pp. 22–32, 2014.
38. C. Giannini, A. A. Shaaban, C. Buratti and R. Verdone, "Delay Tolerant Networking for smart city through drones," in *2016 International Symposium on Wireless Communication Systems (ISWCS)*, Poznan, Poland, 2016, pp. 603–607.
39. Y. Cao, H. Song, O. Kaiwartya, A. Lei, Y. Wang and G. Putrus, "Electric vehicle charging recommendation and enabling ICT technologies: Recent advances and future directions." *IEEE COMSOC MMTC Communications-Frontiers*, vol. 12, no. 6, pp. 23–32, 2017.
40. Y. Cao, X. Zhang, R. Wang, L. Peng, N. Aslam and X. Chen, "Applying DTN routing for reservation-driven EV Charging management in smart cities." in *2017 13th International Wireless Communications and Mobile Computing Conference (IWCMC)*, Valencia, Spain, pp. 1471–1476, 2017.

41. S. Iranmanesh, R. Raad, M. S. Raheel, F. Tubbal and T. Jan, "Novel DTN mobility-driven routing in autonomous drone logistics networks," *IEEE Access*, vol. 8, pp. 13661–13673, 2020.
42. N. Uchida, N. Kawamura, T. Ishida and Y. Shibata, "Proposal of autonomous flight wireless nodes with delay tolerant networks for disaster use," in *2014 Eighth International Conference on Innovative Mobile and Internet Services in Ubiquitous Computing*, Birmingham, UK, 2014, pp. 146–151.
43. N. Uchida, G. Sato, Y. Shibata and N. Shiratori, "Proposal of connectivity support methods with autonomous flight wireless nodes for never die network," in *2015 18th International Conference on Network-Based Information Systems*, Taipei, Taiwan, 2015, pp. 387–392.
44. Y. Li, P. Hui, D. Jin, L. Su and L. Zeng, "Evaluating the impact of social selfishness on the epidemic routing in delay tolerant networks," *IEEE Communications Letters*, vol. 14, no. 11, pp. 1026–1028, 2010.
45. Y. Zhu, B. Xu, X. Shi and Y. Wang, "A survey of social-based routing in delay tolerant networks: Positive and negative social effects," *IEEE Communications Surveys & Tutorials*, vol. 15, no. 1, pp. 387–401, 2013.
46. K. Massri, A. Vitaletti, A. Vernata and I. Chatzigiannakis, "Routing protocols for delay tolerant networks: A reference architecture and a thorough quantitative evaluation." *Journal of Sensor and Actuator Networks*, vol. 5, no. 2, p. 6, 2016.
47. A. Keränen, J. Ott and T. Kärkkäinen, "The ONE simulator for DTN protocol evaluation." in *Proceedings of the 2nd International Conference on Simulation Tools and Techniques*, Rome, Italy, 2009, pp. 1–10.
48. L. P. Portugal-Poma, C. A. C. Marcondes, H. Senger and L. Arantes, "Applying machine learning to reduce overhead in DTN vehicular networks," in *2014 Brazilian Symposium on Computer Networks and Distributed Systems*, Florianopolis, Brazil, 2014, pp. 94–102.
49. A. Kate, G. M. Zaverucha and U. Hengartner, "Anonymity and security in delay tolerant networks," in *2007 Third International Conference on Security and Privacy in Communications Networks and the Workshops - SecureComm 2007*, Nice, France, 2007, pp. 504–513.
50. W. Peng, F. Li, X. Zou and J. Wu, "Behavioral malware detection in delay tolerant networks," *IEEE Transactions on Parallel and Distributed Systems*, vol. 25, no. 1, pp. 53–63, 2014.

CHAPTER 10

# Fog Bank of Vehicles for Next-Generation Connected Traffic

Himanshu Joshi
*Jawaharlal Nehru University*

Kirshna Kumar
*Government of India*

Ahmad M. Khasawneh
*Amman Arab University*

## CONTENTS



## LIST OF ABBREVIATIONS

| | |
|---|---|
| **ART** | Attack resistant trust |
| **CMA** | Contract matching approach |
| **DCs** | Data centres |
| **DSRC** | Dedicated short range communications |
| **MANET** | Mobile ad-hoc network |
| **PROS** | Privacy-preserving-route-sharing |

DOI: 10.1201/b22998-10

| | |
|---|---|
| **PVA** | Parked vehicle assistance |
| **RSU** | Road side unit |
| **SMDP** | Semi-markov-decision-process |
| **V2I** | Vehicle-to-infrastructure |
| **V2V** | Vehicle-to-vehicle |
| **VANET** | Vehicular ad-hoc network |
| **VFB** | Vehicular fog bank |

## 10.1 INTRODUCTION

In the recent years, a surge has been seen in the application of smarter vehicles, and also estimating the rapid growth in their usage in the future with the emergence of Internet of Vehicles (IoV) [1]. These vehicles are not only equipped with better computation facilities but also offer their users a better services experience [2]. These services tend to generate a huge amount of data. Transferring such a huge amount of data on the real time to the centralized computation facilities will sooner or later lead to the bandwidth saturation [3]. Even after a lot of enhancement, these centralized paradigms face a number of challenges such as single point failure, connectivity and transmission latency [4]. Hence, a new computing paradigm called vehicular fog computing is proposed. Vehicular Fog Banks (VFB) use the nearby moving vehicles, parked vehicles or RSUs to carry out computations. The applications of VFBs are not only limited to providing better computation facilities to the vehicles but can also help in downloading content, better and faster network connectivity because of low-latency requirement, geographical distribution, heterogeneous nature of the VFBs and proximity to the end user [5]. However, researchers have not fully explored various challenges faced by VFBs such as proper resource allocation, efficient energy consumption, security challenges and dynamic natures of fog banks [6].

In future technologies for smart cities, autonomous connected vehicles, smart traffic lights, smart roads, smart bus stops and smart stations are the core components towards realizing the next-generation vehicular traffic environments. Thus, the next-generation smart traffic environments in cities will generate big traffic data via the sensors attached throughout the traffic. Data mining and knowledge discovery on big traffic data can be realized towards producing precious traffic information for intelligent decision-making, automation and business model developments in smart cities environments. In this context, one of the biggest challenges in front of smart cities administrators, e.g., city councils, or third-party service providers is to avail a giant mobile computing infrastructure for big traffic data processing.

Today's vehicles are not standalone mechanical means of transportation, rather also have some form of computing and communication capabilities. These capabilities are being advanced and powerful day-by-day enabled via the growing number of embedded system components attached towards intelligent automation of vehicles. Thus, the vehicles can be considered as on the move computing systems. Towards this end, in this chapter, we investigate an innovative idea "Big Traffic Data centric Vehicular Fog Computing Framework for Smart Cities". Basically, the idea is to harness the computing capability

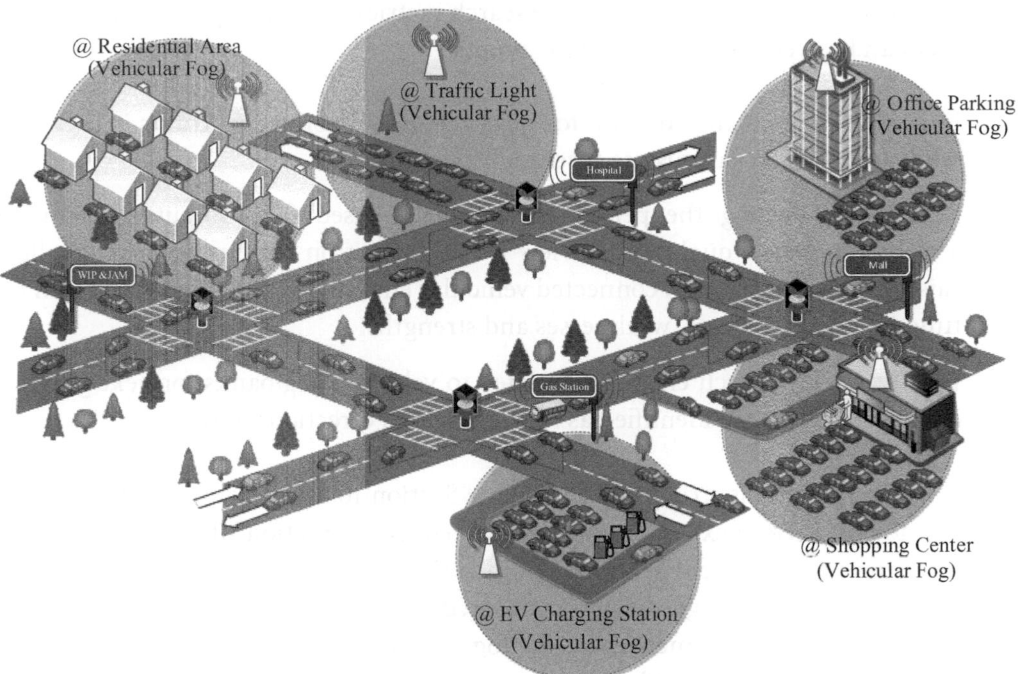


FIGURE 10.1 Vehicular fog bank scenarios.

of geographically close group of vehicles such as at parking space near offices, shopping centres, train stations, residential areas and at on-road junction points (see Figure 10.1, vehicular fogs or fog banks).

Towards realizing the big data centric vehicular fog computing framework, some key research question needs to be answered including following. How to develop a traffic-centric vehicular mobility model to identify and create vehicular fogs? Which type of AI technique is most suited for mobility-centric highly dynamic big traffic data mining and knowledge discovery? How to design and implement AI for traffic automation? What are the distributed authentication- and authorization-centric security models for vehicular traffic environment? How to implement authentication and authorization for joining and releasing of vehicles while vehicular fog formation? How to manage a vehicular virtual machine model from the available vehicular fogs focusing on higher rate of fog generation and deletion? How to develop a cashback model for vehicle as clients of vehicular fog framework in terms of reduction in parking cost, charging cost, road tax, etc.? This chapter study is an initial effort towards highlighting these questions and the significance of vehicular fog bank in the coming years with continuously growing sensor-centric modernization of vehicular traffic environments.

This chapter presents a comprehensive review on VFB for next-generation connected traffic in IoV environment focusing on research issues based studies. The performance metrics-based fog computing techniques for next-generation connected vehicular environments have been categorically described focusing on their limitations and technical contributions while considering various research metrics. Specifically, the contributions of the paper can be presented into following points:

- Firstly, the significance of VFB for next-generation connected traffic is highlighted while focusing on various use case implementations.
- Secondly, following the performance metrics-based fog computing techniques, a technique-based qualitative review is carried out on various research challenges raised in next-generation connected vehicular environments while focusing on main functional components, weaknesses and strengths.
- Finally, open research challenges related to vehicular fog banks for next-generation connected traffic are identified as future research directions in the area.

The remaining chapter is structured as follows: Section 10.2 highlights the significance of vehicular fog banks for next-generation connected traffic. Section 10.3 revisits the related literature surveys on vehicular fog banks in IoV environment. In Section 10.4, a comprehensive review on next-generation connected traffic in IoV environment is presented focusing on performance metrics-based fog computing taxonomy and model-based discussion. Section 10.5 identifies open research issues and challenges related to next-generation connected traffic in IoV environment, followed by conclusions described in Section 10.6.

## 10.2 SIGNIFICANCE OF FOG COMPUTING FOR NEXT-GENERATION CONNECTED TRAFFIC

In this section, the significance of fog computing for next-generation connected traffic is described, while considering two use case implementations. Firstly, fog banks for the moving vehicles and secondly, fog banks of the parked vehicles.

### 10.2.1 Case Study: Fog Bank of Moving Vehicles

The use case significance for fog bank of moving vehicles is further described while considering two sub-scenarios.

#### *10.2.1.1 Network as a Service*

In Intelligent Transport system which aims to achieve productivity and safety by integrating various information, communication is either Vehicle to Vehicle (V2V) or Vehicle to Infrastructure (V2I). For moving vehicles due to constant movement, it becomes very difficult to communicate with the infrastructure. So only vehicle-to-vehicle communication is possible. In cities vehicles move with very less speed but due to continuous movement the connectivity to the server is very poor. So, vehicles can use each other's network to form a fog bank by connecting to the nearest vehicle and therefore connecting to more mobile access points. This allows vehicles to directly communicate with each other rather than first sending the data to the RSUs and then accessing the data. This saves a lot of power consumption and transmission delay time [7].

#### *10.2.1.2 Computation as a Service*

Vehicles communicate either to other vehicles or the infrastructure both come under the category of Dedicated Short Range Communications (DSRC) devices. The range within DSRC works is 5.9 GHz band with the bandwidth of 75 MHz and approximate range of 1000 m [8]. Using DSRC vehicles can leverage the benefit of other vehicles and access their computation capabilities when the traffic is slow moving or there is a traffic jam. During a traffic congestion, vehicles with embedded computers can form a fog bank which can communicate to the RSU which can easily connect to the cloud servers making it a heterogeneous network, thus forming a heterogeneous cloud. This fog bank can not only solve its computation problem but can also perform computation tasks for remote clouds [9].

### 10.2.2 Case Study: Fog Bank of Parked Vehicles

The use case significance for fog bank of parked vehicles is further described while considering two sub-scenarios.

#### *10.2.2.1 Network as a Service*

It becomes very difficult to communicate between moving vehicles due to poor connectivity and is even more difficult to communicate with the RSUs due to constantly changing locations and limited power of the RSUs. Thus, parked vehicles can complement moving vehicles and RSUs in communicating to the cloud or remote servers. The study conducted in the area of Montreal, QC, Canada [10], shows that about 70% of the vehicles are parked for 23 hours a day [11]. With almost all the new vehicles equipped with DSRC devices it becomes obvious to use parked vehicles as a static node while connecting to the network thus creating a hybrid network of both static and mobile nodes. Liu et al. [12] proposes the idea of Parked Vehicle Assistance (PVA) in which parked vehicles can be used as relay nodes which can further improve the connectivity. The study showed that by adding a small number of parked vehicles to the network the connectivity is greatly improved.

#### *10.2.2.2 Computation as a Service*

Role of parked cars is not limited to providing better connectivity but also can be used to provide faster and more efficient computation services not just to moving vehicles but also to the users in their vicinity. Since each vehicle has a limited computational power, it is very difficult to perform large computations by an individual vehicle [13]. Therefore, fog bank of parked vehicles can be formed to perform larger and complex computations. One such application is content downloading using the capabilities of parked cars. Malandrino et al. [14] in his research concluded that parked vehicles improve content downloading efficiency by 25%–30% in comparison to RSUs and also decrease almost half the load of RSUs keeping in view the battery constraint. Data generated from moving vehicles can also be processed by the parked vehicles rather than sending it to the cloud for processing. For example, the videos generated from the moving vehicles can be processed by the fog bank of parked vehicles rather than sending it to the cloud for analysis [15].

## 10.3 LITERATURE SURVEY

A survey on fog computing consisting of its architecture, applications, challenges and opportunities has been suggested in IoT environment [1]. This survey helps the research community and industry for synthesizing and identifying the need of fog computing in upcoming era. The phases needed to design, resource management, implement and realize the practical fog computing system is sequentially reviewed for the support of general IoT landscape [2]. The prominent focus of the survey is to determine the implementation aspects needed for building a practical large-scale fog computing infrastructure in IoT environment. The systematic review has been performed on security requirements in Industrial IoT and novel paradigm of fog computing to address and improve the security of IIoT [3]. The architecture and protocols to build a fog system with the set of evaluation criteria and open research issues has been surveyed to enhance fog usages in various applications [4]. In addition, a review on lesson learned and discussion on future prospects is presented as to play fog technology as a key part in emerging technologies such as tactile internet. Various resource allocation techniques have been summarized for addressing the existing issues such as energy consumption, bandwidth and latency in fog computing environments [5]. In addition, the overview of various trending network applications, main aspects for designing these networks, open research challenges and efforts have also been highlighted. A survey on features, architecture, applications, privacy and security requirements has been performed while including the study on the key role of fog nodes in transient storage, decentralized computation and real-time services [6].

Ongoing research efforts, open challenges and several state-of-the-art approaches concerned to security and privacy issue in fog computing have been reviewed in current research scenarios [16]. A detailed literature review and categorization on various state-of-the-art techniques related to integration of fog computing and blockchain technology has been provided with the open challenges of this integration [17]. In addition, these techniques have been categorized based on domain, year of publication, utilized algorithms, blockchain roles and utilization of blockchain in fog computing architecture. A framework for fog/edge computing-based IoT has been suggested while covering the services, enabling technologies, research issues and the proximity of fog/edge devices [18]. A comprehensive literature review on context aware scheduling has been provided in fog computing while presenting detailed classification, context aware parameter and performance metrics analysis, and various open research issues [19].

## 10.4 PERFORMANCE METRICS-BASED FOG COMPUTING TECHNIQUES FOR NEXT-GENERATION CONNECTED TRAFFIC

In this section, various performance metrics-based fog computing techniques for next-generation connected traffic are qualitatively reviewed based on taxonomy shown in Figure 10.2. The techniques proposed by researchers have been categorized based on three

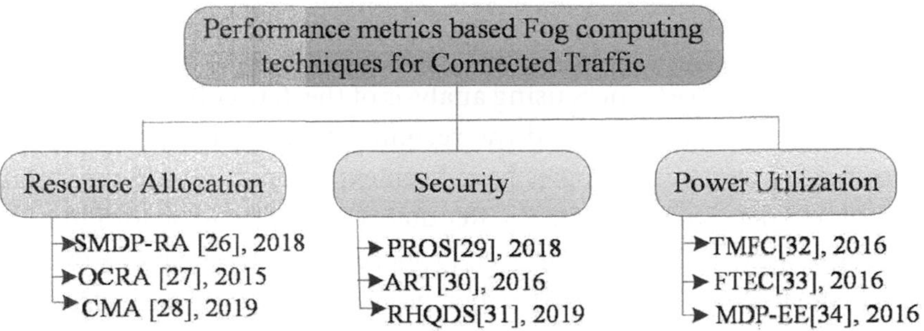


FIGURE 10.2 Fog computing research taxonomy for connected traffic.

performance metrics of next-generation connected traffic. These three performance metrics include resource allocation, security and power utilization.

### 10.4.1 Resource Allocation Techniques

A Semi Markov Decision Process model for Resource Allocation (SMDP-RA) [20] has been proposed in vehicular fog computing system to find the optimal technique for resource allocation. Vehicular fog banks consist of vehicles from different manufacturers thus different vehicles depending on their manufacturers are equipped with different computational capabilities. Since the individual vehicles have limited computational ability, so in order to perform the complex tasks the task is broken down and distributed in the fog among the vehicles. This notion gives rise to the idea of resource allocation. An Optimal Computation Resource Allocation (OCRA) [21] technique has been suggested while utilizing a Semi Markov Decision Process to allocate either Vehicular Cloud (VC) or Remote Cloud (RC) to handle vehicle's service requests with the objective to maximize the long-term expected total reward. Iteration algorithm is considered to develop the optimal scheme that describes which action has to be taken under which state. Another two-stage approach for optimizing resource allocation has been proposed while using a Contract-Matching Approach (CMA) which combines contract theory and matching theory [22]. Contract-based matching theory problem is solved using Karush-Khun-Tucker conditions and the complex task assignment problem is first transformed into a two-sided matching problem with vehicles on one side and user equipment on the other side. Then with the help of an iterative pricing-based stable matching algorithm that carries out 'propose' and 'price-rising' procedures a stable matching is achieved on the dynamically updated preference list.

### 10.4.2 Security Techniques

A Privacy-preserving Route-Sharing (PROS) [23] technique has been proposed in which the user's data is pre-processed by the fog nodes before reaching the service providers then with the help of existing systems and security models a PROS scheme is introduced which improves the anonymous authentication, limits the participation number and guarantee fairness by leveraging the rate-limiting pseudonyms, improves upon the privacy-preserving quality test which helps in forming of groups and protects the privacy of the groups and finally utilizes geo-indistinguishability for requesting the route service and at the same time protecting location privacy. An attack-resistant trust management scheme (ART) [24] has been proposed that is able to detect and cope with malicious nodes in the fog bank and also detect the data trustworthiness using analysis of the data collected and node trustworthiness using the functional trust and the recommendation trust. The benefits of blockchain and smart contract technologies have been exploited to secure vehicular fog banks from unauthorized data sharing [25]. The Reputation-based scheme for High Quality Data Sharing among vehicles (RHQDS) has been proposed, with the help of a three-weight subjective model considering interaction frequency, event timeliness and trajectory similarity for precise reputation management among vehicles.

### 10.4.3 Power Utilization Techniques

A Flow and Time-based Energy Consumption (FTEC) model that has been conducted also focuses on the comparison of energy consumption between the nano data centres (nDCs) of the fog systems and the centralized data centres (DCs) of the cloud computing systems [26]. The results also proved that there is a significant drop in energy consumption by nDCs due to their proximity to the end user. In the comparative study conducted between fog computing and traditional cloud computing by theoretically modelling a fog computing (TMFC) architecture Sarkar and Misra [27] concluded that the average energy consumption of fog computing architecture is 40.48% less than the conventional cloud computing model.

Though the above studies demonstrate fog computing as a greener paradigm, they do not take into consideration factors such as mobility and scalability of nodes in the fog bank which makes the energy consumption management an important area of research. A Markov-Decision-Process (MDP)-based Energy Efficiency (MDP-EE) [28] technique has been designed in fog environment while considering optimal energy charging and transferring problem as a MDP. This technique enables decision making of whether to pay or receive energy from fixed gateway or not and when this gateway comes in contact with the user it decides whether to transfer energy to the user or not but this scheme to has a threshold limit. Still a good amount of work needs to be done in developing new energy efficient protocols and models for VFBs.

Summary of related literatures on resource allocation, security and power utilization techniques for vehicular fog banks is described in Table 10.1, which comprises characteristics/protocol, issues, contributions, techniques, simulation tools utilized, metrics and research limitations of each protocol reviewed.

## 10.5 OPEN RESEARCH CHALLENGES

The advent of vehicular fog banks brings with itself the benefit of using vehicles which otherwise would have been left underutilized, but it brings with itself several issues. Though researchers have addressed several key issues, challenges that still persist are listed below:

- *High mobility support*
  Continuously moving vehicles pose a challenge of establishing a secure and a reliable connection among the vehicular nodes due to constantly changing location of the vehicular nodes. It is necessary to develop the related protocols and the models that

TABLE 10.1 Summary of Existing Techniques for Vehicular Fog Banks

| | Characteristics | | | | |
|---|---|---|---|---|---|
| **Protocols** | **Issues** | **Contributions** | **Techniques** | **Metrics** | **Limitations** |
| SMDP-RA | Not Efficient resource allocation due to heterogeneity | Optimal resource allocation in VFBs having heterogeneous devices | Semi Markov Decision Process | Expected reward | Heterogeneity of service request not considered |
| OCRA | Not efficient utilization of the resources | Maximize total long-term expected reward | Infinite horizon Semi-Markov Decision Process (SMDP) | Expected reward | Limited number of vehicles tested |
| CMA | Inefficient Incentives and task matching | More stable task matching collection | Contract- Matching Approach (CMA) | Contract efficiency | Security is not taken into consideration |
| PROS | Maintaining privacy of the users in the group | Privacy preserved route planning | Privacy-Preserving-Route-Sharing (PROS) | User & Group privacy, Fairness | Limited to vehicular nodes |
| ART | Issue of trustworthiness of nodes and the data shared by them | Enhanced trustworthiness of data & nodes | Attack-Resistant-Trust management scheme (ART) | Precision & Recall | Does not consider data sparsity |
| RHQDS | Issue of trust in fog banks due to their heterogeneity | Authorized and reputation-based data sharing | Consortium Blockchain & Smart contract for secure data storage and sharing | Reputation value & Detection rate | Limited only to secure data sharing |
| TMFC | No theoretical model of fog computing to investigate present till now | Powerful comparison between fog & cloud computing | Mathematical formulation of each component | Service latency and energy consumption | Limited only to comparison between existing components |
| FTEC | Limited investigation of energy consumption by Nano servers | System design & applications that can minimize power consumptions by nDCs | Flow & Time-based energy consumption model | Traffic & Power consumption | Study limited only to power consumption by nDCs |
| MDP-EE | Issue of Efficient energy utilization | Optimal energy management for the mobile energy gateway | Markov Decision Process (MDP) | Maximum energy capacity & Energy transfer rate | Architecture has a threshold limit. |

can efficiently manage the incoming and the outgoing nodes for enabling vehicular fog banks. Both qualitative and quantitative analysis of the available data needs to be done in order to develop more accurate modelling that can be implemented for the changing environments. Systems can also be trained using Machine learning models using the available data sets on the flow of traffic to predict and manage the dynamic fog banks beforehand.

The high mobility support-centric research theme will benefit from handoff management research and development for mobile ad-hoc networks in fog or cloud enabled Internet of Things (IoT) environment. It is due to the similar networking scenario in terms of mobility support while any mobile IoT devices can join and leave to the nearby vehicular fog networks. Similar to vehicular fog banks, IoT-centric fog banks are also newer area of research where there are no established models for implementing high mobility-centric network handoff management support. However, some existing models for supporting mobility in IoT-centric fog banks can be utilized in vehicular fog banks with specific changes considering higher speed and linear movement of vehicles on roads.

- *Hybrid nature of fog banks*
  Nodes in the fog banks are not limited to vehicular nodes but also consist of RSUs, which are connected to the centralized systems. So, the fog banks consist of a multitude of devices making it a heterogeneous system. Every tier in the fog bank consists of a different set of devices operating in different capacities. Moreover, the devices from different manufacturers have different computation and storage capabilities making it more complex even to operate within the same tier. This heterogeneous nature and interoperability system makes the vehicular fog bank a highly hybrid system. Certain standards have to be introduced that can standardize the coordination among different nodes and at different levels that can securely and speedily transfer and process the data in such a geographically distributed system.

  In this hybrid vehicular fog bank research theme, optimal deployment of RSUs, and reliable vehicle-to-infrastructure for enabling RSUs connection the nearby vehicular fog banks. This direction of research has the potential in enabling more vehicles as fog nodes to the fog bank considering not only parked vehicles but also on-the road vehicles. It will also simplify access to the vehicular fog services as consumers and as fog nodes towards enlarging the impact and usefulness of vehicular fog banks in day-to-day computing activities. Vehicle to infrastructure communication research will benefit hybrid nature of fog bank. Various new technologies and protocols have been developed in the last few years for enabling infrastructure communication from on-road vehicles. However, these protocols require updating considering fog bank requirements in terms of reliability and durable communication scenarios in fog bank architecture.

- *Fog bank security and privacy*
  As mentioned above fog banks are highly mobile and hybrid making them an easy target of malicious actors. It becomes even more difficult to locate the threat actors in these heterogeneous environments. Newer encryption and decryption algorithms need to be developed that can work at such a low-latency and can handle real-time data. Proper authentication and authorization mechanisms also need to be developed

for any node joining or leaving the vehicular fog bank. Every security standard developed must be developed keeping in mind the elasticity of the fog bank.

The traditional fog- or cloud-centric security and privacy techniques cannot be directly applied in vehicular fog bank architecture considering the constraints vehicular networking environment in terms of higher mobility and disruptive connectivity to internet. Therefore, cutting-edge light weight security techniques are more applicable in the vehicular fog bank environment. However, the light weight security solutions needs to be verified and validated for different vehicular fog bank scenarios for security performance testing before applying them in real environment. Bench marking of these new age security techniques need to carry out with respect to the cryptography-based traditional security.

- *Efficient energy consumption*

  Due to its proximity to the end users, nodes in the fog banks are energy efficient compared to cloud computing paradigm [27]. But there is a battery constraint when it comes to RSUs or parked vehicles–centric fog bank nodes considering their faster battery drainage and limited battery capacity. So, in order to maximize the service duration of the vehicular fog bank operation, proper task allocation on the basis of energy consumption needs to be done by a service management component in the vehicular fog bank framework.

  Energy efficiency research theme in vehicular fog bank research will benefit from the green computing and green communication–centric research in next-generation wireless systems in IoT environment. Green computing is also a very traditional research theme in cloud or edge computing environment. However, energy efficiency in traditional cloud or edge is quite different than in vehicular fog bank environment considering the power resource availability in cloud environment and limited battery power availability in vehicular fog bank environment. The green computing techniques of cloud or edge environment need to revisit for implementing in vehicular fog bank scenario. IoT-centric green computing and communication techniques are more suitable for vehicular fog bank environment considering limited battery capacity similarity of tiny IoT devices with RSUs and vehicles' battery capacity. However, these green computing and green communication techniques also need to be revisited considering mobility scenario in vehicular environment which is not so usual for sensor-enabled IoT environment.

## 10.6 CONCLUSIONS

Usage of vehicular fog banks for connected traffic are evolving and expanding with the clock of time. The researchers have not fully explored the various research issues such as resource allocation, security and power utilization for connected vehicular traffic instead of rapidly increased usage of vehicular fog banks in several areas in last few years. In this chapter, a comprehensive taxonomy-based structure and a qualitative study of associated publications on performance metrics-based fog computing techniques for next-generation connected vehicular environment have been described. Various performance metrics and approaches employed related to fog computing for next-generation connected traffic have been analysed. Future research directions have also been highlighted. This chapter will enhance the overall understanding of vehicular fog banks research directions in all relevant environments and encourage further researchers to take up and address the issues in vehicular fog bank scenarios.

## REFERENCES


1. R. K. Naha, et al., "Fog computing: Survey of trends, architectures, requirements, and research directions," *IEEE Access*, vol. 6, pp. 47980–48009, 2018.
2. I. Martinez, A. S. Hafid, and A. Jarray, "Design, resource management, and evaluation of fog computing systems: A survey," *IEEE Internet of Things Journal*, vol. 8, no. 4, pp. 2494–2516, 2021.
3. K. Tange, M. De Donno, X. Fafoutis, and N. Dragoni, "A systematic survey of industrial internet of things security: Requirements and fog computing opportunities," *IEEE Communications Surveys & Tutorials*, vol. 22, no. 4, pp. 2489–2520, 2020.
4. C. Mouradian, D. Naboulsi, S. Yangui, R. H. Glitho, M. J. Morrow, and P. A. Polakos, "A comprehensive survey on fog computing: State-of-the-art and research challenges," *IEEE Communications Surveys & Tutorials*, vol. 20, no. 1, pp. 416–464, 2018.
5. M. Mukherjee, L. Shu, and D. Wang, "Survey of fog computing: Fundamental, network applications, and research challenges," *IEEE Communications Surveys & Tutorials*, vol. 20, no. 3, pp. 1826–1857, 2018.
6. J. Ni, K. Zhang, X. Lin, and X. Shen, "Securing fog computing for internet of things applications: Challenges and solutions," *IEEE Communications Surveys & Tutorials*, vol. 20, no. 1, pp. 601–628, 2018.
7. H. Wu, R. Fujimoto, and G. Riley, "Analytical models for information propagation in vehicle-to-vehicle networks," in *Proceedings of 2004 IEEE 60th Vehicular Technology Conference (VTC2004-Fall)*, Los Angeles, CA, vol. 6, pp. 4548–4552, 2004.
8. T. S. Abraham and K. Narayanan, "Cooperative communication for vehicular networks," in *Proceedings of 2014 ICACCCT*, Ramanathapuram, India, May 8–10, 2014, pp. 1163–1167.
9. X. Hou, et al., "Vehicular fog computing: A viewpoint of vehicles as the infrastructures," *IEEE Transactions on Vehicular Technology*, vol. 65, no. 6, pp. 3860–3873, 2016.
10. C. Morency and M. Trépanier, *"Characterizing Parking Spaces Using Travel Survey Data,"* CIRRELT, Montreal, QC, Canada, 2008.
11. T. Litman, *"Parking Management: Strategies, Evaluation and Planning,"* Victoria Transport Policy Inst., Victoria, BC, Canada, 2006.
12. N. Liu, M. Liu, W. Lou, G. Chen, and J. Cao, "PVA in VANETs: Stopped cars are not silent," in *Proceedings of 30th IEEE INFOCOM, Mini-Conference*, Shanghai, China, April, 2011, pp. 431–435.
13. M. Shiraz, A. Gani, R. H. Khokhar, and R. Buyya, "A review on distributed application processing frameworks in smart mobile devices for mobile cloud computing," *IEEE Communications Surveys and Tutorials*, vol. 15, no. 3, pp. 1294–1313, 2013.
14. F. Malandrino, C. Casetti, C.-F. Chiasserini, C. Sommer, and F. Dressler, "The role of parked cars in content downloading for vehicular networks," *IEEE Transactions on Vehicular Technology*, vol. 63, no. 9, pp. 4606–4617, 2014.
15. C. Zhu, G. Pastor, Y. Xiao, and A. Ylajaaski, "Vehicular fog computing for video crowdsourcing: Applications, feasibility, and challenges," *IEEE Communications Magazine*, vol. 56, no. 10, pp. 58−–63, 2018.
16. M. Mukherjee, et al., "Security and privacy in fog computing: Challenges," *IEEE Access*, vol. 5, pp. 19293–19304, 2017.

17. H. Baniata and A. Kertesz, "A survey on blockchain-fog integration approaches," *IEEE Access*, vol. 8, pp. 102657–102668, 2020.
18. B. Omoniwa, R. Hussain, M. A. Javed, S. H. Bouk, and S. A. Malik, "Fog/Edge Computing-based IoT (FECIoT): Architecture, applications, and research issues," *IEEE Internet of Things Journal*, vol. 6, no. 3, pp. 4118–4149, 2019.
19. M. S. U. Islam, A. Kumar, and Y. C. Hu, "Context-aware scheduling in fog computing: A survey, taxonomy, challenges and future directions," *Journal of Network and Computer Applications*, vol. 180, 2021, doi: 10.1016/j.jnca.2021.103008.
20. C. C. Lin, D. J. Deng, and C. C. Yao, "Resource allocation in vehicular cloud computing systems with heterogeneous vehicles and roadside units," *IEEE Internet Things Journal*, vol. 5, no. 5, pp. 3692–3700, 2018.
21. K. Zheng, H. Meng, P. Chatzimisios, L. Lei, and X. Shen, "An SMDP- based resource allocation in vehicular cloud computing systems," *IEEE Transactions on Industrial Electronics*, vol. 62, no. 12, pp. 7920–7928, 2015.
22. Z. Zhou, P. Liu, et al., "Computation resource allocation and task assignment optimization in vehicular fog computing: A contract-matching approach," *IEEE Transactions on Vehicular Technology*, vol. 68, pp. 3113–3125, 2019.
23. M. Li, L. Zhu, Z. Zhang, X. Du, and M. Guizani, "PROS: A privacy-preserving route-sharing service via vehicular fog computing," *IEEE Access*, vol. 6, pp. 66188–66197, 2018, doi: 10.1109/ACCESS.2018.2878792.
24. W. Li and H. Song, "ART: An attack-resistant trust management scheme for securing vehicular ad hoc networks," *IEEE Transactions on Intelligent Transport System*, vol. 17, no. 4, pp. 960–969, 2016.
25. J. Kang, et al., "Blockchain for secure and efficient data sharing in vehicular edge computing and networks," *IEEE Internet of Things Journal*, vol. 6, no. 3, pp. 4660–4670, 2019, doi: 10.1109/JIOT.2018.2875542.
26. F. Jalali, K. Hinton, R. Ayre, T. Alpcan, and R. S. Tucker, "Fog computing may help to save energy in cloud computing," *IEEE Journal on Selected Areas in Communications*, vol. 34, no. 5, pp. 1728–1739, 2016, doi: 10.1109/JSAC.2016.2545559.
27. S. Sarkar and S. Misra, "Theoretical modeling of fog computing: A green computing paradigm to support IoT applications", *IET Networks*, vol. 5, no. 2, pp. 23–29, 2016.
28. Y. Zhang, D. Niyato, P. Wang, and D. I. Kim, "Optimal energy management policy of mobile energy gateway," *IEEE Transactions on Vehicular Technology*, vol. 65, no. 5, pp. 3685–3699, 2016, doi: 10.1109/TVT.2015.2445833.

CHAPTER 11

# Edge Intelligence Empowered Resource Optimization in IoV

Kai Jiang and Yue Cao

*Wuhan University*

Huan Zhou

*China Three Gorges University*

CONTENTS



## LIST OF ABBREVIATIONS

| | |
|---|---|
| **AI** | Artificial intelligence |
| **DDQN** | Double deep Q-Network |
| **DNN** | Deep neural networks |
| **DNS** | Domain name system |
| **EC** | Edge computing |
| **EI** | Edge intelligence |
| **IoV** | Internet of vehicles |
| **MBS** | Macro base station |
| **NAT** | Network address translation |
| **NFV** | Network function virtualization |
| **QoS** | Quality-of-service |

DOI: 10.1201/b22998-11

| | |
|---|---|
| **RL** | Reinforcement learning |
| **RSU** | Roadside units |
| **SDN** | Software-defined network |
| **URLLC** | Ultra-reliable and low latency communication |
| **V2I** | Vehicle-to-infrastructure |
| **V2V** | Vehicle-to-vehicle |
| **V2X** | Vehicle-to-everything |
| **VANETs** | Vehicular Ad-hoc networks |
| **VNFs** | Virtual network functions |

## 11.1 INTRODUCTION

With the progress of wireless multi-access technology and the popularity of vehicles on the roads, the traditional technology-driven transportation system is evolving into a more powerful data-driven intelligent era [1,2]. As a fundamental paradigm in the Fifth Generation (5G) networks, the Internet of Vehicles (IoV) has emerged, which tries to enhance the existing information exchange capabilities by the interconnection between vehicles and all related entities. The IoV has enabled a wide variety of reliable vehicular services and improved passengers' driving safety and comfort [3].

Nonetheless, these flourishing vehicular services generally have severe Quality-of-Service (QoS) requirements involving high-reliability, delay-sensitive, bandwidth-hungry, and compute-intensive applications, which makes cloud-based processing architectures infeasible, because long transmission distance and limited channel bandwidth may lead to extra traffic load, access latency, and unreliability in IoV [4]. Indeed, the accompanying challenges driven by the transmission distance in cloud-based processing have expedited the development of Edge Computing (EC) in IoV. Specifically, EC is emerged as a critical component for future IoV by its physical proximity to the information-generation vehicles [5]. Instead of entirely relying on the cloud, it can sink the cloud's processing capabilities toward the network edges via intermediate Roadside Units (RSU), and provide real-time response while enabling more intelligent services with high performance. Such patterns not only accommodate the diverse requirements of vehicular services but also promise several benefits, including reduced delay and resource consumption, lower uncertainty, privacy protection, more accessibility, and context awareness [6,7].

### 11.1.1 Motivation

As the capabilities of the EC servers are relatively limited, optimizing resource allocation on demand is generally an essential factor for achieving the benefit of vehicular services. By the uncertain and highly dynamic property in IoV, vehicular networks enable particular constraints and requirements on adaptive resource optimization methods [8]. Conventional model-based methods for resource optimization may not be very appealing. Admittedly, such methods always end up with the nonlinear mathematical programming formulation that is undoubtedly NP-hard, while the feasible set of the problem is non-convex. Based on this, model-free resource optimization methods are urgently required to deploy and execute vehicular applications in IoV [4,9].

Fortunately, the development of Artificial Intelligence (AI) has experienced a fall, a leap, and again a fall, until spectacular leaps were made in the past decade. Driven by breakthroughs in hardware upgrading and a series of neural networks, AI can expand technological innovations (such as empowering machine learning, automation, and simulation capabilities) with its superiority for data analysis and extracting insights, especially for dynamic IoV [10]. Driving by the trend, pushing the AI frontiers to the vehicular network edge under this context has given rise to an emerging interdisciplinary, namely, Edge Intelligence (EI). Notably, EI is widely believed to be the next stage of the development of EC. It is not the simple integration of EC and AI, but the complementation and mutual benefits [11]. Instead of entirely relying on the cloud, EI can sink the cloud's processing capabilities at the edge and provide real-time response while enabling more intelligent vehicular services with high performance. Based on this, EI is considered a promising solution to process massive edge data and realize adaptive resource optimization in IoV [10]. However, it is undeniable that although EI has attracted widespread academic concerns recently, the successful implementation of EI empowered resource optimization in IoV is still in its infancy. Therefore, a comprehensive review in this young field from a broader perspective is urgently needed.

### 11.1.2 Structural Organization

The overall structure is organized as follows. Section 11.2 describes the background, including the concept of IoV and EI. Section 11.3 introduces the EI empowered IoV architecture, based on which we make a holistic overview of its related key enablers and specific optimization themes. Section 11.4 elaborates the processes of model training and inference in IoV, together with its particular embodiment of resource optimization. Furthermore, we strive to articulate open challenges and promising research directions in Section 11.5. The conclusion is summarized in Section 11.6.

## 11.2 BACKGROUND

This section describes the background, including the concept of IoV and EI.

### 11.2.1 IoV

The past commercialization process indicates that Vehicular Ad-hoc Networks (VANETs) have deviated from the expected commercial value for a long time since their practical application has been limited by the unstable network quality, user equipment incompatibility, and scanty data processing capabilities [12,13]. Based on this context, conventional VANETs need to accelerate the pace of technological innovation, and IoV has emerged as the evolution of VANETs.

As a fundamental paradigm of 5G networks, IoV enhances the existing vehicular communication capabilities by integrating with the Internet of Things. With involving vehicles as carriers, IoV can realize information interaction with all related entities based on the Vehicle-to-Everything (V2X) technology, where V2X is the umbrella term of different communication modes, including Vehicle-to-Vehicle (V2V), Vehicle-to-Infrastructure (V2I), vehicle-to-pedestrian, and vehicle-to-network [13]. IoV prompts diverse applications with extremely tight service requirements to safeguard traffic safety, relieve traffic congestion, and support vehicular multimedia services. Besides, IoV can also facilitate traffic management, road infrastructure repair, logistics, transportation, etc. Undoubtedly, IoV enables safer, efficient, convenient travel and delivery systems that fundamentally change our daily lives. Nevertheless, IoV generally poses challenges by dynamic network topology, small effective network diameters, fast time-varying channels, and frequent network disconnection. Therefore, an essential open question is "how to support future IoV services via unified air interfaces while satisfying the diverse performance requirements" [9,14].

### 11.2.2 EI

EI, which is given rise by combining the advantages of both EC and AI, is considered an eye-catching emerging interdisciplinary and has gained wide attention lately. More generally, EI can be understood as AI-empowered EC. It is not the simple integration of EC and AI, but the complementation and mutual benefits [11]. On the one hand, AI provides EC with technologies and methods, and EC can unleash its potential and scalability with AI. On the other hand, EC prospers AI with more abundant application scenarios and practices, and AI can expand its applicability with sufficient training resources from network edges [15].

Essentially, EI can enable edge nodes to perform model training and inference locally, frequently avoiding communication with the cloud platform. The emergence of EI is highly nontrivial to system efficiency, service response, scheduling optimization, and privacy protection in IoV. From a technical perspective, the AI models can extract insights from practical edge environments in IoV and endow edge with high-quality asymptotical solutions iteratively [10]. In the last decade, the methods represented by deep learning and Reinforcement Learning (RL) have gradually become the most popular AI techniques in EI. Deep learning can extract features and detect edge anomalies from the data automatically. In contrast, RL, including multi-agent RL and deep RL (the combination of Deep learning and RL), which means realizing objectives via multiple steps and suitable decisions, plays a crucial growing role in the real-time decision-making of edge networks [14,16].

## 11.3 KEY ENABLERS AND OPTIMIZATION THEMES

### 11.3.1 Key Enablers

#### *11.3.1.1 Network Slicing*

Flouring vehicular services always pose heterogeneous network requirements. Network Slicing (NS) is introduced as a cost-effective way to deal with the "one-size-fits-all" network phenomenon. It separates the physical network into multiple segments, allowing self-contained logical network instances (i.e., network slices) with various network characteristics created above the common shared physical infrastructure [17]. As a virtualized and independent end-to-end network, each slice can be optimized and specialized in satisfying demands by particular applications, which further supports customized network functionality, hierarchical abstraction, and resource isolation.

Generally, to create and customize resource combinations for vehicular services, deploying a network-level resource allocation framework with proper V2I access control is necessary. Thus, multiple EC slices with different QoS requirements can be deployed at the appropriate locations to improve the overall resource utilization in IoV. The innate heterogeneity of IoV makes NS indispensable for Ultra-Reliable and Low Latency Communication (URLLC), for which it can support services running across a radio access network [18]. In this way, connections and services could be provided with flexible and dedicated resource allocation and excellent performance guarantees. In other words, not only can the latency be reduced substantially, but also traffic prioritization of vehicular service subscribers is supported by applying NS in IoV.

#### *11.3.1.2 Software-Defined Network*

Software-Defined Network (SDN) is a promising solution to simplify data forwarding and optimize resource allocation, which decouples the control layer's function from the data layer and migrates it to a centralized controller. Recently, SDN is always integrated with EC to propel logic centralized control more reliable and effective in IoV. It draws much attention for enabling the flexibility, scalability, and programmability to resolve IoV bottlenecks, involving QoS guarantee, service customization, and scalability issues inherited from the characteristics of VANETs [5].

Specifically, deploying more RSUs to cope with the ever-increasing vehicles generally leads to extensive costs in IoV. SDN is therefore envisioned as a critical enabler since it offers potential ways for global network configuration and optimization, adaptive resource allocation with cost-effectiveness, and the integration of heterogeneous elements in IoV. Generally, the centralized SDN controller is deployed at underlying edge servers connecting RSUs. It will collect global network information, including traffic load, vehicle density, location and mobility, service types, local resource allocation decisions, and so on. Leveraging the information, the controller can make network-level configurations for resource optimization and access control while deploying adaptive routing protocols for service flows [19]. As shown in Figure 11.1, through the southbound application-program interface, the controller can make network-level optimization for resource slicing and access control with the information gathered from the physical network, which satisfies demands from vehicles and improves the overall resource utilization efficiency. Nevertheless, since the controller needs global control views, applying SDN without expensive signaling overhead in highly dynamic network topologies is a problem worth considering.

#### *11.3.1.3 Network Function Virtualization*

Network Function Virtualization (NFV) is supplementary to SDN technology to decouple network functions, such as load balancing, Domain Name System (DNS), Network Address Translation (NAT), firewall, and video transcoding, from underlying hardware servers. The purpose of NFV is to program network functions as software instances by virtualization technologies [20]. This virtualization makes it possible to transfer network functions from standard general-purpose hardware units to commodity computing platforms, potentially providing the same services as traditional mobile networks.

Generally, the NFV is operated on edge servers connecting RSUs to achieve computation-oriented service customization and provisioning, of which the programmable software instances are often referred to as Virtual Network Functions (VNFs) at edge servers.

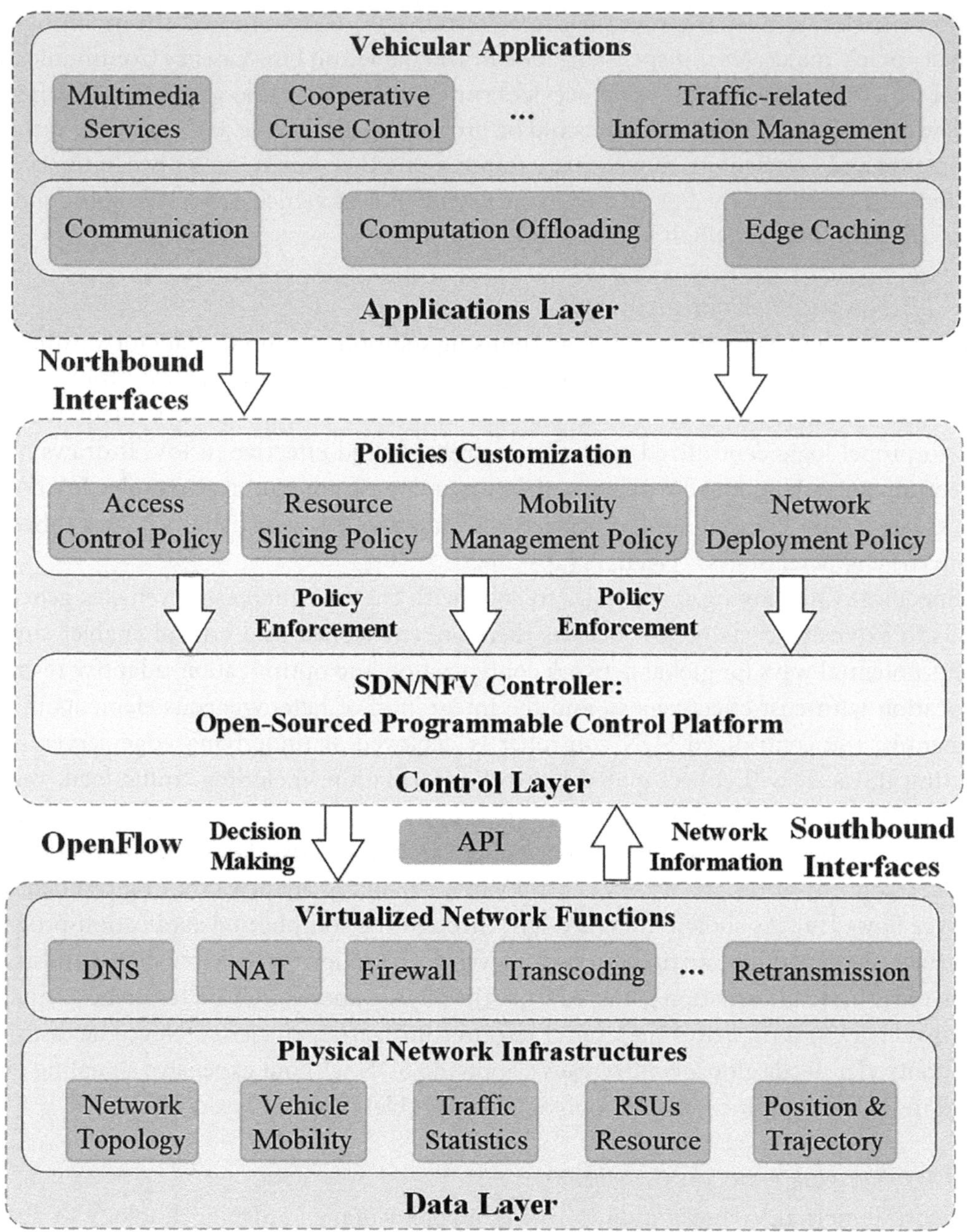


FIGURE 11.1 Heterogeneous SDN/NFV architecture supporting IoV.

By enabling NFV, vehicular services with diverse requirements can be supported, and the EC operators' capital investment and operating costs get decreased significantly in IoV [12]. Indeed, SDN and NFV are complementary and mutually beneficial. As shown in Figure 11.1, under different IoV scenarios, SDN and NFV integrated architecture can facilitate flexible and automated traffic routing management, network-level resource slicing, and efficient resource allocation via proper channel access control from vehicles.

### 11.3.2 Specific Optimization Themes

The emerging EI has been widely considered in addressing resource optimization in IoV based on three themes: computation offloading, resource allocation, and edge caching.

#### *11.3.2.1 Computation Offloading*

Current vehicles usually suffer from the limitation of computation resources, which may be an inevitable bottleneck to process various computation-intensive vehicular services locally in IoV. Therefore, it is feasible to empower vehicles with enough resources offered by transferring computation-intensive tasks to external platforms like cloud and EC servers, known as task offloading. One of the most prominent and widely discussed features of EC is computation offloading, which can extend cloud services to the edges of networks. By computation offloading, the workload of vehicles is enabled to be processed in proximity by leveraging RSUs. Such a pattern meets the expansion requirements of the computation capabilities of vehicles and improves the QoS of diverse vehicular services with lower transmission consumption. Notably, there usually exist two offloading models at the edge. One is entirely offloading, of which the whole process or task is migrated to the EC server. Another is called partial offloading, in this scenario, the task or process is divided into different sub-tasks, and some of them may migrate to various EC servers for collaborative execution [21]. Besides, by involving vehicles as carriers, more resources can be utilized by V2V communication. The computation tasks can either be offloaded to EC servers or be delivered to cooperative vehicles in IoV. Thus, various aspects need to be considered to improve overall network efficiency, such as whether the task needs to be offloaded or processed locally? Where and how to offload the task?

#### *11.3.2.2 Resource Allocation*

Except for computation offloading, adaptively allocating resources is also an essential factor for facilitating the performance of IoV. Resource allocation is the technique of scheduling the resources to find the optimal association that enhances the utilization of the available resources with satisfying QoS requirements, of which its objective is to minimize the cost consumption in task processing. During the process of resource allocation, various resources in IoV are coordinated based on the dynamic system conditions, service types, vehicle density, and heterogeneity of vehicular services. Meanwhile, the pre-defined resource allocation strategies determine the execution sequence of computation tasks. Various resources in IoV are supported to be jointly optimized to reduce the task execution delay and resource consumption [4]. In addition, for multi-RSU scenarios, the problem of resource competition and channel interference between vehicles requires special consideration. Most of the previous studies seldom consider the hidden dynamics of IoV when allocating resources for task processing. To be more precise, they only focus on the performance in quasi-static systems. Thus, an exciting research field in IoV is to design an adaptive resource allocation method while processing the computation task.

#### *11.3.2.3 Edge Caching*

The booming vehicular applications require vehicles to access vast amounts of Internet data (e.g., live traffic information download and navigation map update) while satisfying the QoS requirement, which makes cloud-based processing architectures infeasible, because long transmission distance and limited channel bandwidth may contribute to significant traffic load and access latency in IoV. Fortunately, large-scale data analysis shows that different contents often require different priorities [22]. Only a few popular contents account for the majority of downloads, while the access demand for most of the rest is relatively small. This request pattern promotes the implementation of edge caching in IoV. Expressly, edge caching has provided an alternative to reduce duplicate content transmissions inside the future IoV, which caches frequently used contents near vehicles by pushing cloud functions to intermediate RSUs. Hence, edge caching enables vehicles to access the popular content directly from the caching-enabled RSUs instead of repeatedly downloading it from remote cloud servers. In this way, the redundant traffic and transmission resource consumption at both backhaul and core networks can be significantly alleviated, and the QoS is improved as well. Meanwhile, V2V communication will also enable the storage unit of vehicles to share content according to the traffic load and contact probability [23].

## 11.4 EI EMPOWERED RESOURCE OPTIMIZATION IN IoV

EI empowered resource optimization has been emerged with tremendous opportunities in IoV. Nowadays, its key value propositions have been exemplified in many scenarios. In this part, the processes of model training and inference in IoV will be elaborated, together with its particular embodiment of resource optimization.

### 11.4.1 Model Training in IoV

In general, EI empowered resource optimization relies on the efficient model training and inference along the edge-cloud continuum in IoV [15]. Now, three modes are classified based on the deployment location of model training, as shown in Figure 11.2.

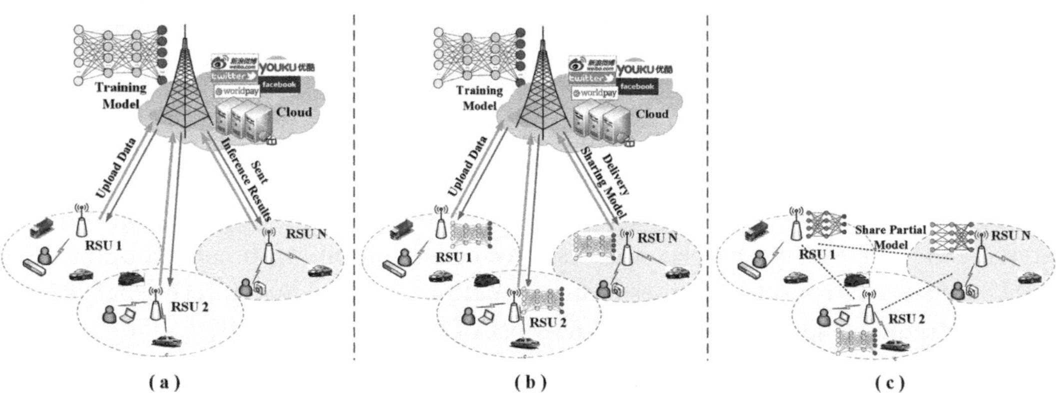


FIGURE 11.2 The processes of model training and inference in IoV. (a) Centralized training, Centralized inference; (b) Centralized training, Decentralized inference; (c) Decentralized training, Decentralized inference.

- *Centralized mode:* This trained model is deployed on the cloud platform, from which the data pre-processing, model training, and message brokering are performed by the cloud. Specifically, the training model is implemented through cloud-edge collaboration, and its performance heavily depends on the quality of network connections. During the training phase, RSUs collect the data within the coverage and upload it to the cloud in real time, of which these data are generated by vehicles, involving vehicular services, sensors, wireless channels, and traffic flow information. After analyzing and storing the data, the model is continually trained in centralized training clusters via data aggregated. Notably, although the centralized mode promises the potential for searching the optimal global solution of the system, the corresponding training complexity often grows exponentially with global network states. Meanwhile, as the deployed model on cloud platforms is spatially far from vehicles, the uploaded data has to pass through multiple networks in IoV. Data delivery and unpredictable network connections may cause prohibitive transmission delays and overhead. In addition, this mode is also vulnerable due to vehicle data resources' high concentricity, which explicitly leads to the leakage and loss of sensitive traffic data.
- *Distributed mode:* Today's vehicular services always have intense requirements on URLLC, which may make centralized mode no longer feasible for the developments of IoV. As an independent node, each RSU should have its own inferences based on the corresponding local states, which could be significantly different among different RSUs. Meanwhile, IoV expects to benefit from EI while keeping privacy. Thus, the model can be trained in a distributed manner to avoid these dilemmas. There is no centralized node (cloud) in the distributed mode, and all the edge nodes (RSUs) perform equal roles [10]. Notably, all the RSUs have a neural network mark, implying that each of them can train the model independently with local data, which further preserves private information locally. However, the training process on an RSU is easy to perform over-fitting due to the correlation among the restricted data source, while the inferences of different RSUs are usually mutually influential in IoV. Thus, to obtain the global training model by sharing local training improvement, multiple RSUs have to undertake model training or data analysis synergistically, and the training set is generated by themselves. In this way, the global model can be trained without the cloud platform's intervention. Furthermore, federated learning is applied to the sensitive data area under the distributed training mode [24]. Unlike the conventional distributed training that focuses on consuming data at the edge, federated learning focuses more on privacy protection.
- *Hybrid mode:* Generally, considering the latency and consumption, it is unrealistic to train a comprehensive model on an RSU. Thus, these modes should have the potential for compatibility. To better facilitate the advantages of multi edges coordination, hybrid mode, which combines the centralized and distributed modes, is considered to adopt in IoV. Note that the hybrid mode is not restricted to practical deployment and is possible to break the bottleneck of a single training mode on performance improvement. In this mode, the RSUs may train the model by either decentralized

updates with each other or centralized training with the cloud platform. Specifically, each RSU trains partial parameters and aggregates them to a central node for the global model upgrade. The private data is only gathered in various RSUs, resulting in privacy preservation, weaker than the decentralized mode but more robust than the centralized mode.

### 11.4.2 Model Inference in IoV

The efficient model inference is equally critical to the resource optimization in IoV. Model inference happens after model training, which means implementing the trained model. As the trained model category described above, the model inference can be executed either on the cloud or on RSUs. There are also three inference modes correspondingly, as shown in Table 11.1. Specifically, the typical inference modes also include centralized and distributed. In the former, the model training and inference are both finished on the cloud. The inference results will be sent to each RSU separately. While in the latter, each RSU can perform its model inference locally. Notably, the cloud can either maintain one training model for a centralized inference of all RSUs in the centralized training mode or send the trained sharing model to the RSU for distributed inferences. Technically, supervised learning, unsupervised learning, and single-agent RL are commonly used for centralized inference, while multi-agent RL is the standard method for distributed inference.

### 11.4.3 Case Study

Now, we strive to discuss a particular embodiment of resource optimization in IoV. As shown in Figure 11.3, an EI-supported IoV architecture with a Macro Base Station (MBS), $N$ RSUs, and $I$ vehicles is considered, of which each RSU is equipped with an EC server, and the MBS is connected to the remote cloud naturally. Here, each RSU in a cluster form can utilize its finite computation and cache resource to enable the task processing and content access requirements at the edge, such that vehicles can request popular contents or offload workloads frequently within the coverage range of RSU. Besides, the system is assumed to operate in a fixed length of time slots $t \in \{0,1,2,\ldots,\tau\}$, and is valid for both V2I and V2V communication, where each vehicle with processing capacities can also be assumed as an edge node. Therefore, under the cloud-edge collaboration, the requesting vehicle can concurrently offload its tasks and download desired contents with the connected RSUs, vehicles, or remote cloud.

We intend to explore the optimization of computation offloading and edge caching, where computing, caching, and resource allocation decisions are considered jointly in this dynamic system. Due to the hidden dynamics in practical IoV, the system conditions, the offloading and caching states, the resource capabilities, and vehicular mobility intensity

TABLE 11.1 Model Training and Inference Modes in IoV

| Model Training | Model Inference | Cloud (Central Node) | RSU |
|---|---|---|---|
| Centralized | Centralized | Training + Inference | N/A |
| Centralized (sharing model) | Distributed | Training | Inference |
| Distributed | Distributed | N/A | Training + Inference |

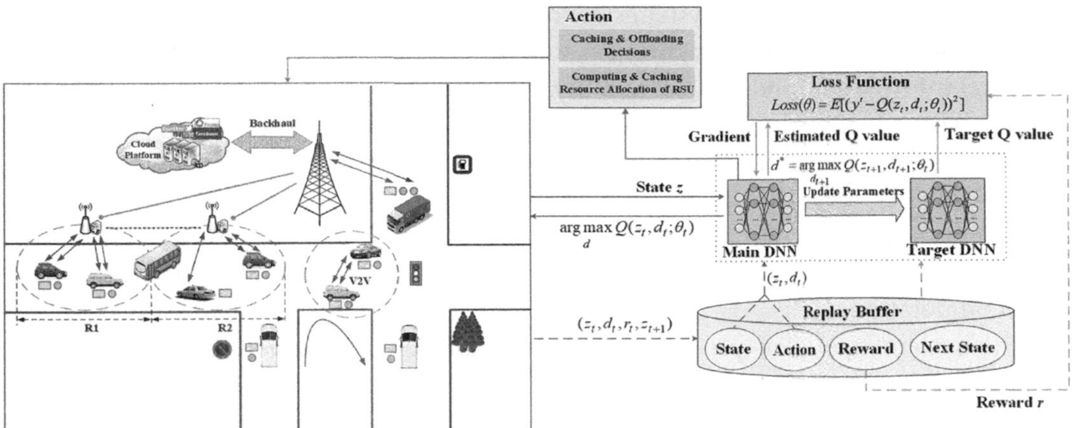


FIGURE 11.3 A particular embodiment of resource optimization in IoV.

are generally uncertain. Then, our objective is to determine the subset of the nearby edge nodes (RSUs, vehicles) and their resources to respond to the requesting vehicle under the dynamic variants and the enormous action-state space. As a branch of AI, RL has exhibited its particular potential for the method design in IoV. In RL, the agent can well capture the hidden dynamics of IoV and imitate the best features, thus learning a series of intelligence behaviors. Thus, we exploit an RL-based method, named Double Deep Q-Network (DDQN), to tackle the corresponding problem, which efficiently learns asymptotically optimal decisions through repeated interactions.

Next, we will provide a detailed description of our algorithm. DDQN is proposed as a promising solution to avoid the curse of dimensionality by utilizing Deep Neural Networks (DNNs). Its basic idea is to adopt different action functions, thereby separating actions' selection and evaluation. Unlike the run-of-mill tabular method that represents and updates state-action combinations and related Q-values in a constructed dimension table, DDQN continually approximates the optimal Q value depending on the update of DNN parameter $\boldsymbol{\theta}$. Meanwhile, we exploit DDQN with an experience replay mechanism, for which the agent will store its transition experience tuples (including current state, executed action, reward, and next state) into the experience replay buffer at each time slot. Generally, experience replay can break the temporal correlation among data training and enhance sample efficiency by using a few previous experiences for current training and update. Besides, to disrupt the pertinence and also ensure the approximation efficiency, dueling exists by maintaining two neural networks with the same structure but different parameters. The temporal difference target Q value is calculated by the target neural network, while the main neural network can acquire the estimated Q-value. During the training process, the stored experience tuples are randomly sampled to train the parameters of these two neural networks. Therefore, for each time slot, the inputs of the main network are the current state from the IoV environment and the training experience tuples from the replay buffer. In contrast, the outputs are the strategies that MBS, RSUs, and vehicles will adopt. Notably, the weight parameters $\dot{\theta}_j$ of target DNN are updated periodically by the counterpart $\theta_j$ of the main DNN via $\dot{\theta} = \zeta\theta + (1-\zeta)\dot{\theta}$ with $\zeta \ll 1$.

In this DDQN model, the system state consists of the available RSU and vehicles, available caches, caching and computing resources at RSUs and vehicles, the transmission channel information, and vehicle mobility. Accordingly, for system action, the agent must decide whether and where the requested content can be cached or the computation task should be offloaded, and how many coded packets should be cached. Hence, the system action consists of the existing caching and computing resources at RSUs and vehicles for the requesting vehicle. Finally, as the value of the reward is negatively correlated to the value of the objective function, we aim to minimize the cost of communication, storage, and computation.

The system operation process is exhibited above. First, each RSU will traverse its resource state and collect the local traffic flow information to obtain the state statistic. Then, the centralized system controller aggregates local states and dispatches them to the main network and experience replay buffer. Based on the acquired state and replay tuples, the agent attempts to learn its own policy through the main network and target network. Specifically, the offloading, caching decision, and resource allocation of the system need to be pre-trained with random policy for a relatively long period. Then, the main neural network and target neural network update their weight parameters $\theta_j$ and $\dot{\theta}_j$ based on the current Q value and samples from the replay buffer, respectively. After that, the trained networks are loaded. The fixed Q-target is used to generate the target Q value $\dot{y}$ as $\dot{y}_j = r_j + \varepsilon Q\left(z_{j+1}, \arg\max_{d_{j+1}} Q\left(z_{j+1}, d_{j+1}; \theta_j\right); \dot{\theta}_j\right)$, and transmitted it to the main network. After receiving $\dot{y}$, the main network performs gradient guiding updates to minimize the loss function, while a variant of stochastic gradient descent is operated until the Q value converges toward the optimal value.

In addition, the BS, RSUs, and vehicles determine and execute the best policies of computing offloading, resource allocation, and content caching derived from the trained DDQN. Henceforth, with the state transition probability, the environment will grant a corresponding immediate reward to the agent in the next state. Technically, the reward value of an agent should satisfy all constraint conditions of the joint optimization to ensure the validity of the results. If the system utility of the current policy is greater than the existing maximal system utility, the reward will be updated, and the environment will update its state based on the current policy. Conversely, punitive negative will be incorporated into the reward if constraints are violated in the proposed problem. For the long-term consideration, there exists a cumulative discounted utility in the system. When the cumulative discounted utility converges, the optimal policy of computation offloading, content caching, and resource allocation are successfully trained. Finally, we also need to be aware that only when the IoV topology undergoes drastic changes, the training in DDQN should be re-executed based on the network dynamics.

## 11.5 OPEN CHALLENGES AND FUTURE DIRECTIONS

Although IoV has paved a path toward intelligent transportation and brought tremendous benefits to daily lives, some challenges still need to be resolved. This section strives to articulate open challenges and promising research directions.

### 11.5.1 Dynamic and Open System

With the increasing density of road traffic, a significant area of concern is the dynamic conditions of networks and the openness of wireless channels. On the one hand, the service environment in IoV is always highly volatile. Vehicles submit new service requests, frequent handovers between edge servers, and vehicle mobility both influence the network topology dynamics interconnections across different edge nodes. IoV requires real-time judgments and decisions to promptly respond to the system dynamics and openness, which poses strict requirements on the training time and cost of the AI model. On the other hand, wireless channels' data rate and access duration will constantly change with the transmission distance between vehicles and RSUs. Meanwhile, interference and fading in wireless channels inevitably reduce communication quality, which retards the training model's learning accuracy and convergence speed. Therefore, EI-supported IoV architecture needs to consider the robustness in various uncertain conditions.

### 11.5.2 Network Architecture Support

The steady growth of vehicles on the roads will drive the vigorous development of vehicular services in IoV. Most of these services are resource-intensive and delay-sensitive, which require massive resources to operate high-dimensional system parameter configuration (e.g., excuse proper model training and inference). Therefore, given the practical dynamic property in vehicular wireless channels, contention-based access conditions, and the limited coverage area of RSUs, an advanced network architecture is desired to make the model training and inference more resource-friendly. Furthermore, network architecture support can also relieve the complexity of collaborative resource optimization and ensure URLLC between vehicles and RSUs via efficient resource management and task scheduling of edge nodes. However, considering the diverse requirements of vehicular services, there still exist hassles to break the barriers among various communication technologies and realize the network architecture controllability.

### 11.5.3 Lightweight Training Models

AI models with powerful data processing ability are required for edge services optimization in IoV, of which they are deployed on RSUs and enable a series of advanced network functions via multi-layer neural networks. Since a single technique is sometimes restricted to performing the model inference, integrating multiple AI techniques is desirable for the over-fitting or under-fitting problem in model training [25]. This integration further complicates the network structures. Therefore, the inevitable challenge to be overcome is how to support AI models in resource-constrained edge environments. Along with a different line, we must construct resource-efficient lightweight AI models with finite resources. Compression techniques such as knowledge distillation, quantization, and weight pruning are feasible to promote lightweight AI models in IoV. However, although these techniques can resize AI models at the edge, it is always accompanied by a loss of precision. Indeed, these static compression techniques usually fail to tailor the dynamic hardware configuration and loads, while the emotional compression technique can be expected to apply with complicated conditions in the future. Finally, advanced technologies such as lightweight virtualization and serverless computing should also be explored to enable efficient model deployment over resource-constrained edge environments in IoV.

### 11.5.4 Collaboration and Incentive Mechanism

With the explosive growth of computation tasks, the resource bottleneck problem of the EC server becomes increasingly prominent. Collaboration between the cloud and edges is desirable to realize efficient task scheduling and service configuration in IoV. It turns out that this synergy combines the advantages of EC and cloud computing, such that seamless and smooth services can be accessed across heterogeneous collaborative platforms anywhere and anytime. Also, to fully exploit the dispersive resource, dividing a computation-intensive task into small sub-tasks and partially offloading them to different edge nodes for collaborative execution is a research hotspot worth expecting in the future [26]. Besides, as for the cooperative distributed system, another issue that requires attention is the incentive mechanism. As the edge ecosystem is a grand open consortium that involves different service providers and vehicles, the operation of services often requires collaboration and integration across service providers in IoV. Thus, stimulating cooperation among multiple entities by applying appropriate incentive mechanisms should be explored. Finally, a challenge for the multi-agent game in the distributed system is to balance fairness and optimization. Incentive mechanisms can also relieve that.

### 11.5.5 Security and Privacy

Vehicles need to collect many traffic data through data transmission among vehicles and RSUs, of which the majority of them are processed at the edge of the network. Vehicle mobility may interrupt the data transmission process and further affect the system stability, thereby blocking the stable supply of channel access and introducing uncertain security threats to the distributed system. This openness imposes intense requirements on distributed trust such that the services provided by scattered RSUs are trustworthy [27]. More seriously, the uploaded data are privacy-sensitive and vulnerable during the model training. Hackers' attacks on nodes tampering and communication channels will lead to severe privacy invasion. Thus, data desensitization and privacy protections are crucial to ensuring user authentication and access control, data integrity, and mutual platform verification in IoV. Despite many advanced techniques, such as federated learning, have been proposed for privacy-friendly distributed training, model parameters or original data sets can still be inferred and reconstructed [24]. Existing feasible solutions include adding noise to protect the original data, communication authentication, homomorphic encryption, and differential privacy. Besides, considering the coexistence of trusted edge nodes with malicious ones, reliable routing distribution protocols for service delivery with predicting vehicles' movement is also a potential area to explore.

## 11.6 CONCLUSION

This chapter aims to provide a study with relative concepts and prospects of this young field from a broader perspective. We first describe the background, including the concept of IoV and EI. Then, we introduce the EI-supported IoV architecture, based on which we make a holistic overview of its related key enablers and specific optimization themes. Subsequently, the processes of model training and inference in IoV are elaborated, together with its particular embodiment of resource optimization. Finally, we strive to articulate open challenges and promising research directions, which may facilitate the transformation of EI empowered resource optimization in IoV from theory to practice.

## REFERENCES


1. Z. Ning, P. Dong, X. Wang, J. P. C. Rodrigues, and F. Xia (2019). Deep reinforcement learning for vehicular edge computing: An intelligent offloading system. *ACM Trans. Intell. Syst. Technol.*, 10(6), 1–24.
2. X. Chen, L. Jiao, W. Li, and X. Fu (2016). Efficient multi-user computation offloading for mobile-edge cloud computing. *IEEE/ACM Trans. Netw.*, 24(5), 2795–2808.
3. Y. Chao, Y. Liu, X. Chen, and S. L. Xie (2019). Efficient mobility-aware task offloading for vehicular edge computing networks. *IEEE Access*, 4(7), 26652–26664.
4. H. Zhou, K. Jiang, X. Liu, X. Li, and V. C. M. Leung (2022). Deep reinforcement learning for energy-efficient computation offloading in mobile-edge computing. *IEEE Internet Things J.*, 9(2), 1517–1530.
5. K. Zhang, S. Leng, Y. He, S. Maharjan, and Y. Zhang (2018). Mobile edge computing and networking for green and low-latency internet of things. *IEEE Commun. Mag.*, 56(5), 39–45.
6. L. Huang, S. Bi, and Y. J. A. Zhang (2020). Deep reinforcement learning for online computation offloading in wireless powered mobile-edge computing networks. *IEEE Trans. Mobile Comput.*, 19(11), 2581–2593.
7. N. Abbas, Y. Zhang, and A. Taherkordi (2018). Mobile edge computing: A survey. *IEEE Internet Things J.*, 5(1), 450–465.
8. T. He, N. Zhao, and H. Yin (2018). Integrated networking, caching, and computing for connected vehicles: A deep reinforcement learning approach. *IEEE Trans. Veh. Technol.*, 67(1), 44–55.
9. H. Zhou, N. Cheng, J. Wang, J. Chen, Q. Yu, and X. Shen (2019). Toward dynamic link utilization for efficient vehicular edge content distribution. *IEEE Trans. Veh. Technol.*, 68(9), 8301–8313.
10. Z. Zhou, X. Chen, E. Li, L. Zeng, K. Luo, and J. Zhang (2019). Edge intelligence: Paving the last mile of artificial intelligence with edge computing. *Proc IEEE*, 107(8), 1738–1762.
11. S. Deng, H. Zhao, W. Fang, J. Yin, S. Dustdar, and A. Y. Zomaya (2020). Edge intelligence: The confluence of edge computing and artificial intelligence. *IEEE Internet Things J.*, 7(8), 7457–7469.
12. J. A. Guerrero-ibanez, S. Zeadally, and J. Contreras-Castillo (2016). Integration challenges of intelligent transportation systems with connected vehicle, cloud computing, and internet of things technologies. *IEEE Wireless Commun.*, 22(6), 122–128.
13. J. Contreras-Castillo, S. Zeadally, and J. A. Guerrero-Ibanez (2018). Internet of vehicles: Architecture, protocols, and security. *IEEE Internet Things J.*, 5(5), 3701–3709.
14. C. Wang, M. Renzo, S. Stanczak, S. Wang, and E. G. Larsson (2019). Artificial intelligence enabled wireless networking for 5G and beyond: Recent advances and future challenges. *IEEE Wireless Commun.*, 27(1), 16–23.
15. K. Jiang, C. Sun, H. Zhou, X. Li, M. Dong, and V. C. M. Leung (2021). Intelligence-empowered mobile edge computing: Framework, issues, implementation, and outlook. *IEEE Netw.*, 35(3), 132–138.
16. N. C. Luong, D. T. Hoang, S. Gong, D. Niyato, and P. Wang (2019). Applications of deep reinforcement learning in communications and networking: A survey. *IEEE Commun. Surv. Tutor.*, 21(4), 3133–3174.

17. E. Wang, D. Li, B. Dong, H. Zhou, and M. Zhu (2020). Flat and hierarchical system deployment for edge computing systems. *Future Gener. Comput. Syst.*, 105(2), 308–317.
18. Z. Mlika and S. Cherkaoui (2021). Network slicing with MEC and deep reinforcement learning for the internet of vehicles. *IEEE Netw.*, 35(3), 132–138.
19. S. Wang, X. Zhang, Y. Zhang, L. Wang, J. Yang, and W. Wang (2017). A survey on mobile edge networks: Convergence of computing, caching and communications. *IEEE Access*, 5(7), 6757–6779.
20. G. Premsankar, M. D. Francesco, and T. Taleb (2018). Edge computing for the internet of things: A case study. *IEEE Internet Things J.*, 5(2), 1275–1284.
21. Z. Ning, P. Dong, X. Kong, and F. Xia (2019). A cooperative partial computation offloading scheme for mobile edge computing enabled Internet of Things. *IEEE Internet Things J.*, 6(3), 4804–4814.
22. X. Wang, C. Min, T. Taleb, A. Ksentini, and V. C. M. Leung (2014). Cache in the air: Exploiting content caching and delivery techniques for 5G systems. *IEEE Commun. Mag.*, 52(2), 131–139.
23. X. Li, X. Wang, K. Li, Z. Han, and V. C. M. Leung (2017). Collaborative multi-tier caching in heterogeneous networks: Modeling, analysis, and design. *IEEE Trans. Wireless Commun.*, 16(10), 6926–6939.
24. X. Wang, C. Wang, X. Li, V. C. M. Leung, and T. Taleb (2020). Federated deep reinforcement learning for internet of things with decentralized cooperative edge caching. *IEEE Internet Things J.*, 7(10), 9441–9455.
25. T. Taleb, et al. (2017). On multi-access edge computing: A survey of the emerging 5G network edge cloud architecture and orchestration. *IEEE Commun. Surv. Tutor.*, 19(3), 1657–1681.
26. X. Chen, W. Li, S. Lu, Z. Zhou, and X. Fu (2018). Efficient resource allocation for on-demand mobile-edge cloud computing. *IEEE Trans. Veh. Technol.*, 67(9), 8769–8780.
27. R. Khan, P. Kumar, D. N. K. Jayakody, and M. Liyanage (2020). A survey on security and privacy of 5G technologies: Potential solutions, recent advancements, and future directions. *IEEE Commun. Surv. Tutor.*, 22(1), 196–248.

CHAPTER 12

# A Trend in Smart Charging, Vehicle-to-Grid and Route Mapping

Aanchal Khatri

*Sat Jinda Kalyana College*

Geetika Aggarwal

*Nottingham Trent University*

Vishal Khatri

*Guru Gobind Singh Indraprastha University*

Harish Chander

*Sat Jinda Kalyana College*

CONTENTS



DOI: 10.1201/b22998-12

## LIST OF ABBREVIATIONS

| | |
|---|---|
| **BEP** | Binary evolutionary programming |
| **CC** | Constant current |
| **CPM** | Charging point manager |
| **DSO** | Distributed system operator |
| **EV** | Electric vehicles |
| **EVCS** | Electric vehicle charging scheduling |
| **GHG** | Greenhouse gases |
| **G2V** | grid to vehicle |
| **MDP** | markov decision process |
| **OCST** | optimal charging starting time |
| **PEV** | Plug-in hybrid electric vehicle |
| **RES** | Renewable energy resource |
| **ToU** | It uses time-of-use |
| **TWh** | terawatt-hours |
| **V2G** | Vehicle-to-grid |
| **VGI** | Vehicle-to-grid integration |

## 12.1 INTRODUCTION

With increase in modernization, the transport sector is burdening the environment and the infrastructure. Some 350 million registered vehicles recorded in India, in which more than 98% depends on combustible fuel like petroleum, diesel and natural gas. As a matter of fact, transport sector solely consumes some 62.3% of oil in the word in 2011, which is the cause of both urban and regional air pollution. The emission of fossil fuel by the transport sector emits some 6.8Gt of CO2 and other greenhouse gases (GHG), which causes air pollution and global warming [1]. The reliance on the fossil fuel for transport and other uses led to various issues: one of the foremost is climate change and air pollution in the metropolitan cities which causes health problems for the peoples but also at the same time affect infrastructure and nature [2]. Further, it burdens the economy, as the fluctuation and price rise in fuel prices led to inflation in the country, which directly put peoples under financial burden. The geopolitical instability of oil-producing countries uses oil as an economic weapon against other rival countries which affects region peace and put supply side constraint. In this situation, electric vehicles (EV) [3] and plug-in hybrid electric vehicle (PHEV) come into play, as they provide sustainable alternatives to combustible engine vehicles. EV can be defined as light vehicles that consume power from the battery with a capacity of at least 4 KWh and which can be charged by plug in from external source. Plug-in electric vehicles (PEV) offer various advantages over conventional vehicles like, reduce consumption of fossil fuel leading to less economic burden on the user and the country fiscal, improve air quality, and can be charged from various primary energy sources. Further, integration of PEV with electric grid which also comprises of renewable energy resource (RES) led to reduction in GHG emission and growth in renewable energy.

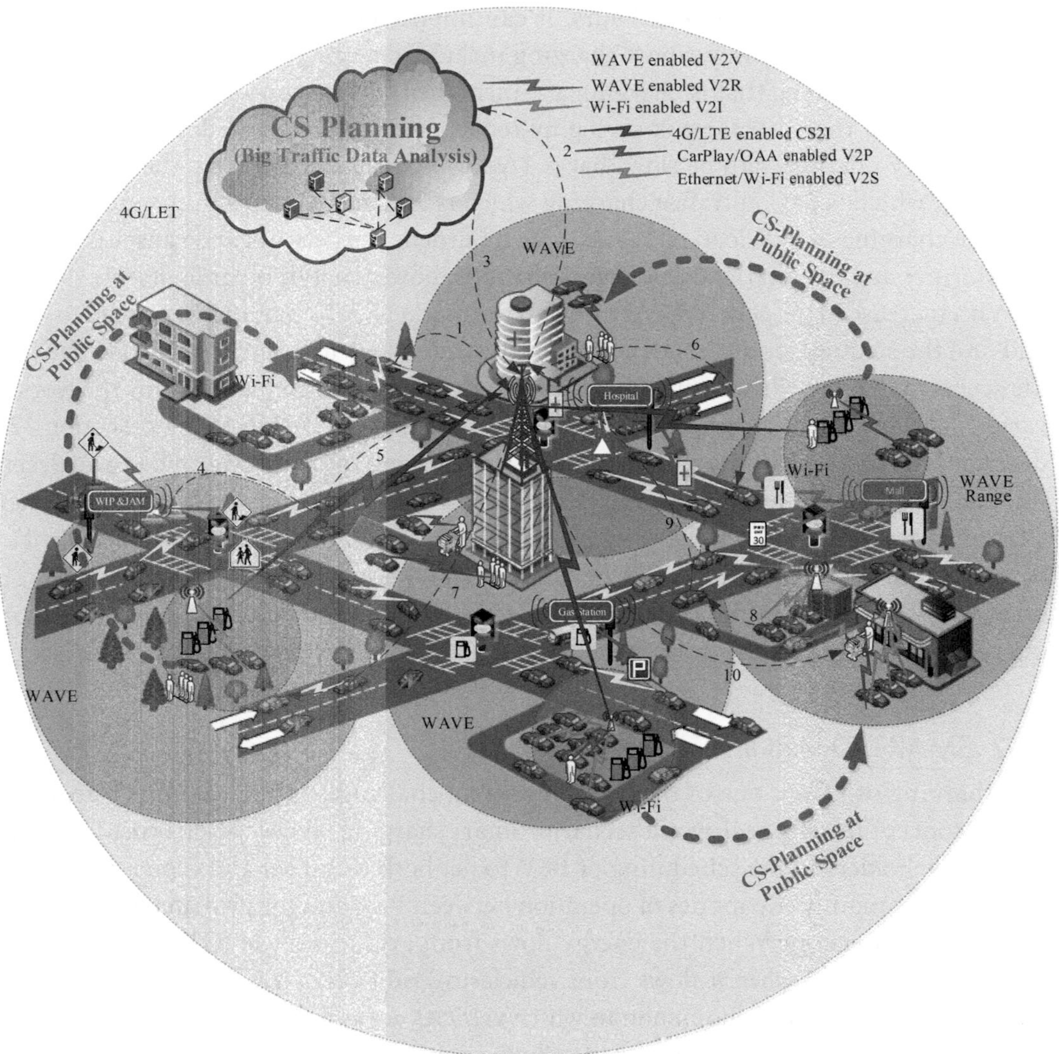


FIGURE 12.1 Electric vehicles charging infrastructure in smart city environment.

Considering all these benefits, the government of various countries are encouraging the use of electric vehicles (see Figure 12.1), with various subsidies and tax benefits like tax deduction, tax exemption, parking fee exemption, etc. After rapid growth in 2020, the electric vehicles stock went up to 10 million mark that is more than 40% up from the previous one. Electric vehicles account for some two-thirds of new car registration in 2020 [4]. PEV is powered by electric battery with a capacity of 5–15 and 25–40 KWh for battery electric vehicles. The charging of PEV can be done in three ways: off-peak charging, smart charging (valley filling), and smart charging (peak filling). To avoid the off-peak charging, it can be delayed during the peak hours as soon as it connects to electric vehicle supply equipment. To help reduce demand in peak hours and better smooth electricity demand, the utility coordinator can impose dynamic price rates which are adjusted according to demand and supply during peak hours. It can impose peak demand charge in times of peak hours. In the same way, smart charging includes charging at various off-peak hours, leading to valley filling that is charging at times when there is low demand for electricity in the grid and the other is peak saving by reducing energy demand during peak hours. Further, PEV charging can be slow that is 1AC, which is standard plug, the other is Fast charging, which is 3AC, ultrafast charging which is 3AC or DC done by an external charger. The charging of PEV load increases, which burdens the network and causes issues like transformer and line overloads, voltage deviations, power supply, other faults, etc.

With increase in the penetration of PEV in the market day by day, there is a need to study and find the solution for the issues it face, for the technology to be implemented in the real-life scenario in more efficient manner. The objective of this chapter is to review the work of various authors in this promising technology which is organized as follows. Section 12.2 describes the smart charging and its different architectures that are centralized and decentralized. Further, the authors' work is divided on the basis of which architecture of smart charging they followed, and accordingly Sections 12.3 and 12.4 are explained. Section 12.5 describes the route mapping and scheduling techniques used by authors for easing out undue pressure on the infrastructure. In the last part, the chapter is concluded by mentioning relevant points and also with the recommendation for future work.

## 12.2 SMART CHARGING

The charging of PEV is one of the most important challenges in respect of its integration of this concept into the real-world scenario. Smart charging allows various customers and other stakeholders in the scheduling of PEV to get both technical and economic benefits [5]. There are mainly two modes of operation between PEV and the grid, in terms of direction of flow of energy. When the energy flows from grid-to-vehicle (G2V), it is called as charging mode, and when it flows from vehicle-to-grid (V2G), it is called as discharging mode. V2G is the new phenomenon in which vehicles act as a supplement of energy to the grid, which acts as storage source or distributed sources [6]. It can be possible as the PEV are stranded 90% of time which will help in providing ancillary services like frequency control, load balancing, etc. Extensive research has proven that vehicle-to-grid integration (VGI) can be optimized by charging the vehicles at off hours and stored in the batteries and later on in the peak utilization hours the vehicles can transfer the electricity to the grid. In a way to control charging, supply equipment for electric vehicles, energy management facility, and price signals are used. There are different measures taken by coordinators to control the demand for electricity at peak hours. Though V2G is a useful technology for the grid with intermittent renewable energy source (RE), there is a fluctuation in RE provided by solar, wind energy, etc. Though it suffers from many shortcomings like premature degradation of batteries and various energy losses during operation [7].

In order to make this technology feasible, PEV management can be done using aggregator. Aggregator function is used to connect and manage various PEVs for the management and integration into electrical devices into the system. The control done by the aggregator will depend on the number of devices connected and the flexibility provided to the users.

The Aggregator will manage its load based on the number of devices connected. Further, it provides the ancillary services, by controlling frequency in the instantaneous increase or decrease of battery usage. Finally, the smart charging can be done using two different architectures: centralized and decentralized defined in the next section.

## 12.3 V2G CENTRALIZED CONTROL ARCHITECTURES

In the centralized control, aggregator acts as the manager for managing directly the charging of all PEV via charging point manager (CPM). Apart from the technical support, it also supports buying and selling the electricity market. Based on the historical data of PEV demand, aggregator will provide the demand forecast to distributed system operator (DSO) for approval as shown in Figure 12.2. DSO will provide approval based on the total requirement of distribution network and according to its capacity.

After receiving the approval, aggregator will purchase the bidding in day-to-day market. The transmission network operator will perform the required changes in the profile of the transmission network according to the needs of aggregator. If everything goes right, the aggregator will provide the activated charging points to all PEVs in their region.

Further aggregator also provides the ancillary services like secondary and tertiary services, when the transmission operator demands. If any uncertain circumstance happens, then aggregator will take corrective actions to return to safe state. Also, aggregator forms a database consisting of electric vehicle identification id (PEV id), Charging point id, the battery charging level, and the user choice of using charging point. In this way, the aggregator will help manage the overall charging and the network of devices. Every PEV is fitted with control unit which controls the mode in which it is working, i.e., either in charging or discharging based on the state of the network.

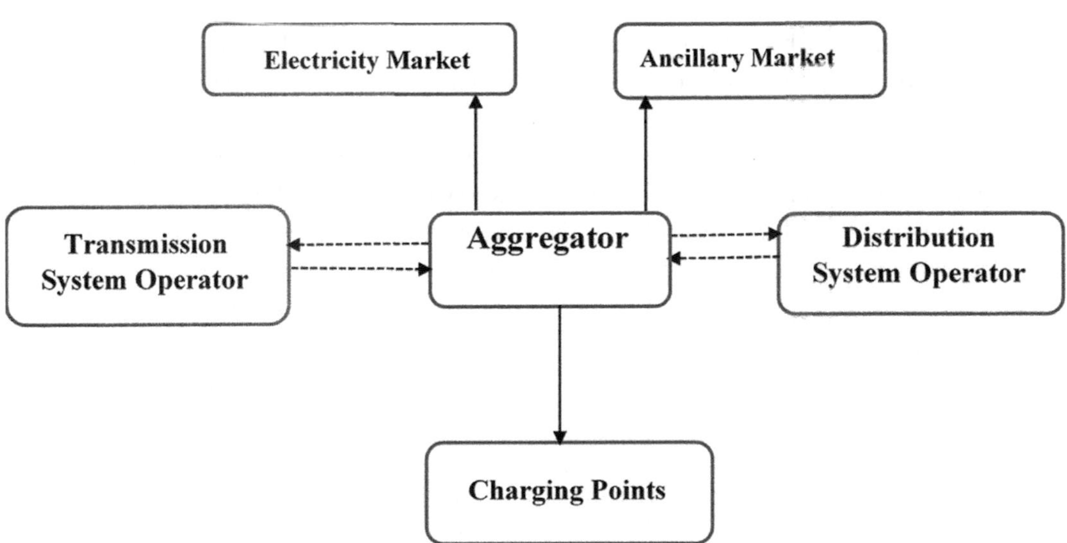


FIGURE 12.2 Vehicle-to-grid centralized control architectures.

Soares et al., in the work [8], defined a methodology for managing the EV charging for the quasi-real-time grid in which aggregator will control the electricity market and distribution system operator (DSO) will look into the technical restriction imposed by the electric components. The method used for the aggregator is to reduce penalty and provide positive reward for how well the forecast of electricity requirement of the PEV is provided. On the other hand, DSO manages the grid by controlling voltage fluctuation and reducing the overload of branches in the grid. In order to assess the performance, data set provided by aggregator is used which is based on the PEV movement and its power requirement from time to time. It uses Markov chain algorithm to simulate the movement of PEV. In this, fluctuating voltage and overloading of specific branches have been solved. In the case of centralized architecture, energy loss minimization can also be another objective considered by many authors in the literature [9,10]. Usman et al. [9] studied that the uncontrolled and random charging can cause reduced performance and on the other hand overload the network resources. In the paper, charging algorithm based on optimal charging starting time (OCST) is suggested in which the charging of EV is managed by its departure time and also schedule the night charging in the time of low network load. It uses time-of-use (ToU) tariff scheme in which tariffs are calculated based on the time of usage. Binary evolutionary programming (BEP) is used for finding the optimal solution for less tariff and OCST. The result has shown the cost-effective tariff with reduced power loss and improved network voltage. In work [11], the author proposed an intelligent charging system based on meta-heuristic function for scheduling PEV loads. The problem is optimized using simulated annealing so that stationary and ramping operation cost is reduced. The result has shown that it reduced the cost by 5%–16%, compared to system with no control. In Ref. [12], the problem related to overload of lines and transformers are analysed using meta-heuristic function. As the PEVs are charging, it analysed the network; if there is no problem, it keeps on charging, but if there is a problem, then it will find out whether it is due to line congestion or voltage deviation. In these circumstances, it stops the charging by some percentage of PEV, adding them to the waiting queue of PEV to be charged. When the congestion in the network comes down, it will allow the waiting PEV to start charging.

From the above, it can be concluded that centralized architecture suffers from many shortcomings, due to its centralized nature. The first one is overall dependence on the management of aggregator. Therefore it is necessary to have a backup system for the management in case of failure. Secondly, as the size of PEV increases, the data to be handled by aggregator increases enormously, which burden it with processing. Finally, it has constrained on the security side. As the aggregator has all the information of the charging points and probable everyday routes.

Considering the above constraints, it is important to move to decentralized architecture.

## 12.4 V2G DECENTRALIZED ARCHITECTURE

In the decentralized approach, the control of how much to charge and where to charge resides with the individual PEV. The aggregator has no role in decision-making. There are different ways to influence the behaviour of PEV like price or control signals. The individual optimizes according to the charging cost.

The PEV participates in the market and its price will change based on the demand and supply constraints. Further, aggregator will set the prices based on the charging profiles of PEVs. In the literature, researcher studied the effect of charging cost on overall charging profiles. In paper [13], Moghaddama presented a model for bidding in which various aggregators in the distribution network participate to buy and sell the energy and other ancillary services. The problem is considered as Nash equilibrium problem. The objective function is calculated based on the cost of energy purchased from the day ahead market and the revenue incurred. Further, the uncertainties are associated with day ahead energy price, supply of the energy and the real-time price. The data is analysed using case study. The work proposed in Ref. [14] used scheduling of PEV for optimal charging. Further for filling the valley, global optimization problem is formulated. PEV updates the charging autonomously according to their own demands and further control signals sent by the distributor. To work this algorithm, each PEV profile is communicated to the utility.

The algorithm is entirely based on cost optimization and lacks in other parameters faced by the distribution network like line congestion and voltage frequency regulations. In abnormal situations, it is necessary to carry out the control technique in which DSO changes the price signal which is called as nodal price strategy. Further, in case of other ancillary services, secondary and tertiary services are difficult to manage as it lacks the centralized control. The primary frequency can be managed using the droop control method.

Individual PEV can control the voltage regulation by either reducing their demand in peak hours or by injecting the energy into the grid using V2G option. In paper [15], the aggregator aggregates the total requirement of PEV in the distribution network. The algorithm for smart charging is designed for charging the price and creating threshold for regulations. Further, it is combined with profit maximization algorithm based on system load and price constraints. The simulation result shows that it has substantially increased the aggregator profit and on the other hand reduced system load.

In Luo et al. [16], motivated from the concept of dealing the internet traffic, the concept of congestion pricing is implemented for the decentralized architecture. The fluctuation in the supply of wind power is balanced by adjusting its charging through V2G and discharging through G2V which is based on real-time virtual market price signal and its need at the time of charging whether at peak hours or not. The mismatch of supply and demand of wind power is reflected in the differential price signal. The charging at peak hours defined by urgency level led to different allocation of compensation duties to different PEV users. Another approach for the efficient architecture based on stochastic model is proposed in paper [17]. The author presented the scheduling of PEV charging, which observes the network load and voltage of various buses. When PEV is plugged into specific bus, the smart agent will schedule the charging based on two stochastic probabilities, access and suspend probabilities. The access probability is for accessing the network and to continue the charging, and the other is to cease charging in the case of error in the network. This mechanism ensured that the PEV will reach optimum level of charging before leaving the station. The strategy has reduced the load to the grid in duration of charging periods and gained high probability of charge completion. Further, compared to other decentralized approaches, this approach is implemented without any exchange of information between grid and PEV which led to reduction in cost.

In the subsequent section, E-mobility based on route mapping and scheduling is discussed.

## 12.5 ROUTE MAPPING

With increase in the intensity of electric charging vehicles, there is undue pressure on the charging infrastructure and the distributed network. To solve this problem the concept of scheduling was evolved. The literature focuses on many solutions for this problem, which includes the shortest path first scheduling, etc. Liu et al. [18] have suggested electric vehicle charging scheduling (EVCS) based on multi-objective optimization problem based on the amount of charging energy, station selected for charging, and availability of charging slot at station. For solving the problem various operators are defined. The result has shown that the algorithm works effectively in real-world scenario. In paper [19], the proposed scheme for the scheduling is based on the charging station selection for the heterogeneous vehicles, which is one with different battery capacity, brands, etc. The work is mainly based on the vehicles making reservations and those which are parked in the area. Further, it gains the knowledge from the arrival time and the time vehicles take to charge at the station. In this way, author proposed a scheme for effective scheduling which is evaluated at Helsinki city area, which has shown smooth charging performance with low communication cost. Zhang et al. in the paper [20] proposed delay optimal charging scheduling based on Markov decision process (MDP). It reduced the waiting time of EV based on the arrival time, charge level, RE stored in the battery and power price in the grid. The policies for charging and no charging are defined as radical and conservative, which are analysed. In a way, the research is moving towards finding various constraints in this novel technology.

In paper [21], Chen et al. has proposed a model for high-efficiency dynamic transit system as shown in Figure 12.3. The proposed model is multi-objective optimization problem which optimizes on two criteria, one is energy-efficient optimal routing and charging infrastructure availability of EV fleet. It is based on finding the best possible route to the destination from the starting point which satisfies the need of all passengers and at the same time reduces the distance and cost of travel. Further, it also reduces the charging and fuel cost which is constrained by the charging infrastructure. The optimization problem is solved using mixed integer quadratically constrained programming.

Yao et al. [22] proposed an optimization problem for the joint charging and routing of EV is the objective of paper which is based on electric vehicle travel time, charging cost, charging time, pricing and the revenue generated from the customer. This problem is proposed as mixed integer programming problem (MIP), defined as NP hard due to coupling between routing and where to charge and for what time. The paper proposed a two-stage linear programming problem which generates result in polynomial time. Another approach for the joint charging and routing problem which is not based on the shortest path problem is proposed in Ref. [23]. For the deterministic and stochastic network, a dynamic programming based en route charging navigation is proposed. A simplified charge control algorithm is defined which defines the charging control decision. Further, state recursion algorithm is defined for finding accurate navigation.

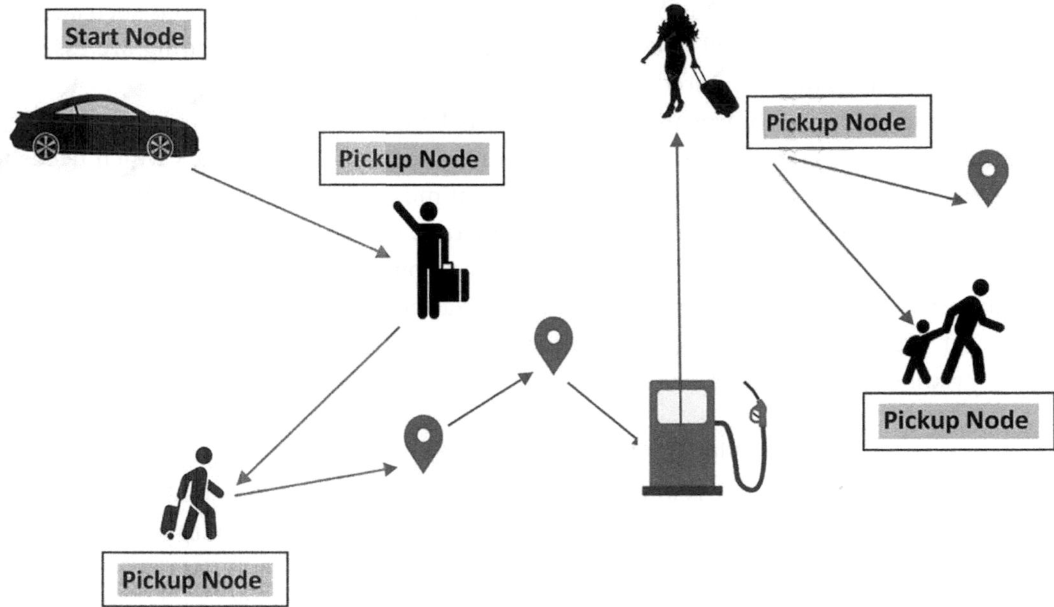


FIGURE 12.3 Routing of dynamic transit system.

The uncertainty of grid power supply led towards RE for the charging in the PEV. The technology is promising, but it still demands the research and innovation. In the paper [24], for the heterogeneous charging facility consisting of different charging capacity, price signal and services type. In this, every PEV can choose to take the route that is charge scheduling problem and the amount of charge or discharge of battery to be done on the chosen path. The problem is formulated as Joint routing and charge scheduling optimization problem, which is NP-hard problem. The problem is solved using approximate algorithm which is implemented on the large size internet of electric vehicles (IoEV). The algorithm defined the optimized route and efficient charging points in the distributed settings which improved the computational complexity and also protected the autonomy of PEV user. Scholars [16,24,25] have worked towards integrating RE source for transmission and providing the effective backup in the time of peak hours.

## 12.6 OPEN RESEARCH ISSUES IN E-MOBILITY

### 12.6.1 Charging Station Recommendation

Charging station recommendation research is one of the potential research themes for E-mobility landscape considering the growing number of EVs in the consumer market leading to higher traffic for charging at EV charging stations [19]. Initially, charging station recommendation was investigated considering travel distance optimization-based travel path recommendation for EVs. However, the growing number of EVs on roads significantly impacts the dynamic traffic environment in terms of both the dense traffic on roads and increased competition at charging stations. Therefore, design and development of charging station recommendation techniques using novel traffic predictions such as modern machine learning–based heuristics techniques including deep learning are essential research direction to be explored in the future E-mobility landscape.

Charging station recommendation research is benefited from the mathematical optimization research considering the multiple number of traffic parameters which needs to be optimized at the same time. The mathematical modelling of dynamic EV traffic environment is essential for implementing mathematical optimization techniques to predict dynamic traffic parameters for precise and better recommendation of charging stations to EV drivers. More specifically, appropriate optimization function derivation is very important and first step towards implementation optimization for E-mobility traffic environment prediction and recommendation.

### 12.6.2 Charging Station Placement Optimization

Charging station placement is another potential research theme in E-mobility landscape due to the geographical significance of charging stations for both charging service provider and EV consumers or drivers [20]. The geographical optimization of charging station depends on the number of traffic parameters, and surrounding traffic environment apart from future traffic growth in the particular area. An appropriate charging station planning can significantly improve EV driver experience in terms of accessibility of the charging station. A well-planned charging station placement considering future traffic growth can significantly increase the economic benefits to charging service providers as the traffic growth could result in future charging revenue.

Charging station placement optimization research is also highly relevant to government city administrators due to the strategic value of the well-planned EV charging services in a particular city. This direction of research is quite related to the electricity grid consideration or availability of electricity resource and connectivity. This could ensure blackout issue can be effectively managed in a particular charging station utilizing different electricity resource connectivity. Considering the multiple optimization parameters involved, charging station optimization is multi-variable optimization problem which should be solved using modern heuristic solutions including deep learning and various versions of genetic algorithms.

### 12.6.3 EVs as Energy Resource

EVs should also be considered as potential energy resource particular with the growing number of EVs in the market [21]. EVs can be used to charge our personal devices while on the move in traffic while in the parking group of EVs energy network can be used to enable as electricity grid kind of service. This is a very potential use case of EVs and has significant economic impact particularly in developing countries where electricity coverage and availability are not always guaranteed.

However, to enable EVs as energy resource, distributed energy network modelling is needed so that faults in one energy point do not impact other energy networks nearby. Wireless charging connectivity research would benefit EV energy resource research as the technical advancement can significantly ease energy connectivity in both the direction from EVs to energy network and energy network to EVs.

### 12.6.4 Cyber Security in EV Energy Network

The growing EV market penetration and the charging networks around the world pushing towards green mobility has significantly increased the cyber security threat on EV energy networks [22]. The cyber security threat of EV charging energy network is quite different from the traditional cyber security challenges considering the catastrophic impact of a malicious security attack on energy network could not only in the EV rather on electricity network. Considering the longer connection duration of EV charging schedules while charging at stations, security-related malicious agents have longer time to activate the attacks on the charging networks. As the charging networks are interconnected, a security risk in individual EV could easily reach any other EVs in the network, as well as it can reach the electricity charging grid network.

With the advancement of home smart charging points for EVs, the cyber security in EV energy network has become more complex, and serious threat as it directly connects EV energy networks to home energy networks. A malicious security agent can reach any home via EV charging points and can control even home appliances online. Towards addressing this security concerns hardware-centric cyber security techniques are potential candidates for enhancing the capabilities of cryptography-centric traditional techniques. This is due to the longer duration of charging point hardware connectivity while charging EVs either at smart home charging points or at charging stations.

## 12.7 CONCLUSION

For making the living more sustainable in the modern society, electric mobility is the promising technology. The PEV incorporation into the electric grid on the one hand provides benefits but on the other hand puts undue pressure on the infrastructure. In this way, the solution lies in intelligent charging control management for which centralized and decentralized architecture have been explored. From the review, it can be concluded that the decentralized architecture works best when the number of vehicles is limited. On the other hand, as the number of vehicles will increase, centralized control architecture works well, as all the information concerning network is available at one central point for managing. Further, for integration of PEV real-life scenario, a lot of work has to done towards managing voltage deviations, overloads, etc. Many authors suggested solutions but lack on the analysis part in the power system simulation in finding the actual results. In our further studies, these issues require more consideration and efforts from researchers.

## REFERENCES

1. International Energy Agency, Key World Energy Statistics, 2021, Available online at: https://www.iea.org/reports/key-world-energy-statistics-2021.
2. UN Environment Programme, Emissions Gap Report, 2021, Available online at https://www.unep.org/resources/emissions-gap-report-2021.
3. International Energy Agency, PEV Definition, Available online at: https://www.iea.org/fuels-and-technologies/electric-vehicles.
4. International Energy Agency, Annual Global Electric vehicle Outlook, 2021. Available online at https://www.iea.org/reports/global-ev-outlook-2021.

5. M. Broussely, "Battery Requirements for HEVs, PHEVs, and EVs: An Overview", *Electric and Hybrid Vehicles*, 2010, Vol. 10(1), pp. 305–345.
6. L. IoanDulăua and D. Bică, "Effects of Electric Vehicles on Power Networks", *Procedia Manufacturing*, 2012, Vol. 46, pp. 370–377.
7. A. Ahmad, Z. A. Khan, M. S. Alam, and S. Khateeb, "A Review of the Electric Vehicle Charging Techniques, Standards, Progression and Evolution of EV Technologies in Germany", *Smart Sciences*, 2018, Vol. 6, pp. 36–53.
8. F. J. Soares, P. M. Rocha Almeida, and J. A. Pec, "Quasi-Real-Time Management of Electric Vehicles Charging", *Electric Power Systems Research*, 2014, Vol. 108, pp. 293–303.
9. M. Usman, W. U. K. Tareen, A. Amin, H. Ali, I. Bari, M. Sajid, M. Seyedmahmoudian, A. Stojcevski, A. Mahmood, and S. Mekhilef, "A Coordinated Charging Scheduling of Electric Vehicles Considering Optimal Charging Time for Network Power Loss Minimization", *Energies*, 2021, Vol. 14, pp. 5336.
10. S. M. Mousavi and D. Flynn, "Controlled Charging of Electric Vehicles to Minimize Energy Losses in Distribution Systems", *IFAC-PapersOnLine*, 2016, Vol. 49(27), pp. 324–329.
11. K. Valentine, W. G. Temple, and K. M. Zhang, "Intelligent Electric Vehicle Charging: Rethinking the Valley-Fill", *Journal of Power Sources*, 2011, Vol. 196, pp. 10717–10726.
12. J. A. P. Lopes, F. J. Soares, and P. M. R. Almeida, "Integration of Electric Vehicles in the Electric Power System", *ProcIEEE*, 2011, Vol. 99, pp. 168–83.
13. S. Z. Moghaddama and T. Akbarib, "Network-Constrained Optimal Bidding Strategy of a Plug-In Electric Vehicle Aggregator: A Stochastic/Robust Game Theoretic Approach", *Energy*, Vol. 151, 2018, pp. 478–489.
14. L. Gan, U. Topcu, and S. Low, "Optimal Decentralized Protocols for Electric Vehicle Charging", *IEEE Transactions on Power Systems*, 2011, Vol. 6(1), 940–951.
15. E. Sortomme and M. A. El-Sharkawi, "Optimal Charging Strategies for Unidirectional Vehicle-to-Grid", *IEEE Transactions on Smart Grid*, 2011, Vol. 2(1).
16. X. Luo, S. Xia, and K. W. Chan, "A Decentralized Charging Control Strategy for Plug-In Electric Vehicles to Mitigate Wind Farm Intermittency and Enhance Frequency Regulation", *Journal of Power Sources*, 2014, Vol. 248, pp. 604–614.
17. W. Zhang, D. Zhang, B. Mu, L. Y. Wang, Y. Bao, J. Jiang, and H. Morais, "Decentralized Electric Vehicle Charging Strategies for Reduced Load Variation and Guaranteed Charge Completion in Regional Distribution Grids", *Energies*, 2017, Vol. 10(2), pp. 147.
18. W.L. Liu, Y. J. Gong, W. N. Chen, Z. Liu, H. Wang, and J. Zhang, "Coordinated Charging Scheduling of Electric Vehicles: A Mixed-Variable Differential Evolution Approach", *IEEE Transaction on Intelligent Transportation Systems*, 2019, Vol. 21(12), pp. 5094–5109.
19. Y. Cao, T. Jiang, O. Kaiwartya, H. Sun, H. Zhou, and R. Wang, "Toward Pre-Empted EV Charging Recommendation through V2V-Based Reservation System", *IEEE Transaction on Systems, Man, Cybernatics: Systems*, 2019, 51(5), 3026–3039.
20. T. Zhang, W. Chen, Z. Han, and Z. Cao, "Charging Scheduling of Electric Vehicles with Local Renewable Energy under Uncertain Electric Vehicle Arrival and Grid Power Price", *IEEE Transactions on Vehicular Technology*, 2013, 63(6), 2600–2612.
21. T. Chen, B. Zhang, H. Pourbabak, and A. Kavousi-Fard, "Optimal Routing and Charging of an Electric Vehicle Fleet for High-Efficiency Dynamic Transit Systems", *IEEE Transactions on Smart Grid*, 2016, 9(4), 3563–3572.
22. C. Yao, S. Chen, and Z. Yang, "Joint Routing and Charging Problem of Multiple Electric Vehicles: A Fast Optimization Algorithm", *IEEE Transactions on Intelligent Transportation Systems*, 2021, 21(1), 1–10.
23. C. Liu, M. Zhou, J. Wu, and C. Long, "Electric Vehicles En-Route Charging Navigation Systems: Joint Charging and Routing Optimization", *IEEE Transactions On Control Systems Technology*, 2017, 27(2), 906–914.

24. X. Tang, S. Bi, and Y. A. Zhang, "Distributed Routing and Charging Scheduling Optimization for Internet of Electric Vehicles", *IEEE Internet of Things*, 2018, 6(1), 136–148.
25. L. Yao, Z. Damiran, and W. H. Lim, "Optimal Charging and Discharging Scheduling for Electric Vehicles in a Parking Station with Photovoltaic System and Energy Storage System", *Energies*, 2017, Vol. 10, pp. 550–570.
26. D. Dallinger and M. Wietschel, "Grid Integration of Intermittent Renewable Energy Sources Using Price-Responsive Plug-In Electric Vehicles", *Renewable and Sustainable Energy Reviews*, 2012, Vol. 16, pp. 3370–3382.

CHAPTER 13

# Vehicle-to-Vehicle-Enabled EV Charging Scenario Iteration

Shuohan Liu

*Lancaster University*

Xu Xia

*Telecom Research Institute*

Jixing Cui and Yue Cao

*Wuhan University*

Qiang Ni

*Lancaster University*

CONTENTS



DOI: 10.1201/b22998-13

## LIST OF ABBREVIATIONS

| | |
|---|---|
| **AC-DC** | Alternating current to direct current |
| **BSS** | Battery switching station |
| **CS** | Charging station |
| **DC-DC** | Direct current to direct current |
| **EV** | Electric vehicle |
| **EV-C** | EV as energy consumer |
| **EV-P** | EV as energy provider |
| **G2V** | Grid to vehicle |
| **GC** | Global controller |
| **ICV** | Internal combustion vehicle |
| **KM** | Kuhn-munkres algorithm |
| **MIP** | Mixed integer optimization problem |
| **NP** | Nondeterministic polynomial |
| **PL** | Parking lot |
| **RSU** | Road side unit |
| **V2G** | Vehicle to grid |
| **V2V** | Vehicle to vehicle |
| **V2V-Pair** | V2V charging pair |

## 13.1 INTRODUCTION

Vehicles become an important transportation method in modern society. However, the traditional Internal Combustion Vehicles (ICVs) use non-renewable energy such as fossil oil, which produces harmful gases and causes serious environmental damage. Meanwhile, ICVs request large energy consumption, and their inefficient energy conversion rate causes energy waste. Facing the potential energy shortage in the future, Electric Vehicles (EVs) are introduced and have been widely used. EVs are environmentally friendly. Furthermore, EVs can alleviate the energy shortage and improve energy usage efficiency.

However, compared with the traditional ICVs, EVs require a longer charging time to refill their driving range. EVs refill battery by travelling towards Charging Stations (CSs), Battery Switching Stations (BSSs) or other places with charging facilities for energy replenishment. Therefore, in most previous studies, considering current EV charging technology and market supplement, EVs' energy replenishment is mainly divided into plug-in charging mode at CSs [1] and battery switching mode at BSSs [2]. The plug-in charging mode is widely used in daily areas including private transportation, electric taxi. Meanwhile, the battery switching mode takes the batteries as replaceable components. EVs replace used batteries with fully charged batteries at the BSSs, which are mainly used in public transportation (uniform battery model).

However, the above plug-in charging mode and battery switching mode have certain spatial and temporal limitations. For example, a large amount of EVs drive towards a same CS/BSS will cause charging congestion, which results in the problem of insufficient service providing. Due to the location and management constraints, CSs/BSSs request huge operating costs. Therefore, it is not realistic to solve charging congestion by deploying a great amount of CSs and BSSs. In addition, the peak charging period also restricts the energy replenishment modes above. To solve this, with the improvement of battery technology, a more flexible Vehicle-to-Vehicle (V2V) charging mode [3,4] is introduced. It allows EVs' energy transfer from EVs with energy supplies to EVs with energy requests. The V2V charging mode is beneficial to both charging and energy providing EVs as it provides convenient energy transfer and large revenues in the energy transfer. At the same time, the V2V charging mode can also be used to balance the load of the grid, which reduces the adverse impact of the grid during peak hours.

In Figure 13.1, the structure and problems in the V2V charging mode are listed. In Section 13.2, the scalability of V2V charging mode is discussed, including three different V2V charging communication manners. In Section 13.3, charging service modes are compared. Here cooperative Grid-to-Vehicle (G2V) and V2V charging and pure V2V charging modes are detailed respectively. The charging problems under the EVs' on-the-move mode are discussed in Section 13.4, including the V2V charging Pairs (V2V-Pair) matching and Parking Lots (PLs) selection schemes. The problem under the EVs' parking mode is discussed in Section 13.5. The core factors affecting the V2V charging efficiency are energy

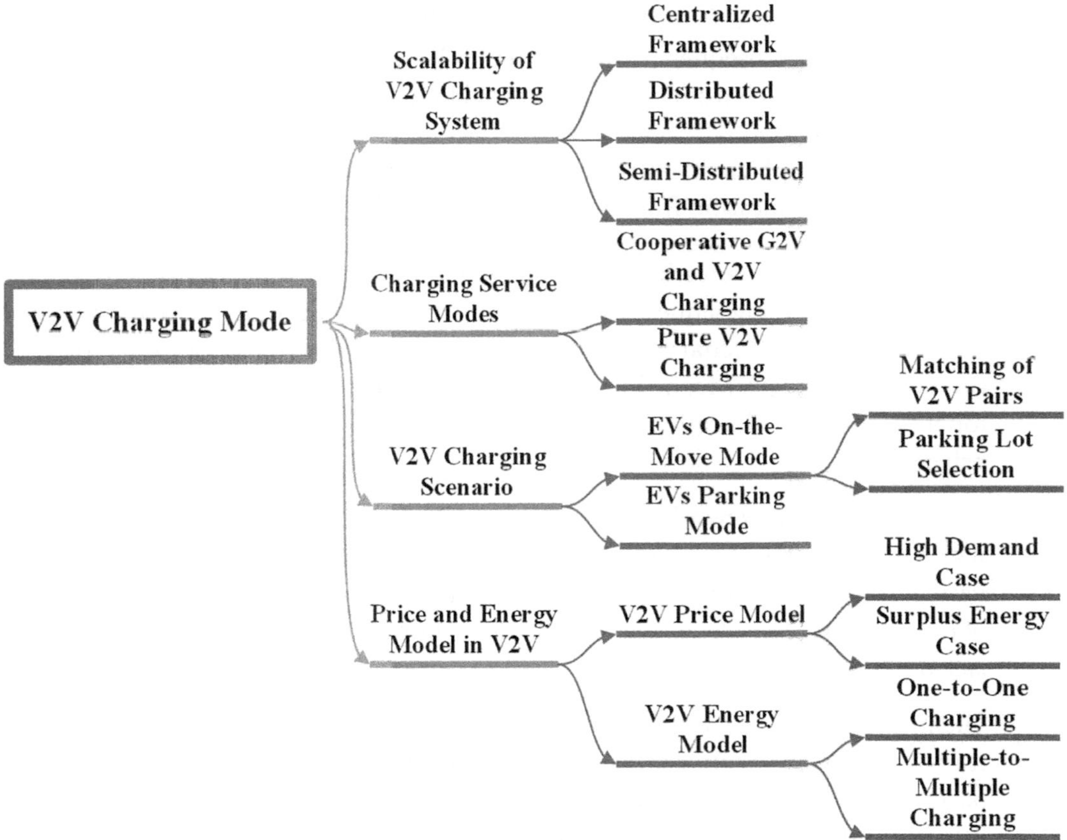


FIGURE 13.1 Overview of the V2V charging mode.

supply and energy price. The V2V charging logics are different under the buyer's market (V2V charging oriented by EVs as energy consumers) and under the seller's market (V2V charging oriented by EVs as energy providers). Therefore, the price models in V2V charging mode are discussed in Section 13.6. Followed by Section 13.7, the energy models in V2V charging mode are discussed. Here, the V2V charging process is discussed into two groups based on one-to-one V2V charging or multiple-to-multiple V2V charging.

## 13.2 SCALABILITY OF V2V CHARGING SYSTEM

In the V2V charging network, information is communicated between vehicles, Global Controller (GC) and Road Side Units (RSUs). Vehicles provide charging communication functions through mobile technologies (including Wi-Fi, Bluetooth, 4G and 5G networks). The information communication includes EV charging information and PLs operating information:

- PL's location and occupation status which includes the number of parked EVs and the available time for V2V converter.
- The expected energy required from EVs as energy Consumers (EV-Cs), the expected energy provided by EVs as energy Providers (EV-Ps).
- EVs' arrival time at a PL and the expected V2V charging time when it reaches the PL. It affects the subsequent PL allocation for EVs. Maximizing the use of limited PL improves the V2V charging efficiency (avoid long V2V charging waiting.)

Thanks to V2V and vehicle-to-infrastructure communications network technology, the V2V charging communication framework is realized and divided into three frameworks:

- The centralized framework relies on a cloud-based GC, which is used to globally control V2V charging requests, such as V2V-Pair matching and V2V charging scheduling. Here, the V2V charging requests contain the status of EVs (location, trip destination and energy). Meanwhile, the GC monitors the local occupation status of PLs for V2V charging allocation (PL-Selection). However, this framework brings many privacy concerns.
- The distributed framework relies on the locally cached information of PLs and EVs without considering global information. Here, EV-Ps and EV-Cs are matched as V2V-Pairs by communication among EVs. This distributed framework protects the privacy of information and reduces the computational burden of the GC. However, due to the lack of global deployment, this framework is more likely to cause charging congestion.
- The hybrid (semi-distributed) framework enhances the robustness of computing by transferring long-distance tasks to the GC. Meanwhile, RSUs near EVs deal with the basic information aggregation and mining tasks. In this framework. EVs can obtain information from RSUs, which reduces communication cost and protect privacy.

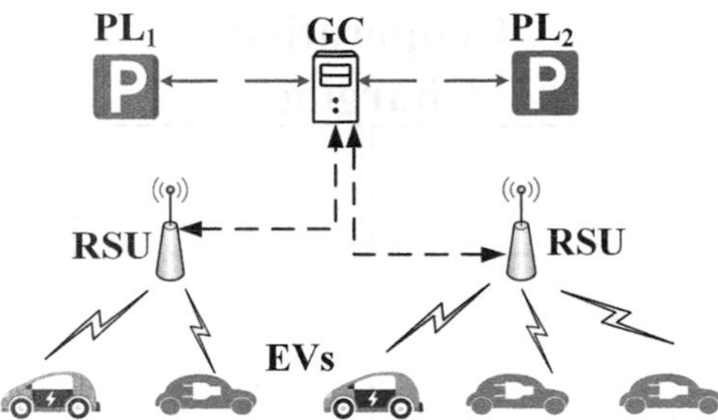


FIGURE 13.2 Semi-distributed framework.

Figure 13.2 demonstrates a semi-distributed V2V framework. The GC communicates with PLs and RSUs through 4G/5G cellular network. Here, RSUs temporarily cache and update occupation status at PLs with a fixed frequency. Mobile EVs send V2V charging requests to RSUs. RSUs reply EVs with V2V-Pair matching result and PL-Selection based on cached information.

It is worth noting that reporting EVs' V2V charging reservation (considered as ancillary services) has a positive effect on V2V charging result. The introduction of reservation information can effectively predict the occupation status of V2V charging converters at the PL and avoid that PL selected has no converter available when EVs reach the PL. However, V2V charging reservation information produces a relatively large communication cost. Although a high update frequency can bring higher accuracy, frequent updates also cause a burden on the communication network.

## 13.3 CHARGING SERVICE MODES

### 13.3.1 Cooperative V2V and G2V Charging

EVs are able to obtain charging services from the grid by mature plug-in charging mode. Here, EVs refill energy at CSs, home or other places equipped with charging facilities. Under the current battery and electric technology, the charging time of EV is much longer than the refill time of traditional ICVs. When the grid is processing with a large amount of parallel EVs charging requests, charging congestion and price fluctuations will happen.

Meanwhile, EVs are mobile uncertain, which brings a great charging challenge to the grid in CSs deployment. Therefore, previous V2V charging research takes V2V charging as a supplement to the G2V mode as the V2V charging mode is flexible in location selection (use PLs as commonplaces for energy transfer). Here, EV-Ps transfer energy to EV-Cs in the form of V2V-Pairs via V2V charging converters. Through the V2V charging service, the local energy balance can be adjusted to smooth the energy price fluctuation. The cooperation of V2V and G2V technologies adequately alleviates the charging difficulty of EVs. Here, it possibly involves the revenue comparison and energy optimization between V2V and G2V [5–8]. When the grid energy price is high, EV-Cs could choose to obtain energy from cheaper EV-Ps. At the same time, EV-Ps could also store extra energy from low energy prices for subsequent energy transfer.

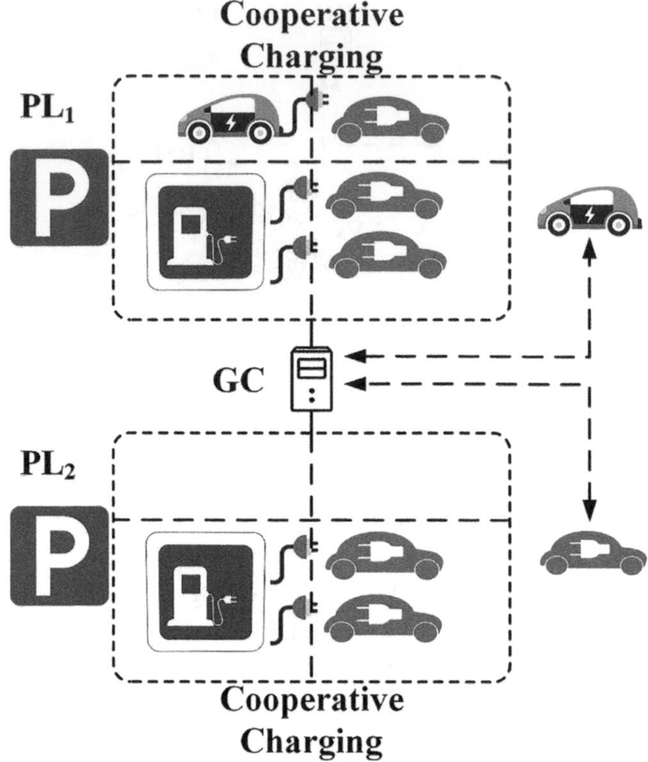


FIGURE 13.3 Cooperative V2V and G2V charging.

Figure 13.3 introduces a cooperative charging mode underlying a centralized framework. PLs are the places where EVs park and supplement energy. Here, PLs can provide EVs with traditional plug-in charging from the grid. Meanwhile, PLs also allow energy transfer via V2V charging converters, which facilitate the EVs with V2V request for energy transfer. This framework relieves the pressure of the power grid or helps EVs earn additional revenue (for EV-Ps) and minimize charging cost (for EV-Cs).

To encourage EVs to use renewable energy other than non-renewable energy, the work in Ref. [6] formulates the V2V-Pair matching as a constrained mixed integer linear programming, which aims to minimize the non-renewable energy consumption of EVs through the grid. EVs are encouraged to charge renewable photovoltaic power and store them for V2V charging purposes. In this work, a concept of virtual EVs is proposed to encourage EVs to consume the remaining renewable energy among themselves. The work in Ref. [7] proves that the cooperative V2V and G2V mode maintains the load balance of the grid and regulates the price of energy.

### 13.3.2 Pure V2V Charging

In addition, some works take V2V charging mode as an alternative to the plug-in charging mode for EVs [7,9–12]. These works are EVs oriented and aims to achieve high efficiency of pure V2V charging through optimization of V2V-Pairs matching and PL-Selection.

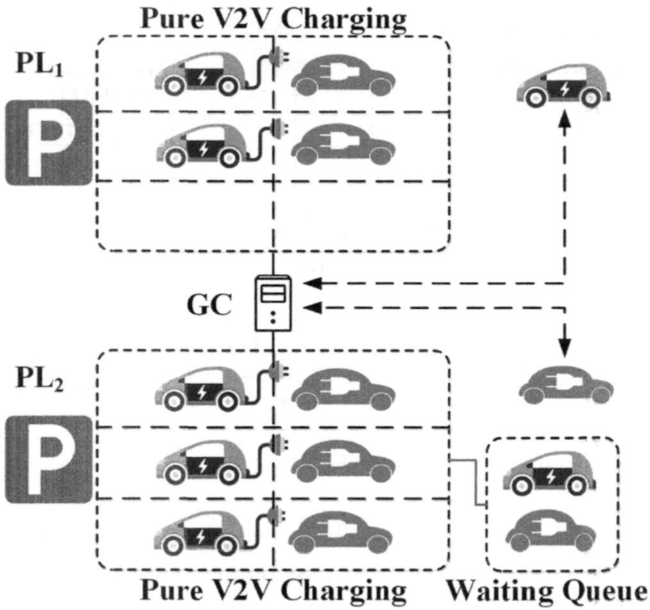


FIGURE 13.4 Pure V2V charging.

Figure 13.4 demonstrates a framework that EVs' energy replenishment only takes place between EVs. Here, EV-Ps and EV-Cs are matched as V2V-Pairs and travel towards PLs for V2V charging. The GC aggregates the V2V charging requests from EVs and monitors the local occupations status of PLs (including V2V-Pairs currently charging and waiting at PLs). Then the GC allocates PL-Selection for V2V-Pairs to optimize the overall EV drivers' charging experience.

## 13.4 V2V CHARGING FOR ON-THE-MOVE EVs

The V2V charging mode provides adaptability for EVs on-the-move. Usually, in the V2V charging mode, EVs are divided into EV-Ps and EV-Cs. Through the energy transfer devices (such as DC-DC converters [13]), energy can be transferred from EV-Ps to EV-Cs [14–17].

Major problems that exist in the V2V charging mode include the matching of V2V-Pairs and the PL-Selection for V2V-Pairs [7,9,18,19]. The core factor that affects V2V-Pair matching is charging utilities of EV-Ps and EV-Cs [20,21]. Maximizing the charging utility of EVs in V2V charging is conducive to the promotion of the V2V charging mode. Meanwhile, it is necessary to consider where to operate the V2V charging for a matched V2V-Pair [5,7,22].

In Figure 13.5, a typical procedure for on-the-move EVs' V2V charging is structured as follows:

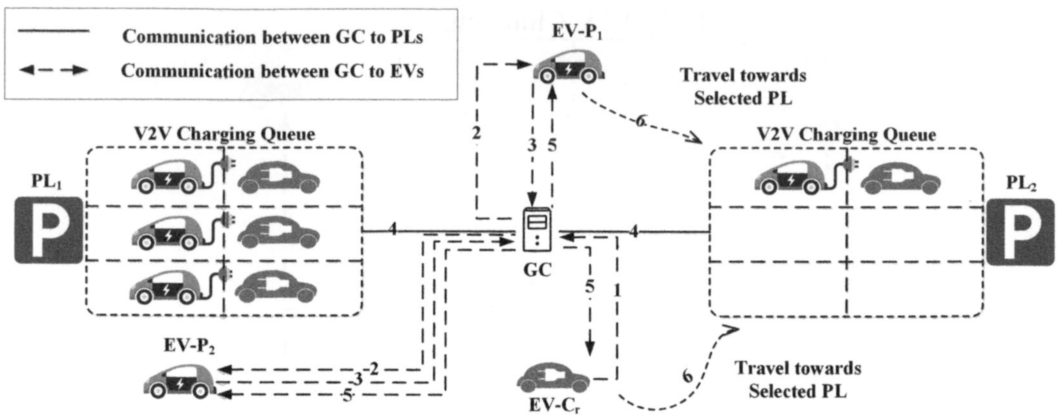


FIGURE 13.5 V2V charging for on-the-move EVs.

- **Steps 1–2**: If an EV-C (EV-$C_r$) is on-the-move and its State of Charge (SoC) is below the threshold, it sends its V2V charging request (including SoC, location, energy request amount) to the GC. The GC communicates to other EV-Ps to collect their status (location, energy surplus amount).
- **Step 3**: EV-Ps (EV-$P_1$ and EV-$P_2$) report their status to the GC and wait for V2V-Pair matching result. Then the GC matches an optimal V2V-Pair for V2V charging (in terms of location, travelling time and energy surplus).
- **Step 4**: The GC communicates with PLs ($PL_1$ and $PL_2$) and aggregates their local occupation status (including current charging V2V-Pairs, waiting queue).
- **Steps 5**: With the aggregated information from EV-Ps and PLs, the GC selects the appropriate PL to EV-$C_r$ and its match EV-P (in terms of the minimized V2V trip duration or cost). Then the GC sends the matching result and PL-Selection back to the EVs.
- **Steps 6:** Once the PL-Selection and V2V-Pair matching results are received by the EVs. For example, in this figure, $PL_2$ is selected as it can provide immediate V2V charging, EV-$P_1$ is matched as the V2V-Pair for EV-$C_r$ as it is closer in distance to EV-$C_r$ and selected PL. Then EV-$C_r$ and EV-$P_1$ will travel towards $PL_2$.

### 13.4.1 V2V-Pair Matching

Under the V2V charging mode, when an EV-C requests for charging, it requires a suitable EV-P to match for a maximized charging utility and minimized charging cost. In the V2V-Pair matching process, the distance between EVs and the energy provided by EV-Ps are considered [7]. The work in Ref. [9] proposes an efficient KM-based matching algorithm for optimal V2V-Pair assignment. However, the above works do not guarantee that every EV will obtain a stable EV-P to match. Therefore, the marriage algorithm is involved in V2V-Pair matching [19,20]. In the work [20], it proposes a matching algorithm that considers

EVs' charging utilities (regards to the energy cost and price). Here, the V2V charging utility of an EV-C is given by its charging payment minus its extra travelling cost (energy cost travelling towards the parking place). The V2V charging utility of an EV-P is given by its energy providing profit minus its travelling, time and amortized cost. Based on charging utility analysis, this work proposes a stable matching algorithm that could be regarded as a one-to-one marriage model for V2V-Pairs.

During V2V-Pair matching, the following constraints should be noted:

- EV-Ps need to ensure that their remaining energy should be higher than the energy demand of EV-Cs when they arrive at PLs.
- If the subsequent travel of EV-Ps is considered, the residual energies of EV-Ps need to meet the minimum requirement of completing their travels.
- EV-Ps' moving to PLs and energy providing processes will produce additional energy loss, which needs to be considered in the energy calculation.

The work in Ref. [18] proposes a V2V charging system in the city scenario. Here, the system records EVs status information (charging energy offered, location and expected duration) and completes the matching through the maximum weighted bipartite graph. Meanwhile, in the work [19], EV-Cs are allocated to the optimal energy supplier (CS or EV-P) to minimize the total charging cost. Here, in the matching process, this work considers the EVs' charging time, distance and charging revenue/cost. The matching problem in this work is described as a Mixed Integer optimization Problem (MIP) which is NP-hard, then this work develops an algorithm based on stable marriage algorithm.

Figure 13.6 demonstrates the stable V2V matching in the work [21], which proposes an EV-C oriented optimal marriage matching solution. Here, EV-Cs and EV-Ps have their preference list (regarding V2V charging utilities). It needs to be mentioned, there are two different possibilities under V2V mode: buyer's market (pair matching is oriented by EV-Cs) and seller's market (pair matching is oriented by EV-Ps). The two cases regarding EV-Cs oriented market and EV-Ps oriented market will be detailed in Sections 13.4.1 and 13.4.2.

Table 13.1 briefly introduces V2V-Pair matching algorithms applied in related works.

### 13.4.2 Charging Place Selection

Due to the EV mobility, V2V-Pairs need to select a PL (commonplace that provides converters for V2V energy transfer) after pair matching. When the GC conducted a feasibility analysis of V2V-Pairs' PL-Selection, the core factors are based on trajectory driving time, energy consumption, charging time and charging comfort quality.

Considering the energy cost and charging service quality, in the works [7,9], the optimal PL-Selection is obtained based on the aggregated information such as the distance from EVs to the PLs, the occupation status of the PLs (current occupation status and potential reservation status at PLs) and the charging efficiency (V2V charging waiting time) [23]. Uneven allocation of V2V charging will cause charging congestion problem, and PLs will have potential waiting queues for EV-Cs and EV-Ps. Thus, by efficiently allocating

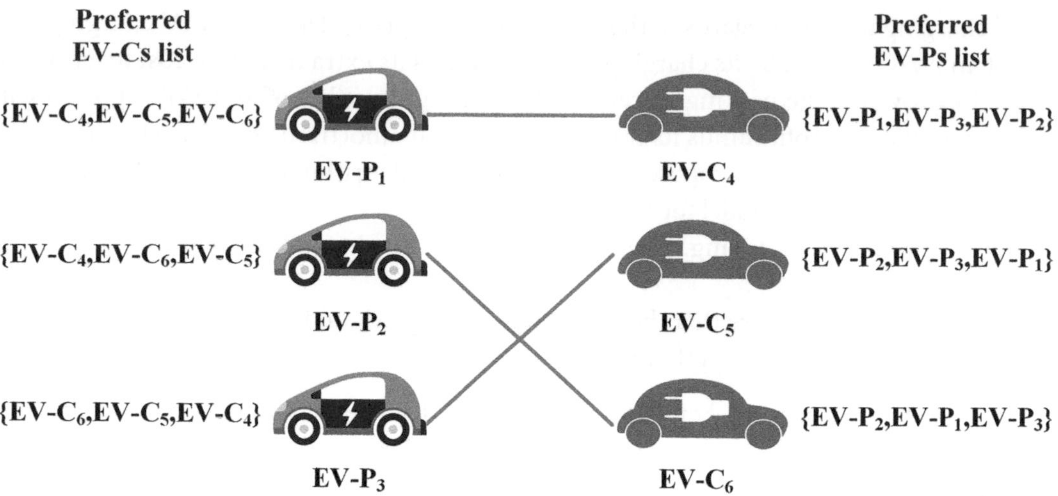


FIGURE 13.6 Stable match algorithm.

TABLE 13.1 Matching Algorithm Applied in V2V-Pair Matching

| Research Works | Matching Algorithm |
|---|---|
| [7] | First arrive first service rule combined with an energy transfer satisfaction principle |
| [9] | KM-Based matching algorithm for the optimal V2V-Pair |
| [18] | Apply maximum weighted bipartite matching algorithm to find the optimum matching |
| [19] | The deferred-acceptance algorithm matching game to seek a stable matching |
| [20] | 1. Maximize charging utility matching algorithm with a constructed bipartite graph according to Kuhn-Munkres algorithm<br>2. Propose a stable matching algorithm based on marriage algorithm |

PL-Selection would maximize the utilization of parking and charging resources at PLs, the V2V charging effect can be effectively improved.

## 13.5 V2V CHARGING FOR PARKED EVs

In addition to the V2V charging for on-the-move EVs, some works focus on solving the V2V charging under the EVs' parking mode. Since EVs have parked at PLs, the scheduling problem of EVs and the related local optimization problem of PLs occur.

The work in Ref. [10] proposes a localized V2V trading system based on blockchain, which maximizes the overall charging efficiency through an iterative double auction mechanism of charging and energy providing. As the auctioneers, local aggregators conduct double auctions among EVs according to the bidding price. Meanwhile, EVs can recharge from both EV-Ps and the grid. The work in Ref. [24] jointly considers the V2V charging and plug-in charging modes and proposes an offline optimal MIP formula to solve the energy transfer scheduling problem between EVs. To solve the coupling constraint problem, the charging loads of PLs are considered in the work [3]. The optimization problem is decomposed into a group of sub-problems by the Lagrange relaxation method. Each sub-problem corresponds to a single EV charging scheduling problem. In addition, EVs usually have their own travel plans and do not park at PLs for a long period, so it is very important to arrange V2V charging scheduling within the parking duration limit.

## 13.6 V2V CHARGING PRICE MODEL

The V2V charging mode can be used as a supplement to the traditional plug-in charging mode [5,6,25]. Here, EV-C drivers benefit from the charging V2V mode that reduces their charging costs and obtains a flexible charging strategy. To achieve an efficient V2V charging, the objectives of the V2V charging price model are:

- Maximizing the revenues of EV-Ps
- Minimizing the cost of EV-Cs

Unlike plug-in charging mode or battery switching mode that energy price depends on the grid or a single CS/BSS. In the V2V charging mode, the energy price is affected by the energy supply in the market via EV-Ps [26]. Based on the comparison of charging and providing energy/number of EVs, the V2V charging market is divided into high demand case and surplus supply case.

### 13.6.1 High Demand Case

When the number of EV-Ps or the total energy supply is less than the number/energy demand of EV-Cs, it is named high demand case. Here, the V2V charging price depends on EV-Ps. Under this case, EV-Ps aim to maximize their revenues. However, considering the charging price in the grid, the maximum charging price of EV-Ps should be no higher than the grid energy price. Meanwhile, due to the limited energy supply by EV-Ps, in some cases, EV-Cs need to charge additional energy from the grid [6].

The work in Ref. [12] discusses the constraints of local energy trading and focuses on the V2V charging cost-effectiveness and possibility. Furthermore, the work in Ref. [22] maximizes the energy providing revenues through the oligopoly game and finds the equivalence of the energy providing price to minimize the costs of EV-Cs. The work in Ref. [8] proposes a dynamic price model based on energy storage price. EVs are encouraged to use photovoltaic power generation for charging in outdoor and daytime, and this work achieves a higher energy consumption rate among EVs through energy distribution, which reduces the dependence on the grid and non-renewable energy.

### 13.6.2 Surplus Supply Case

When the energy provided by EV-Ps is greater than the energy required by EV-Cs, the price is friendly to EV-Cs. Therefore, EV-Ps must reduce the V2V energy providing price to attract EV-Cs. The work in Ref. [22] minimizes the V2V charging cost through the oligopoly game and finds the equivalence of the energy providing price to maximize the revenues of EV-Ps. The work in Ref. [5] also simulates under surplus supply case, which

TABLE 13.2 Price Model Applied

| Research Works | Price Model Approaches |
|---|---|
| [5] | Minimize energy exchanged unit cost with help of Lagrange multipliers |
| [6] | Mixed integer programming in offline optimization |
| [8] | Dynamic pricing scheme based on the average price of the stored energy in EVs |
| [12] | Reduce yearly energy bill in neighbours with V2V charging |
| [22] | Apply oligopoly game to minimize charging cost in surplus supply case and maximize the energy providing revenues in high demand case |

proposes a bidding model. Here, the extremum point is found by Lagrange multiplication to maximize the revenues of EV-Ps.

Table 13.2 introduces price models in related works.

## 13.7 V2V CHARGING ENERGY MODEL

A V2V charging converter mainly consists of a front-end AC-DC converter and a back-end DC-DC converter [13]. Thus, different power transmissions can be achieved between two EVs. The direct V2V charging method (one-to-one V2V charging mode) is a more flexible approach that abandons the connection with the grid (G2V and vehicle-to-grid). A one-to-one V2V charging allows directly energy transfer between EVs. The other traditional indirect method uses the power grid (or condensers) as an intermediate point. Here, EV-Ps transfer energy to condensers for storage, then condensers transfer stored energy to EV-Cs, which allows multiple-multiple V2V charging.

### 13.7.1 One-to-One V2V Charging

EVs are matched as one-to-one V2V-Pair through matching algorithms which are detailed in Section 13.3.1. Here, an EV-P transfers its energy towards another EV-C though a V2V charging converter.

A basic scheme of one-to-one V2V charging allows energy transfer between two EVs parked at a PL by bidirectional charging technology [13–17] and bidirectional V2V charging converter connecting EVs. Furthermore, the work in Ref. [11] proposes a mobile V2V charging facility for one-to-one V2V charging approach which allows energy transfer on the move. This system allows the robot's charging arm to reach out when the EVs are driving on the highway.

Figure 13.7 demonstrates a simple one-to-one V2V charging model (direct V2V charging services) at PLs between V2V-Pairs. One EV-P transfers its surplus energy to another EV-C via off-board converter. Here, several V2V-Pairs are able to parallelly transfer energy in one-to-one V2V charging manner at PLs depending on the number of converters.

### 13.7.2 Multiple-to-Multiple V2V Charging

Different from the above one-to-one scheme, EVs can also achieve multiple-to-multiple V2V charging approach. This approach solves the limitation of fixed V2V-Pairs matching requirement. Through the cooperation of Vehicles-to-Grid (V2G) and G2V technologies, PLs (or other places that provide V2V charging service) are used as additional energy

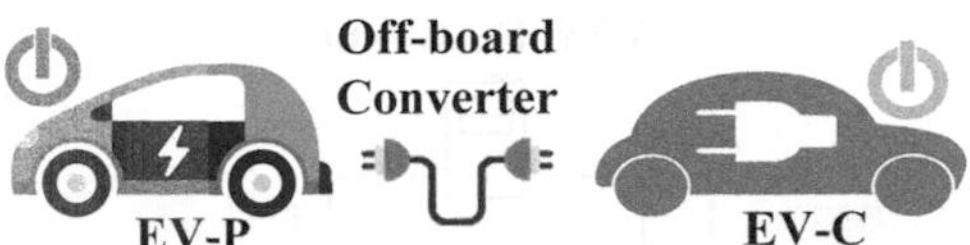


FIGURE 13.7 One-to-one V2V charging.

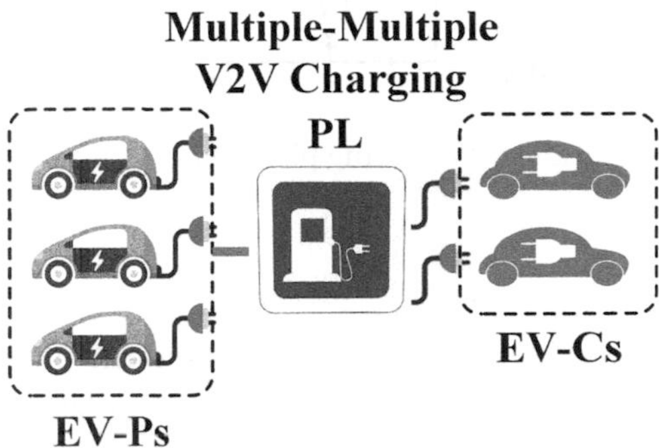


FIGURE 13.8 Multiple-to-multiple V2V charging.

storage devices. EVs with surplus energy can offer their remaining energies to a PL through V2G technology, and this PL can provide the stored energy to one or more EVs with charging requests to realize multiple-to-multiple V2V charging services.

Figure 13.8 demonstrates a multiple-multiple V2V charging model. Here, the EV-Ps transfer energy for revenue at PLs (according to the energy providing price and energy amount and energy loss). Condenser deployed at PLs storages and provides the surplus energy. Thus, PLs are enabled to provide parallel charging services for several EV-Cs based on the number of charging slots.

Here, several constraints need to be considered:

- Due to the energy loss and efficiency in V2V charging, the total energy providing at PLs needs to be greater than the total energy request from EV-Cs.
- The energy providing revenue of a single EV-P should be higher than its energy providing cost (regards to travelling and energy providing cost).
- The arrival time of the EV-Ps (considering the potential energy providing time) needs to be earlier than the arrival time of the EV-Cs.

Under multiple-multiple V2V charging scenario, the optimization objective in the work [6] is to optimize EV-Cs energy charging at a single PL. Here, EVs announce their arrival, departure time and the required charging energy to the PL. Then the PL is used as an aggregator to collect the charging requests and constraints of EVs, and then manages the V2V charging requests to maximize the charging efficiency.

Figure 13.9 depicts the framework of the V2V transaction system [27]. EV-Ps and EV-Cs are parked at PLs where energy transfer is performed through V2V charging

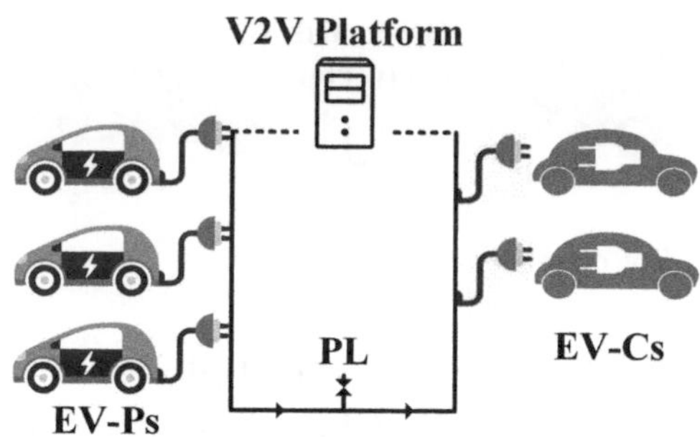


FIGURE 13.9 Multiple-to-multiple V2V trading platform.

converters. Both EV-Ps and EV-Cs report their amount of energy transfer (provide/demand) and expected parking time to the V2V trading platform. Here, the V2V energy trading platform allocates multiple V2V-Pairs according to the price, amount of energy transfer and the parking time of EVs. The difference between this platform and one-to-one V2V charging model is that the PL handles many sequential V2V charging requests through the trading platform. A single EV-C can obtain energy transfer from multiple EV-Ps. In the same way, a single EV-P can also provide energy to multiple EV-Cs in a time sequence.

## 13.8 FURTHER DISCUSSION

### 13.8.1 Security and Privacy

The V2V charging involves a large amount of information interaction. As the V2V charging mode uses intelligent devices, it has potential privacy leakage. Here, the sensitive data needs to be encrypted in V2V charging mode. In addition, the V2V charging mode faces with many EVs, and a wide range coverage. This brings great challenges to security. One solution is designing an encryption scheme with light weight and high security. Another development direction is to design an efficient and scalable key management scheme. For privacy protection, EVs may use pseudonyms to hide their identities.

At the software level, V2V charging involves Peer-to-Peer transaction operations, which is vulnerable to attacks. It is necessary to avoid V2V charging transaction information tampering. Here, attacker tampers with the information sent to EVs via Wi-Fi or Bluetooth without encryption. To solve this, the authentication protocol for V2V charging applications is introduced [18]. Problem also exists under hardware level. In G2V modes, CS operators are usually stable and safe. However, the V2V charging is completed at deployed V2V charging converters. These plug-in power supply devices have certain safety risks, including battery overload, battery overheating caused by EV model mismatch, etc. Therefore, a unified safety standard is urgently needed for the V2V charging to realize the safety functions such as full battery power failure, no-load power failure, overload power failure and emergency stop.

### 13.8.2 Data Communication and Analytics

As EV-Cs and EV-Ps are under move simultaneously, the V2V charging mode has more stringent requirements for communication than G2V modes. The V2V charging mode requests communication with shorter latency. In addition, due to the large communication range, the loss in the communication process should be avoided through an efficient data pre-processing. A potential solution is to analyse EV drivers' behaviours, simulate EV drivers moving models and charging preferences. This could predict EVs' V2V charging requests, and better optimize the energy and price model of V2V charging.

### 13.8.3 Heterogeneity of EV Batteries

Battery matching (battery heterogeneity) is ignored in previous works. It is necessary to establish a service classification strategy under the condition of EVs battery specification heterogeneity. In addition, the V2V charging mode is required to further improve the coupling between differentiated service characteristics and power supply demands. Meanwhile, before V2V-Pair matching, the battery heterogeneity of EVs is required to be clarified. After the establishment of the service classification strategy under the condition of heterogeneous EV battery specifications, the mapping relationship between the change of energy supply demand and the service characteristics need to be considered.

## 13.9 CONCLUSION

This chapter reviewed state-of-the-art works on the V2V charging mode and energy-sharing models. The V2V charging mode benefits from communication frameworks (centralized, distributed and hybrid communication framework). V2V charging communication frameworks efficiently aggregate V2V charging requests from EVs and occupation status at PLs, which lays the foundation in the V2V charging mode. Furthermore, two charging modes were introduced. In the cooperative V2V and G2V charging mode, V2V charging is applied as a supplement to the grid charging for keeping energy load and price balance. The pure V2V charging mode is regarded as an alternative approach to G2V modes as it provides flexible charging service. To solve the problems encountered in the V2V charging mode for on-the-move EVs, V2V-Pairs matching algorithms and PL-Selection schemes were discussed in this chapter. Here, these approaches maximize the EV charging utility participating in V2V charging. For parking EVs at PLs, V2V charging scheduling strategy has become the core of optimization. The energy price and supply affect the participation of V2V charging. EVs face high demand case and surplus supply case in V2V charging energy models; this determines whether the dominant player in V2V charging transactions is EV-Ps or EV-Cs. In addition, considering that a large amount of EVs participate in the V2V charging mode, the energy model is divided into one-to-one and multiple-to-multiple energy transfer models.

## REFERENCES

1. J. Mukherjee and A. Gupta, "A Review of Charge Scheduling of Electric Vehicles in Smart Grid," *IEEE Systems Journal*, vol. 9, no. 4, pp. 1541–1553, 2015.
2. T. Raviv, "The Battery Switching Station Scheduling Problem," *Operations Research Letters*, vol. 40, no. 6, pp. 546–550, 2012.

3. P. You and Z. Yang, "Efficient Optimal Scheduling of Charging Station with Multiple Electric Vehicles via V2V," in *2014 IEEE International Conference on Smart Grid Communications (SmartGridComm)*, Venice, Italy, Nov. 2014, pp. 716–721.
4. C. Liu, K. T. Chau, D. Wu, and S. Gao, "Opportunities and Challenges of Vehicle-to-Home, Vehicle-to-Vehicle, and Vehicle-to-Grid Technologies," *Proceedings of the IEEE*, vol. 101, no. 11, pp. 2409–2427, 2013.
5. R. Alvaro, J. Gonza´lez, C. Gamallo, J. Fraile-Ardanuy, and D. L. Knapen, "Vehicle to Vehicle Energy Exchange in Smart Grid Applications," in *2014 International Conference on Connected Vehicles and Expo (ICCVE)*, Vienna, Austria, Nov. 2014, pp. 178–184.
6. A. Koufakis, E. S. Rigas, N. Bassiliades, and S. D. Ramchurn, "Offline and Online Electric Vehicle Charging Scheduling with V2V Energy Transfer," *IEEE Transactions on Intelligent Transportation Systems*, vol. 21, no. 5, pp. 2128–2138, 2020.
7. G. Li, L. Boukhatem, L. Zhao, and J. Wu, "Direct Vehicle-to-Vehicle Charging Strategy in Vehicular Ad-hoc Networks," in *2018 9th IFIP International Conference on New Technologies, Mobility and Security (NTMS)*, Paris, France, Feb. 2018, pp. 1–5.
8. S. Aznavi, P. Fajri, M. B. Shadmand, and A. Khoshkbar-Sadigh, "Peer-to-peer operation strategy of PV equipped office buildings and charging stations considering electric vehicle energy pricing," *IEEE Transactions on Industry Applications*, vol. 56, no. 5, pp. 5848–5857, 2020.
9. G. Li, Q. Sun, L. Boukhatem, J. Wu, and J. Yang, "Intelligent Vehicle-to-Vehicle Charging Navigation for Mobile Electric Vehicles via Vanet-Based Communication," *IEEE Access*, vol. 7, pp. 170888–170906, 2019.
10. J. Kang, R. Yu, X. Huang, S. Maharjan, Y. Zhang, and E. Hossain, "Enabling Localized Peer-to-Peer Electricity Rrading among Plug-in Hybrid Electric Vehicles Using Consortium Blockchains," *IEEE Transactions on Industrial Informatics*, vol. 13, no. 6, pp. 3154–3164, 2017.
11. P. Chakraborty, R. Parker, T. Hoque, J. Cruz and S. Bhunia, "P2C2: Peer-to-Peer Car Charging," in *2020 IEEE 91st Vehicular Technology Conference (VTC2020-Spring)*, Antwerp, Belgium, May 2020, pp. 1–5.
12. O. M. Almenning, S. Bjarghov, and H. Farahmand, "Reducing Neighborhood Peak Loads with Implicit Peer-to-Peer Energy Trading under Subscribed Capacity Tariffs," in *2019 International Conference on Smart Energy Systems and Technologies (SEST)*, Porto, Portugal, Sept. 2019, pp. 1–6.
13. T. J. C. Sousa, V. Monteiro, J. C. A. Fernandes, C. Couto, A. A. N. Mele´ndez, and J. L. Afonso, "New Perspectives for Vehicle-to-Vehicle (V2V) Power Transfer," in *IECON 2018–44th Annual Conference of the IEEE Industrial Electronics Society*, Washington, DC, USA, Oct. 2018, pp. 5183–5188.
14. X. Mou, R. Zhao, and D. T. Gladwin, "Vehicle to Vehicle Charging (V2V) Bases on Wireless Power Transfer Technology," in *IECON 2018–44th Annual Conference of the IEEE Industrial Electronics Society*, Washington, DC, USA, Oct. 2018, pp. 4862–4867.
15. S. Taghizadeh, P. Jamborsalamati, M. J. Hossain, and J. Lu, "Design and Implementation of an Advanced Vehicle-to-Vehicle (V2V) Power Transfer Operation Using Communications," in *2018 IEEE International Conference on Environment and Electrical Engineering and 2018 IEEE Industrial and Commercial Power Systems Europe (EEEIC / I CPS Europe)*, Palermo, Italy, June 2018, pp. 1–6.
16. B. Roberts, K. Akkaya, E. Bulut, and M. Kisacikoglu, "An Authentication Framework for Electric Vehicle-to-Electric Vehicle Charging Applications," in *2017 IEEE 14th International Conference on Mobile Ad Hoc and Sensor Systems (MASS)*, Orlando, FL, USA, Oct. 2017, pp. 565–569.
17. E. Bulut and M. C. Kisacikoglu, "Mitigating Range Anxiety via Vehicle-to-Vehicle Social Charging System," in *2017 IEEE 85th Vehicular Technology Conference (VTC Spring)*, Sydney, NSW, Australia, June 2017, pp. 1–5.

18. O. T. T. Kim, N. H. Tran, V. Nguyen, S. M. Kang, and C. S. Hong, "Cooperative between V2C and V2V Charging: Less Range Anxiety and More Charged EVs," in *2018 International Conference on Information Networking (ICOIN)*, Chiang Mai, Thailand, Jan. 2018, pp. 679–683.
19. R. Zhang, X. Cheng, and L. Yang, "Flexible Energy Management Protocol for Cooperative EV-to-EV Charging," *IEEE Transactions on Intelligent Transportation Systems*, vol. 20, no. 1, pp. 172–184, 2019.
20. E. Bulut, M. C. Kisacikoglu, and K. Akkaya, "Spatio-Temporal Non-Intrusive Direct V2V Charge Sharing Coordination," *IEEE Transactions on Vehicular Technology*, vol. 68, no. 10, pp. 9385–9398, 2019.
21. M. Wang, M. Ismail, R. Zhang, X. Shen, E. Serpedin and K. Qaraqe, "Spatio-Temporal Coordinated V2V Energy Swapping Strategy for Mobile PEVs," *IEEE Transactions on Smart Grid*, vol. 9, no. 3, pp. 1566–1579, 2018.
22. P. You, Z. Yang, M. Chow, and Y. Sun, "Optimal Cooperative Charging Strategy for a Smart Charging Station of Electric Vehicles," in *2017 IEEE Power Energy Society General Meeting*, Chicago, IL, USA, July 2017, pp. 1-1.
23. Y. Cao, T. Jiang, O. Kaiwartya, H. Sun, H. Zhou, and R. Wang, "Toward Pre-empted EV Charging Recommendation Through V2V-Based Reservation System," *IEEE Transactions on Systems, Man, and Cybernetics: Systems*, pp. 1–14, 2019.
24. A. Koufakis, E. S. Rigas, N. Bassiliades, and S. D. Ramchurn, "Towards an Optimal EV Charging Scheduling Scheme with V2G and V2V Energy Transfer," in *2016 IEEE International Conference on Smart Grid Communications (SmartGridComm)*, Sydney, NSW, Australia, Nov. 2016, pp. 302–307.
25. M. Wang, M. Ismail, R. Zhang, X. S. Shen, E. Serpedin, and K. Qaraqe, "A Semi-Distributed V2V Fast Charging Strategy Based on Price Control," in *2014 IEEE Global Communications Conference*, Austin, TX, USA, Dec. 2014, pp. 4550–4555.
26. R. Alvaro-Hermana, J. Fraile-Ardanuy, P. J. Zufiria, L. Knapen, and D. Janssens, "Peer to Peer Energy Trading with Electric Vehicles," *IEEE Intelligent Transportation Systems Magazine*, vol. 8, no. 3, pp. 33–44, 2016.
27. Y. Xu, S. Wang and C. Long, "A Vehicle-to-Vehicle Energy Trading Platform Using Double Auction with High Flexibility," in *2021 IEEE PES Innovative Smart Grid Technologies Europe (ISGT Europe)*, Espoo, Finland, Oct. 2021, pp. 01–05.

CHAPTER 14

# Introduction and Overview of V2G Security

Sifan Li and Yue Cao

*Wuhan University*

Xu Xia

*Telecom Research Institute*

Qiang Tang

*SG Star Energy (Sichuan) Technology Co. Ltd*

Yonglong Peng

*Keepv Technology (Nanjing) Co. Ltd*

## CONTENTS



## LIST OF ABBREVIATIONS

| | |
|---|---|
| **DDoS** | Distributed denial-of-service |
| **DoS** | Denial-of-service |
| **EVs** | Electric vehicles |
| **IDS** | Intrusion detection system |
| **IPS** | Intrusion prevention system |

DOI: 10.1201/b22998-14

| | |
|---|---|
| **MITM** | Man-in-the-middle |
| **OCSVM** | One-class support vector machine |
| **SDN** | Software-defined network |
| **SIEM** | Security information and event management |
| **SoC** | State of charge |
| **V2G** | Vehicle to grid |

## 14.1 INTRODUCTION

With the continuous depletion of fossil energy, solving the global energy crisis is imminent. Traditional fuel vehicles are the "main force" of greenhouse gas emissions, research on new energy vehicles is to optimize the energy structure, and the development of clean renewable energy. "Electricity" is one of the cleanest energy sources. The concept of electric energy substitution provides a direction for the development of new energy vehicles. At present, all over the world, the era of electric vehicles (EVs) is beginning, and it has become an inevitable trend for EVs to popularize in people's daily life with the advantages of clean and efficiency, green emission reduction, and national policy support. Vigorously developing EVs has become an important measure for countries around the world to deal with climate change, environmental pollution, promote energy conservation, and emission reduction. The wide adoption of EVs is expected to accelerate especially with advancements in reforming the power grid to a smart grid. Due to their environmental benefits (i.e., reduced greenhouse emissions), their role in the grid's ancillary services, dynamic energy pricing, and governmental incentives, the number of EVs in use is expected to grow dramatically in the future. EVs must be charged in order to work properly. Although home charging can also meet daily needs, it takes too long to charge. Therefore, the construction of charging stations or small charging piles is the key to the popularization of EVs.

With the popularity and promotion of new energy vehicles, the demand for charging piles is also increasing, and the network security issue in the interaction between cars and charging piles has also attracted more attention. In Figure 14.1, we have found that the charging pile has the characteristics of many points, wide range and dispersion. Its user, communication network and cloud lack effective security protection mechanisms, and each node is relatively vulnerable to intrusion. The main risks are concentrated in the system security of the charging pile itself, the security of the power grid, the data transmission with the local charging station, the data transmission between the charging station and the operation platform, the stable operation of the operation platform, and the user settlement security and other fields. At the same time, when new energy vehicles interact with charging piles, users are also prone to security problems such as the destruction of viruses and malicious codes, interruption attacks, and the disclosure of sensitive privacy. For example, theft is a serious threat in the application of distributed EV charging piles. Stealing refers to stealing sensitive information, such as user ID and password, and the amount of money stored, by all kinds of illegal users through measures such as seizing packets and eavesdropping. An interrupt attack is to destroy information transmission, resulting in

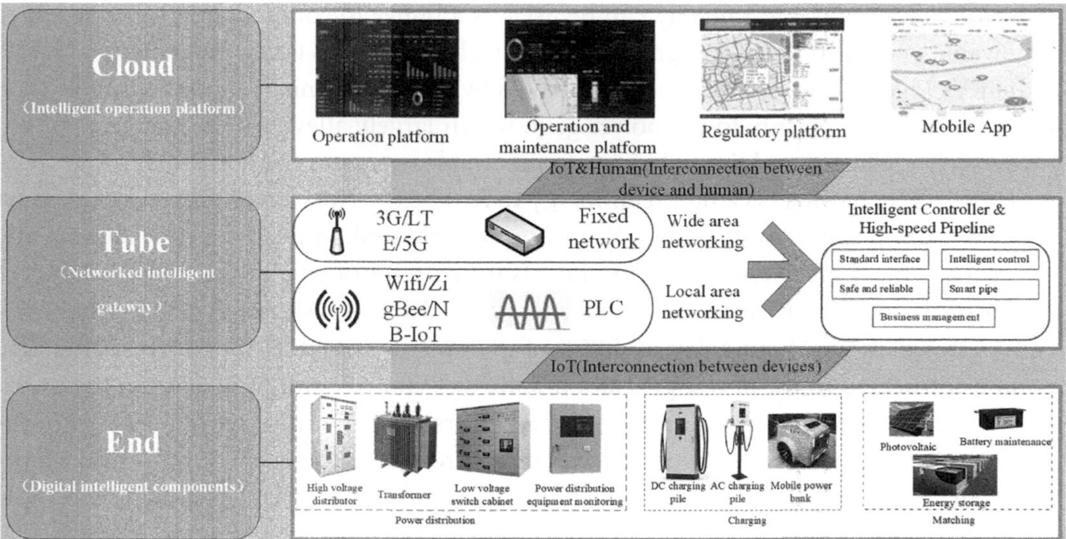


FIGURE 14.1 The frame diagram of charging pile.

users not to obtain system information resources stably and damage to system hard disk and communication line. In the communication of the charging pile, the attackers mainly damage the body of charging piles, such as 4G and 5G modules and the interface of the card reader. Those threats escalate with the wide deployment of public charging stations, and the random mobility of EVs which augment the attack surface along with chances to spread infections across the system. Furthermore, critical data will be shared across the EV-charging system, including location information, charging duration, EV identity, state of charge (SoC), payment info, etc. This gives rise to cyber-attacks targeting user privacy and charging network availability.

The remainder of this chapter is organized as follows: Section 14.2 presents background and related work; Section 14.3 classifies and describes the attacks; Section 14.4 lists literature related to mitigation and protection; Section 14.5 presents future directions; and Section 14.6 ends with a summary and conclusions.

## 14.2 BACKGROUND AND RELATED WORK ABOUT EV CHARGING

Compared with the traditional power grid, smart grids embody the advantages of reliability, efficiency and economy.[1,2] It can not only carry out two-way communication of information flow, power flow with users, and meet the diversified needs of users but also has the ability of power grid fault prediction, isolation, diagnosis, and self-recovery. More importantly, the smart grid can also support the access of energy storage equipment and renewable energy (such as solar, wind) power generation systems. However, due to the characteristics of discontinuity and randomness of renewable energy, it will cause power grid fluctuation when connected. Therefore, other auxiliary systems need to compensate for the access of renewable energy, so as to smooth the fluctuation caused by renewable energy to the power grid and ensure the stability of power grid voltage and frequency. The vehicle-to-grid (V2G) technology can be used as a buffer to compensate for the access of renewable energy to the power grid. The V2G technology uses the stored energy of large-scale EVs to store the electric energy generated by renewable energy, and then stably send the stored electric energy to the power grid, which largely alleviates the problems of low power grid efficiency and limited access to renewable energy to the power grid.

The V2G technology is a new technology to realize the two-way exchange of information flow and power flow between EVs and power grid under the unified dispatching and control of smart grid system. Its core idea is to use a lot of energy stored by EVs as the buffer of the power grid and renewable energy.[3–10] Research shows that more than 90% of EVs travel about 1 hour a day, and 95% of the time is parked. However, the parked EV can be connected to the power grid as a mobile distributed storage device. On the premise of meeting the driving requirements of EV users, the remaining electric energy in the battery of the EV is fed back to the power grid through the V2G network to smooth the load curve of the power grid. In other words, when the load of the grid is too high, the energy stored by the EV is fed (discharged) to the grid; when the grid load is low, the energy storage of a large number of vehicles can be used to store the excess electric energy (charging) of the grid to avoid waste. In this way, EV users can buy electricity from the grid when prices are low and sell it back to the grid when the price is high, thus making a profit.

### 14.2.1 Threats to Smart Grid

At present, the penetration rate of EVs is not high, and the impact of their access on the distribution network is not obvious. However, with the increasing penetration rate of EVs and the tilt of regional policies, the impact of their charging and discharging behaviors on the distribution network will gradually become prominent. The influence of EVs on distribution networks mainly focuses on system reliability, power quality, economic operation, and other aspects.

#### *14.2.1.1 System Reliability*

Eshou and Wenmin[11] pointed out that the centralized charging behavior of EVs under fast charging mode was easy to cause overload in local areas of the distribution network and threaten the reliability of the power system. Qian et al.[3] analyzed the impact of EV charging behavior on daily load curves under disordered charging. The research showed that with the improvement of EV permeability if the charging behavior of EVs was not managed, the load peak of the distribution network will increase.

#### *14.2.1.2 Power Quality*

Power electronic equipment in EV chargers and dischargers is easy to produce harmonics, which affects power quality to a certain extent. Chen et al. analyzed the influence of charging and discharging equipment on harmonics of distribution network, and gave countermeasures[12,13] for EVs that were not controlled charging behavior research. Research had shown that large-scale EV disordered charge would cause certain influence on transformer current harmonic, even harm the life of the transformer. They put forward a method through policy guide to avoid the EV concentrated charging and minimize the impact on the transformer.

#### *14.2.1.3 Economic Operation*

Zhong studied the influence of EV participating in peak regulation on the distribution network. In this operation mode, EV charging behavior did not have a great impact on the power grid, but its discharging behavior may have a certain impact on the system operation mode and the direction of power flow. The study of literature showed that the charging behavior of EV in the disordered charging mode was not economical, and it may also cause negative effects such as a sharp drop in power grid voltage and passing load.

### 14.2.2 Threats to Users

As an important part of the smart grid, the V2G network is a new direction for the development of EV in the future. It can not only bring benefits to vehicle users, but also realize the orderly charge/discharge management of vehicles, and play the role of "cutting peak and filling valley," so as to improve the efficiency, safety, and stability of power grid operation. However, when the EV is connected to the power grid for service, the V2G network will collect, communicate, and share the data of vehicle users, and the attacker is easy to obtain user privacy through the data. For example, the time and frequency of vehicles staying in the hospital may reveal the user's health status; the location of vehicle charging/discharging and transaction bills may expose sensitive information such as users' economic status, home address, and social activities.[14,15] On the other hand, the V2G network will regularly make statistics on user participation. System managers can reward high-quality customers, track malicious vehicles, and optimize power grid regulation parameters according to the statistical results.[16–18] If the V2G network defense capability is weak, the system manager is likely to become an internal attacker under the influence of the enemy to obtain or disclose the user's sensitive information, which will bring great security risks to the V2G network. If the privacy protection of the V2G network cannot be paid enough attention to, the insecure operating environment will make users lose confidence in the V2G network, and many potential users will not be willing to accept the services of the V2G network.[19]

## 14.3 ATTACK CLASSIFICATION

In Figure 14.2, we have found that the attacks on the charging process of electric cars are varied. The attacker will launch SQL injection attack, XSS attack, Worm and steal permission on the smartphone or charging APP used by the user. The main attacks on EV charging stations and infrastructure include physical damage, denial of charging attack, denial of service (DoS) attack, price attack, and data injection attack. The communication phase is also vulnerable to attacks, including man-in-the-middle attacks, DoS attack, RF jamming attack, and tamper attack. Aurora attack, topology attack, load redistribution attack, AGC attack, etc. will affect the normal operation of the smart grid.

Next, the chapter will introduce and list some main attacks, such as physical damage, price information attack, DoS attack, privacy-sensitive information attack, Man-in-the-middle attack, and data injection attack.

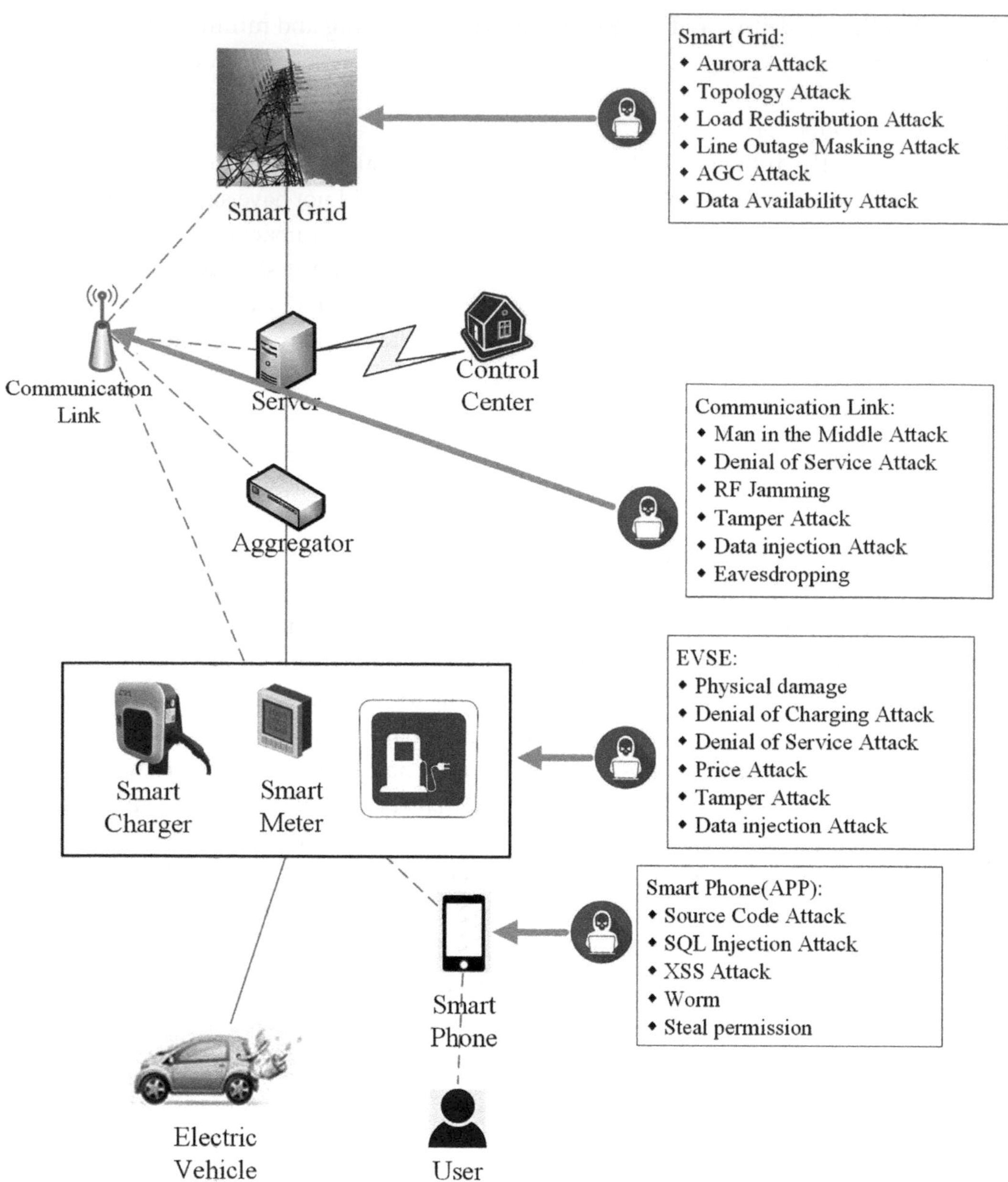


FIGURE 14.2 The security threats of intelligent charging.

## 14.3.1 Physical Damage

The improvement of charging stations, charging piles and charging infrastructure is very important for the safe charging of users. However, many public charging piles and charging stations have suffered some man-made physical damage. Consult the parking lot administrator that the charging pile is damaged from time to time, some owners forgot to pull the charging gun after charging, the results with the charging gun away, some owners will reverse the EV into the charging parking space, accidentally knocked the charging pile. There are also owners for their own parking convenience, with glue blocked the charging port and cut cables, causing the charging pile cannot be used, disguised as a parking space for fuel cars.

Usually, part of the charging pile is installed in an open or semi-open environment, so the charging pile itself is easily affected by the weather and external environment, such as lightning strikes and rain.

### 14.3.2 Price Information Attack

For users, the price of charging is their main concern. If price information is attacked, it will result in property damage to customers or grid operators. Then the harm that price attack brings is incalculable. Because of the volatility of electricity price, power grid suppliers need to transmit electricity price information to users in real time. This makes it difficult to protect electricity price information and gives the attackers an opportunity to exploit it.

However, when designing EV energy management policies, pricing policies are rarely studied with potential price information attacks in mind. For instance, although a bidirectional communications infrastructure can bring many benefits to the smart grid, it can introduce new vulnerabilities. For example, a malicious attacker can attempt to tap into the smart grid's communication system with the aim to cause malfunctions, disrupt the electricity market, or make monetary profits.[20] One easy target for such attacks is the manipulation of the real-time pricing information that is communicated by the public utility to the vehicles.[21] The attacker may disrupt the transmission of the electricity price information to the EV owners, resulting in the loss of the pricing information, which is, in fact, one of the possible DoS attacks on the smart grid. Alternatively, it is possible for the attacker to manipulate the pricing information by injecting incorrect price values so as to compromise the charging policies of the EV owners.

### 14.3.3 Denial of Service Attack

DoS attack is a family of attacks in which an attacker tries to block or prevent legitimate users from accessing a specific network service or resource through the attack source. Attackers can create botnets by compromising vulnerable Internet-connected devices, as well as devices such as charging piles, and launch attacks by controlling servers to control botnets. As a result, the victim receives a large amount of attack traffic from different sources from the damaged device, disrupting its normal activity.

Compared to traditional DoS attacks, IoT DoS attacks are harder to defend against because charging piles have limited storage and computing resources. Attackers can easily forge malicious packets based on design flaws in charging piles firmware or vulnerabilities in communication protocols. On the other hand, in addition to being victims of DoS attacks, charging piles can also serve as powerful DoS attack AIDS. The dumb security implementation of charging piles makes them easy for attackers to control to set up distributed botnets, which is considered to be the primary mechanism for building distributed DoS attacks based on flooding.

Even though the number of charging piles is growing rapidly, their security conditions remain unpleasant. Due to the limited computational power and constrained storage of charging piles, it is extremely hard to implement practical protection mechanisms on them. As a result, simplified lightweight protocols are generally adopted in charging systems, making them susceptible to various attacks as an inadequate level of protection can be easily bypassed by attackers. More seriously, attackers can exploit the vulnerabilities of the charging system and create botnets to launch severe distributed denial-of-service (DDoS) attacks without much effort. A DDoS attack is one in which an attacker attempts to obstruct the legitimate operations of a device and makes the device unavailable to legitimate users via extreme resource consumptions by distributed attack sources. For example, the Mirai virus infected 65,000 devices within the first 20 hours of its release in August 2016, forming a botnet to attack more IoT devices. In November 2016, the smart heating systems of two buildings in Finland were shut down due to a DDoS attack caused by a Mirai botnet, which made the heating controllers continually reboot until the heating systems came into crashes. Similar DDoS events occurred worldwide after more Mirai variations were developed. Such examples indicate that IoT devices can be employed as DDoS weapons to collaboratively create botnets after being compromised by attackers; they can also become DDoS victims when their normal operations are interrupted by IoT DDoS botnets.

Vulnerable charging piles are the potential attack targets to form botnets, which then threaten the IoT and the smart home system security via DDoS attacks. Moreover, the bigger the botnet, the more powerful the attack can be. Even with only a small fraction of the charging piles are compromised to generate DDoS packets, the magnitude of the launched attacks would be enormous and such attacks could easily disrupt any target including powerful servers equipped with heavily protected strategies.

### 14.3.4 Privacy-Sensitive Information Attack

In the V2G network, the user privacy is easily leaked in the process of charging/discharging, battery management, data communication, data management, and communication protocol.

The charge/discharge is the most basic service in the V2G network. When performing the service, the vehicle must provide its valid identity to the charging pile or local aggregator, and the identity can be authenticated before communicating with the power supply provider to obtain the corresponding service. However, data will be continuously monitored by the V2G network during the charge/discharge service, which may expose sensitive information such as the location and time of the vehicle charge/discharge.[22] The attacker can also analyze the movement trajectory of the vehicle by analyzing the status information of the vehicle battery, so as to infer the activity rule of the owner.[23]

When the system monitors the V2G network, a large amount of data will be generated, including user identity, charge and discharge policies, location, and charging information. Data management is the collection, aggregation, storage, and distribution of this data. However, the database manager may become an internal system attacker and attack the database for profit. Therefore, considering how to defend against the internal attackers of the system is also an issue worth studying in V2G network privacy protection.

### 14.3.5 Man-in-the-Middle Attack

Man-in-the-middle attack refers to an attack in which an adversary can establish a connection between a vehicle and a local aggregator or charging point, intercept normal communication data without the knowledge of the communication parties, extract information from the data or tamper with the data to achieve malicious purposes. Adversaries can also link fake charging piles to real ones and sell electricity to users without paying the grid. In addition, man-in-the-middle attacks can expose users' privacy.

### 14.3.6 Eavesdropping Attack

The eavesdropping attack is a kind of passive attack aimed at eavesdropping on network information. Attackers listen to wireless channels to collect data packets, which often contain private information about vehicles, as they travel through the network. Encrypting data is generally used to resist eavesdropping attacks. If there is no decryption key, the attacker cannot identify the true content of the ciphertext data even if he obtains the ciphertext data.

### 14.3.7 Counterfeit Attack

A counterfeit attack is when an attacker forges legitimate entities in a network (such as EV, local aggregators, or charging points) to gain access. If the local aggregator or charging pile cannot identify the legitimacy of the vehicle, then malicious vehicles could gain access to the grid to steal power and information. On the other hand, if a legitimate vehicle is unable to distinguish between local aggregators or charging points that have been controlled by an adversary, its real identity or other sensitive information could be compromised. In general, the way to prevent impersonation attacks is to hide the true identity of the entity.

## 14.4 MITIGATION AND PROTECTION

### 14.4.1 Physical Defense

The purpose of physical security is to protect the charging pile, computer, server, and other equipment and communication facilities in the station from natural disasters, human operation error or error, power failure, equipment theft and destruction, electromagnetic interference, and other environmental accidents. Physical security includes the safety of the station environment, site, and equipment. Therefore, the physical security of the station can be divided into two aspects: the safety of the environment and the safety of the equipment and facilities.

#### *14.4.1.1 Safety of the Station Environment*

The safety protection of the communication network environment, according to the "Electric Conduction Charging System part I: General requirements" and other five national standards, GB2887-89 national standard "computing station site technical Conditions," GB9361-88 "computer station site safety requirements" and ISO/IEC27001 information security management standard design specifications, the station should have fire alarm, temperature and humidity monitoring system, anti-theft alarm system, and electromagnetic disturbance monitoring system. To protect the equipment and facilities in the station from flood, fire, earthquake, high temperature, chemical decay substances, and strong magnetic interference, ensure that the communication network is in a normal and stable running state.

#### *14.4.1.2 Safety of Devices*

The station mainly includes the devices anti-theft, anti-destruction, anti-electromagnetic information leakage, prevent line interception, anti-electromagnetic interference, and power supply protection. Communication equipment and communication lines need to be firmly installed, with a certain ability to resist disaster and damage. These are directly related to the security of equipment information data, and even can lead to the breakdown of the communication network, affecting the reliability of network information security.

Physical security is the basic guarantee of information security. Therefore, in the construction of communication networks, physical security needs should be fully considered, appropriate equipment should be selected, and the environment and equipment conditions of the monitoring stations should be selected to ensure the safe transmission of communication network information.

### 14.4.2 DoS Attack Defense

In order to defend against DoS attacks and make servers steadily serve their legitimate users, researchers came up with many techniques, such as IP traceback, packet marking, entropy variations, and intrusion detection and prevention, to prevent DoS attacks before they occur.

#### *14.4.2.1 IP Traceback Techniques*

For a DDoS attack, higher detection accuracy can be achieved at the closest vicinity of the victim due to the presence of aggregated attack flows, in comparison with remote disjoint attack flows. Similarly, it is desirable to perform filtering of the attack traffics closer to their attackers to avoid influence on other network users. For the purpose of packet filtration, revealing the attack source and path followed by attack packets is essential. Many researchers adopted IP traceback techniques to assist in packet filtration. For example, Burch and Cheswick[24] developed a link testing technique that performed a recursive search from a victim until the attacker was reached. The path of the link testing started from the router closest to the victim and ends until the source router of the attacker was identified. After continuously iterating on the determination of incoming links, the node in the path at each upstream level can be revealed. Snoeren et al.[25] proposed a hash-based IP traceback approach by using practical packet logging, which is aimed at storing packet digests on intermediate routers. Each router stored the digests of the packets passing through it and a digest could be the hashed value of the IP header fields of a packet. The network path is then determined by using the stored information. This technique allowed us to eliminate all but a handful of physical networks that could be the source of the attacking packet stream, which could be used to defend against reflection-based and botnet-based DDoS attacks.

#### 14.4.2.2 *Entropy Variations*

Nychis et al.[26] analyzed the detection capabilities and correlations of different entropy-based metrics, such as flow-header features and behavioral features. Flow-header features involve IP addresses, ports, flow sizes, and the behavioral features were related to the number of packets during communications between nodes. By finding the difference in the entropy values of attack packets at the source or destination IP address and the corresponding port, slow request or response attacks can be detected. In Ref. [27], the entropy was classified into long-term entropy when the number of packets is more than 10,000 and short-term entropy when it was less than 10,000. The short-term entropy was used for early detection while the long-term entropy was employed to classify attacks. Such an approach could be employed to identify protocol exploitation flooding attacks and amplification-based flooding attacks with appropriate thresholds.

#### 14.4.2.3 *Intrusion Detection and Prevention Systems*

Roschke et al.[28] proposed a deployment architecture of intrusion detection systems by taking a cloud computing paradigm, which included a system layer, a platform layer, and an application layer. The intrusion detection system (IDS) is deployed on each layer of the cloud to gather alerts from sensors and the central management unit in the cloud correlates and analyzes the alerts. Rengaraju et al.[29] presented a distributed firewall with an intrusion prevention system (IPS). The firewall was deployed on the switches in software-defined network. These switches monitored the passing traffics for anomaly detection, and suspicious packets were transferred to the firewall for further analysis. Once a malicious activity was identified, the IPS could prevent intrusions by sending alarms, dropping packets, resetting connections, or blocking the traffics from the attackers. Then, the software-defined network controller updated the firewall rules. This IPS system was investigated for naïve flooding attacks.

#### 14.4.2.4 *Machine Learning–Based DoS Defense Techniques*

A variety of machine learning algorithms have been developed in defense of DDoS attacks. Xiao et al.[30] proposed an effective detection scheme based on KNN with correlation analysis to detect DDoS attacks. In Ref. [31], a detection mechanism based on a one-class support vector machine was introduced for application-layer DDoS attacks, specifically for HTTP flooding attacks, SYN flooding attacks, and NTP amplification attacks. As Vishwakarma and Jain suggested in Ref. [32], honeypots with machine-learning-based approaches are effective in detecting botnet-based DDoS attacks in IoT. They used heterogeneous IoT honeypots to capture device malware installation attempts and adopted unsupervised machine learning techniques, such as clustering and anomaly detection to automate the process of detection and prediction of the incoming security threats by extracting features from honeypots. Asad et al.[33] proposed a detection mechanism by relying on artificial neural networks with feedforward and backpropagation algorithms to accurately discover several application-layer DDoS attacks.

A deep bidirectional LSTM-based recurrent neural network model was developed to detect botnet activities within consumer IoT networks by focusing on text recognition at the packet level. Besides, Meidan et al.[34] adopted deep autoencoders called N-BaIoT to detect anomalous network traffics from compromised IoT devices, and nine commercial IoT devices were infected with two widely known DDoS attacks, namely, Mirai and BASHLITE, to prove the model's effectiveness.

The abovementioned machine learning algorithms build mathematical models based on sample data, known as "attack packets," to make predictions for potential DDoS attacks. To achieve effective detection and explore their feasibility, these models were deployed on different devices, such as honeypots, routers, network switches, and firewalls.

### 14.4.3 Privacy Information Protection

In order to ensure the confidentiality, availability, integrity, non-repudiation, and non-repudiation of communication information in the process of charge and discharge of EV. It is necessary to study the privacy protection of charge/discharge from two aspects: privacy protection based on identity information and privacy protection based on data information.

#### *14.4.3.1 Privacy Protection Scheme Based on Identity Information*

The main idea of privacy protection scheme based on identity information is to anonymize the user's identity, so that the fine-grained data received by the charging service operation platform cannot be associated with the actual user, so as to achieve the purpose of privacy protection. It mainly adopts blind signature, pseudonym technology, group/ring signature, zero-knowledge proof, and other technologies. Qiao et al.[35] used the method of embedding tamper proof equipment module in smart meter to protect the privacy and security of EV users and used the means of generating pseudonym in this process. At the same time, an anonymous group authentication scheme based on revocable group signature was proposed to solve the problems of high overhead and user privacy security caused by dynamic access of EVs to public charging stations.

#### *14.4.3.2 Privacy Protection Scheme Based on Data Information*

The main idea of the privacy protection scheme based on data information is to use data aggregation,[36] so that the attacker cannot be associated with the data of a specific user even on the premise of knowing the user's identity. Li et al. use homomorphic encryption to achieve privacy protection demand aggregation and effective response. For data communication from EV users to smart grid operation center, data aggregation is directly performed on the ciphertext of the local gateway without decryption, and the aggregation results of the original data can be obtained in the operation center.[37]

### 14.4.4 Enhancing Network Protection

The charging pile is connected to the network and smart grid, which increases the attack targets and diversity of attackers. There are more and more attacks on network resources initiated by charging piles in recent years. Therefore, it is very important to protect network resources and enhance network protection. The most basic protection measure is to deploy a firewall and IDS.

#### *14.4.4.1 Firewall*

The firewall is a barrier between the public network and the private network. It can detect some low-level attacks and filter malicious traffic, provide safety information services and ensure network security. The new generation firewall must also combine DoS attack detection and network protocol selection technology. The former filters attack traffic and reduce the threat of attack, while the latter deals with injection and spoofing. Transport Layer Security (TLS) and Secure Sockets Layer (SSL) protocols can be combined with Secure Hash Algorithm (SHA), Hash-based Message Authentication Code (HMAC), and other encryption technologies to complete communication to data verification.

#### *14.4.4.2 Intrusion Detection System*

The charging process is very vulnerable to DoS attacks, which may lead to denial of charging. DOS attacks attempt to delay, block, or disrupt communications, and may severely degrade network performance, causing the charging pile to fail. The IDS can detect and mitigate DoS attacks, and it is also the first line of defense to protect smart grid. The IDS monitors the network or system for malicious activities or policy violations. Any detected activity or violation is usually reported to the administrator or collected centrally using the security information and event management system.

## 14.5 FURTHER DISCUSSION

As a substitute for traditional fuel vehicles, the popularity of EV is rising because of their environmental protection characteristics and the support of national policies. In addition, many countries have successively issued relevant policies and plan to realize the electrification of vehicles in the next few years. This is accompanied by the increase and construction of a series of EV supporting facilities such as charging piles and replacement power stations. The EV charging pile has the characteristics of a large investment amount and a long industrial chain. It is an important guarantee for the development of new energy vehicle industry and an important part of the emerging digital economy such as smart transportation and smart energy. On the contrary, there is a lack of protection for the charging process and charging pile of EV, and people's awareness of the protection of charging infrastructure and process needs to be strengthened. On the basis of Internet connection, EV charging pile is related to information security. We should pay attention to potential threat prevention and control. At present, it is necessary to optimize the application of communication technology, improve the comprehensive protection ability to charging pile information security, build a communication simulation platform, and improve the security and comprehensive efficiency of technology applications.

## 14.6 CONCLUSION

The information of charging pile and replacement station is important information transmitted in the communication network. The security and correctness of information, the integrity of infrastructure, and the smooth operation of the network are the basis to ensure the normal operation of the charging and exchange station. Due to the complexity of the station communication network, there are many potential security risks in the network, which makes the equipment, network, and information vulnerable to various attacks or influences. Therefore, network security is very important. If there is no security guarantee, network resources are vulnerable to DoS and MITM attacks, and data resources are prone to abuse, theft, tampering, destruction, and deletion, resulting in immeasurable losses. Taking the charging of EV as the main scenario, this chapter considers the possible infrastructure damage and attack, price information attack, DoS attack, privacy data attack, MITM attack, counterfeiting attack, and eavesdropping attack. Some corresponding protection schemes are listed according to the types of different attacks. Network security is not only the premise of the normal operation of the charging station but also an essential part to ensure the safe and stable operation of the smart grid.

## REFERENCES

1. Zhu Q, Rieger C, Basar T. A hierarchical security architecture for cyber-physical systems. *Resilient Control Systems (ISRCS), 2011 4th International Symposium on*. 2011, Boise, ID, USA.
2. Wang W, Lu Z. Cyber Security in the Smart Grid: Survey and Challenges. *Computer Networks*, 2013, 57(5):1344–1371.
3. Qian K, Zhou C, Allan M, et al. Modeling of Load Demand Due to EV Battery Charging in Distribution Systems. *IEEE Transactions on Power Systems*, 2011, 26(2):802–810.
4. Sousa T, Morais H, Vale Z, et al. Intelligent Energy Resource Management Considering Vehicle-to-Grid: A Simulated Annealing Approach. *IEEE Transactions on Smart Grid*, 2012, 3(1):535–542.
5. Rong J, Lu R, Lai C, et al. A Secure communication protocol with privacy-preserving monitoring and controllable linkability for V2G. *IEEE International Conference on Data Science in Cyberspace*. Guilin, China, IEEE, 2017.
6. Guille C, Gross G. A Conceptual Framework for the Vehicle-to-Grid (V2G) Implementation. *Energy Policy*, 2009, 37(11):4379–4390.
7. Kennel F, Gorges D, et al. Energy Management for Smart Grids with Electric Vehicles Based on Hierarchical MPC. *Industrial Informatics*, 2013, 9:1528–1537.
8. Rottondi C, Fontana S, et al. Enabling Privacy in Vehicle-to-Grid Interactions for Battery Recharging. *Energies*, 2014, 7:2780–2798.
9. Han S, Han S, Sezaki K. Development of an Optimal Vehicle-to-Grid Aggregator for Frequency Regulation. *IEEE Transactions on Smart Grid*, 2010, 1(1):65–72.
10. Zhang Y, Jin L, Zheng D, et al. Privacy-Preserving Communication and Power Injection over Vehicle Networks and 5G Smart Grid Slice. *Journal of Network & Computer Applications*, 2018, 122:50–60.
11. Eshou L, Wen-Min W. Influence and Countermeasure of Electric Vehicle Battery Charging to Power Systems. *East China Electric Power*, 2010, 38(1):109–113.
12. Gomez J C, Morcos M M. Impact of EV Battery Chargers on the Power Quality of Distribution Systems. *IEEE Power Engineering Review*, 2007, 18(10):63–63.

13. Kazerooni M, Kar N C. Impact analysis of EV battery charging on the power system distribution transformers. *Electric Vehicle Conference*. Greenville, SC, USA, IEEE, 2012: pp. 1–6.
14. Ghosh D P, Thomas R J, Wicker S B. A privacy-aware design for the vehicle-to-grid framework. *System Sciences (HICSS), 2013 46th Hawaii International Conference on*. Hawaii, USA, IEEE, 2013.
15. Tseng, H R. A secure and privacy-preserving communication protocol for V2G networks, 2012: 2706–2711. doi: 10.1109/WCNC.2012.6214259.
16. Wang H, Qin B, Wu Q, et al. TPP: Traceable Privacy-Preserving Communication and Precise Reward for Vehicle-to-Grid Networks in Smart Grids. *IEEE Transactions on Information Forensics & Security*, 2017, 10(11):2340–2351.
17. Au MH, Liu JK, Fang J, et al. A New Payment System for Enhancing Location Privacy of Electric Vehicles. *IEEE Transactions on Vehicular Technology*, 2014, 63(1):3–18.
18. Yang Z, Yu S, Lou W, et al. $P^{2$ : Privacy-Preserving Communication and Precise Reward Architecture for V2G Networks in Smart Grid. *IEEE Transactions on Smart Grid*, 2011, 2(4):697–706.
19. Shen J, Zhou T, Lai C F, et al. Hierarchical Trust Level Evaluation for Pervasive Social Networking. *IEEE Access*, 2017, 5:1178–1187.
20. Li H, Zhu H. Manipulating the electricity power market via jamming the price signaling in smart grid. *2011 IEEE GLOBECOM Workshops (GC Wkshps)*. Houston, TX, USA, IEEE, 2012.
21. Bursztein E, Cochran G J, Durumeric C Z, et al. Understanding the Mirai Botnet[C]// USENIX Security Symposium. USENIX Association, 2017.
22. Mustafa MA, Ning Z, Kalogridis G, et al. Roaming electric vehicle charging and billing: An anonymous multi-user protocol. *2014 IEEE International Conference on Smart Grid Communications (SmartGridComm)*. Venice, Italy, IEEE, 2014.
23. Saxena N, Choi B J. Authentication Scheme for Flexible Charging and Discharging of Mobile Vehicles in the V2G Networks. *IEEE Transactions on Information Forensics & Security*, 2016, 11(7):1438–1452.
24. Burch H, Cheswick B. Tracing anonymous packets to their approximate source. Unpublished paper. *Proceedings of USENIX LISA Conference*. California, USA, Dec, 2000.
25. Snoeren A C, Member S, Partridge C, et al. Single-Packet IP Traceback. *IEEE/ACM Transactions on Networking*, 2008, 10(6):721–734.
26. Nychis G, Sekar V, Andersen D, Kim H, Zhang H. An empirical evaluation of entropy-based traffic anomaly detection. *Proceedings of the ACM SIGCOMM Internet Measurement Conference*, IMC. 2008: pp. 151–156. doi: 10.1145/1452520.1452539.
27. Oshima S, Nakashima T, Sueyoshi T. Early DoS/DDoS detection method using short-term statistics. *2010 International Conference on Complex, Intelligent and Software Intensive Systems*, Krakow, TBD, Poland. 2010: pp. 168–173.
28. Roschke S, Cheng F, Meinel C. Intrusion detection in the cloud. *Eighth IEEE International Conference on Dependable, Autonomic and Secure Computing, DASC 2009*, Chengdu, China, 12–14 December, 2009. IEEE, 2009.
29. Rengaraju P, Ramanan V R, Lung C H. Detection and prevention of DoS attacks in Software-Defined Cloud networks. *2017 IEEE Conference on Dependable and Secure Computing*. Taipei, Taiwan, IEEE, 2017.
30. Xiao P, Qu W, Qi H, et al. Detecting DDoS Attacks Against Data Center with Correlation Analysis. *Computer Communications*, 2015, 67:66–74.
31. She C, Wen W, Lin Z, et al. Application-Layer DDOS Detection Based on a One-Class Support Vector Machine. *International Journal of Network Security & Its Applications*, 2017, 9(1):13–24.

32. Vishwakarma R, Jain A K. A honeypot with machine learning based detection framework for defending IoT based botnet DDoS attacks. *2019 3rd International Conference on Trends in Electronics and Informatics (ICOEI)*. Tirunelveli, India, 2019.
33. Asad M, Asim M, Javed T, et al. DeepDetect: Detection of Distributed Denial of Service Attacks Using Deep Learning. *The Computer Journal*, 2020, 63(7):983–994.
34. Meidan Y, Bohadana M, Mathov Y, et al. N-BaIoT: Network-based Detection of IoT Botnet Attacks Using Deep Autoencoders. *IEEE Pervasive Computing*, 2018, 17(3):12–22.
35. Qiao Q. Anonymous group authentication scheme for aggregation Signcryption and V2G in smart grid [D]. Xi'an University of Electronic Science and technology, 2014
36. Liu Y B, Song X L, Xiao Y G. Authentication Mechanism and Trust Model for Internet of Vehicles Paradigm. *Journal of Beijing University of Posts and Telecommunications*, 2017, 40(3):1.
37. Li H, Lin X, Yang H, et al. EPPDR: An Efficient Privacy-Preserving Demand Response Scheme with Adaptive Key Evolution in Smart Grid. *IEEE Transactions on Parallel & Distributed Systems*, 2014, 25(8):2053–2064.

9781032307534